AWIK

5/12/12 Major 8, '72.

Fuzzy
Logic
in
Chemistry

Fuzzy Logic in Chemistry

Edited by

DENNIS H. ROUVRAY
Department of Chemistry
The University of Georgia
Athens, Georgia, U.S.A.

ACADEMIC PRESS

San Diego   London   Boston   New York   Sydney   Tokyo   Toronto

Copyright © 1997 by ACADEMIC PRESS

Academic Press
a division of Harcourt Brace & Company
525 B Street, Suite 1900, San Diego, California 92101-4495, USA
http://www.apnet.com

Academic Press Limited
24-28 Oval Road, London NW1 7DX, UK
http://www.hbuk.co.uk/ap/

Library of Congress Cataloging-in-Publication Data

Fuzzy logic in chemistry / edited by Dennis H. Rouvray.
          p.   cm.
     Includes index.
     ISBN 0-12-598910-5 (alk. paper)
     1. Fuzzy logic.   2. Chemistry--Mathematics.   I. Rouvray, D. H.
   QD39.3.M3F89 1997
   540'.01'5113--dc21   96-50417
                                                       CIP

PRINTED IN THE UNITED STATES OF AMERICA
97   98   99   00   01   02   EB   9   8   7   6   5   4   3   2   1

Contents

4 Fuzzy Classical Structures in Genuine
Quantum Systems   91

ANTON AMANN

5 Fuzzy Measures of Molecular
Shape and Size   139

PAUL G. MEZEY

8 Fuzzy Logic in Computer-Aided Structure Elucidation   283

IVAN P. BANGOV

9 Fuzzy Hierarchical Classification Methods in Analytical Chemistry   321

DAN-DUMITRU DUMITRESCU

Contributors

Numbers in parentheses indicate the pages on which the author's contributions begin.

Anton Amann (91) ETH Hönggerberg, CH 8093 Zurich, Switzerland

Ivan P. Bangov (283) Bio-Rad Sadtler Division, Philadelphia, Pennsylvania 19104

Jürgen Brickmann (225) Institute of Physical Chemistry and Darmstadt Center for Scientific Computing, Technical University of Darmstadt, D-64287 Darmstadt, Germany

Dan-Dumitru Dumitrescu (321) Faculty of Mathematics and Computer Sciences, Babes-Bolyai University, RO-3400 Cluj-Napoca, Romania

George J. Klir (31) Center for Intelligent Systems, Binghamton University, Binghamton, New York 13902

Paul G. Mezey (139) Department of Chemistry and Department of Mathematics and Statistics, University of Saskatchewan, Saskatoon, Saskatchewan, Canada S7N 5C9

Kurt Mislow (65) Department of Chemistry, Princeton University, Princeton, New Jersey 08544

Dennis H. Rouvray (1) Department of Chemistry, University of Georgia, Athens, Georgia 30602

Jun Xu (249) Oxford Molecular Group, Inc., Baltimore, Maryland 21286

Professor Lotfi A. Zadeh

Foreword

When Professor Rouvray asked me to write a foreword to *Fuzzy Logic in Chemistry*, I felt both flattered and challenged: flattered because his request made me aware of the existence of applications of fuzzy set theory, or fuzzy logic as it is commonly referred to today, to chemistry—and challenged because the remoteness of chemistry from my fields of expertise makes it difficult for me to comment in specific terms on the contributions assembled in this volume. This is the first volume, as far as I know, to focus on the applications of fuzzy logic to chemistry.

During the past decade, applications of fuzzy logic have grown rapidly in number, variety, and visibility. What is the explanation for this phenomenon? What are the basic concepts in fuzzy logic that underlie its applications? These are the questions that I will attempt to cast some light on in this foreword.

First, a bit of history. My first paper on fuzzy sets (1965) was motivated by the realization that there is a wide gap between the precision of mathematics and the pervasive imprecision of the real world. At the center of this gap is the fact that almost all concepts in mathematics are sharply defined, whereas almost all real-world classes have unsharp, that is, fuzzy, boundaries.

We all know that mathematics has scored brilliant successes in dealing with a wide variety of real-world problems. But what is also true is that there are many problems in economics, psychology, decision analysis, and other fields that do not lend themselves to precise analysis in the classical spirit. And, what is perhaps more important, there are many problems in which tolerance for imprecision can be exploited—through the use of fuzzy logic—to achieve tractability, robustness, low solution cost, and better rapport with reality. Today, most of the applications of fuzzy logic fall into this category.

In every field of science—including chemistry—a pivotal role is played by the ways of representing and dealing with dependencies between variables. It is standard practice to deal with such dependencies through the use of differential, difference, or algebraic equations. But there are many cases in which the dependencies are too complex or too ill-defined to be amenable to representation by conventional methods. In such cases, fuzzy logic provides an effective way of dealing with dependencies through the use of so-called fuzzy 'if–then' rules. For example, the dependence of a variable Z on variables X and Y may be described as:

> if X is small and Y is small then Z is large
> if X is small and Y is medium then Z is medium
> if X is medium and Y is small then Z is small
> .
> .
> .
> if X is large and Y is large then Z is large.

In such rules, X and Y and Z are linguistic variables whose values, e.g., small, medium, and large, are words rather than numbers. In effect, the values of linguistic variables are labels for fuzzy sets. It is understood that the membership functions of these sets must be specified in context. Usually, the membership functions are assumed to be triangular or trapezoidal.

In fuzzy logic, the use of fuzzy 'if–then' rules is governed by the *calculus of fuzzy rules*, CFR. A major part of CFR is the *Fuzzy Dependency and Command Language*, or FDCL for short. Basically, FDCL is a fuzzy programming language that provides a powerful tool for the representation and manipulation of imprecise or ill-defined dependencies. Two issues play pivotal roles in FDCL. The first is interpolation, and the second relates to the induction of rules from observations.

The problem of interpolation may be described as follows. Assume that we have a collection of fuzzy 'if–then' rules that express the dependency of the linguistic variable Y on the linguistic variables $X_1, \ldots, X_n$, with the ith rule having the form if X_1 is A_{1i} and X_n is A_{in}, then Y is

B_i, where the A_{ij} and B_i are the linguistic values of $X_1, \ldots, X_n, Y$. The question is: What is the value of Y if $X_1, \ldots, X_n$ are assigned linguistic values $A_1, \ldots, A_n$, with the understanding that the A_i are different from the A_{ij}?

FDCL provides a straightforward interpolative answer to this basic question. The interpolation process leading to the value of Y lies at the base of most of the applications of fuzzy logic.

Interpolation serves an important purpose: it greatly reduces the number of rules that are needed to describe a dependency. Thus, in many of the applications of fuzzy logic in control, the number of rules is of the order of 20 and rarely exceeds 90. By contrast, when crisp 'if–then' rules are used, their number may be in the hundreds.

In many of the early applications of fuzzy logic, the A's and B's in the 'if–then' rules had to be calibrated by cut-and-trial to achieve a desired level of performance. During the past few years, however, the techniques related to the induction of rules from observations have been developed to a point where the calibration of rules—by induction from input–output pairs—can be automated in a wide variety of cases. Particularly effective in this regard are techniques centered on the use of neural network methods and genetic computing for purposes of system identification and optimization. Many of the so-called neuro-fuzzy and fuzzy-genetic systems are of this type.

Prior to the development of fuzzy logic, the standard practice in dealing with uncertainty and imprecision was to draw upon the concepts and techniques of probability theory. There are still some who claim that anything that can be done with fuzzy logic can be done equally well or better through the use of probability theory. Such claims reflect a lack of familiarity with fuzzy logic and an unwillingness to develop an understanding of what it offers.

A concept that plays a central role in fuzzy logic—and differentiates it from other methodologies—is the concept of a linguistic variable. As was alluded to earlier, the concept of a linguistic variable enters in the characterization of dependencies through the use of fuzzy 'if–then' rules. More importantly, the concept of a linguistic variable is the point of departure for a methodology that might be called *computing with words*, or CW for short. CW may be viewed as the principal contribution of fuzzy logic.

In CW, the initial data set is assumed to consist of a collection of propositions expressed in a natural language or, more particularly, in the form of a collection of fuzzy 'if–then' rules. The result of computation, that is, the terminal data set, is likewise a collection of propositions expressed in a natural language.

A basic idea underlying CW is that a proposition may be viewed as a fuzzy constraint on a variable. To arrive at the terminal data set, these constraints are propagated from premises to conclusions through the use of the rules of inference in fuzzy logic.

A simple example of CW is the following. Assume that a function, f, $Y = f(X)$, is described in words by the fuzzy 'if–then' rules:

> if X is small then Y is small
> if X is medium then Y is large
> if X is large then Y is small.

The question is: What is the maximum value of f?

Computing with words is not as yet a standard tool in the fuzzy logic toolchest. It is not employed explicitly in *Fuzzy Logic in Chemistry*. Nevertheless, my conviction is that—once it is understood—CW will be employed widely and effectively in the solution of a variety of problems that do not lend themselves to computing with numbers. In many cases, computing with words in place of numbers enhances tractability and lowers solution cost.

Fuzzy Logic in Chemistry is a bold venture into an exploration of the use of nonstandard computing methodologies in chemistry. It opens the door to a much wider use of fuzzy logic and related techniques in the analysis and design of chemical systems.

Fuzzy Logic in Chemistry is a tribute to Professor Rouvray's vision and initiative. Professor Rouvray and the contributors to this volume deserve our thanks and congratulations.

Lotfi A. Zadeh

Preface

How does a new idea or a new paradigm come to replace an older one? If we are to believe Kuhn,[1] the new paradigm ultimately wins out because of its elegance, its consistency, its comprehensiveness, and especially its usefulness. Judged on these criteria, fuzzy methods should have long since replaced the earlier paradigm of binary logic and its concomitant probabilistic reasoning. But things are not that simple.

Although fuzzy methods and fuzzy logic in particular have evident advantages over traditional methods, they have encountered some fierce opposition. Fuzzy logic is clearly much closer to ordinary commonsense reasoning and also provides an inferential system that enables us to obtain specific answers to nonspecific and vague questions. Why such an approach should have provoked so much criticism becomes apparent if we refer again to Kuhn.[1] The road to success, it would seem, is never direct and straightforward when it comes to overthrowing paradigms. Initially, the new paradigm is attacked, reviled, and laughed out of court. When such treatment is no longer tenable, the paradigm is admitted grudgingly, although it is still regarded as something of an outcast. Eventually, after most of the older courtiers have died, the breakthrough is achieved. The new paradigm takes its rightful place and is then accorded widespread acceptance.

How far along the path to widespread acceptance is fuzzy logic? It is certainly now out of the wilderness of rejection and contempt and is viewed as a very upwardly mobile discipline. Has it achieved the breakthrough? There are many indications that it has. Fuzzy logic is now riding the crest of a wave, the like of which has not been seen in its previous history. Specialized journals on the subject have been introduced to cope with the flood of technical papers; to date, over 20,000 such papers have appeared. Added to this are numerous books, both highly technical and less so, articles, and reviews. Even popular books on fuzzy logic have begun to hit the newsstands.[2, 3] Perhaps the clearest indication that fuzzy logic has arrived is the fact that fuzzy logic controllers are currently being employed in a host of commercial applications that range from domestic appliances such as cameras and television sets to elevators and giant power stations. World front-runner Japan is now exporting products with fuzzy logic components to the tune of $40 billion annually. Moreover, worldwide trade in such products is showing exponential growth.[4] The exploitation of applications of fuzzy logic over the past few years in particular has been so spectacular that the term *fuzzy logic* is now rapidly becoming a household term.

What about applications of fuzzy logic in the physical sciences? Here, too, an increase in interest in recent years has been very noticeable. Initially, the focus fell on a wide variety of physicochemical concepts, all of which are inherently vague to some extent unless they are rigorously mathematical in nature and defined in strictly mathematical language. The first such concept to be fuzzified was that of the electron following the enunciation of Heisenberg's Uncertainty Principle. Thereafter, numerous other concepts in which electrons are directly or indirectly involved were also shown to be fuzzy in nature. These included the concepts of molecular structure, chirality, molecular shape, symmetry, reactivity, acidity, aromaticity, inductive and mesomeric effects, and selectivity. In more recent years fuzzy logic has been exploited extensively in research that embraces the areas of molecular engineering and design. This work has involved the classification, clustering, and sorting of molecules by techniques based on fuzzy pattern recognition or the use of fuzzy neural networks. There is currently rapidly growing interest in the role that linguistic variables can play in such computational methodologies; Chapter 6 herein provides a case in point.

With all this burgeoning interest in fuzzy methods, it seemed to me that it was high time that an international conference be organized to explore how far we had come and to consider the role that fuzzy logic might play in the chemical domain in the future. Among other things, such a conference could discuss the fuzzification of chemical concepts, the current use of fuzzy methods in molecular design, and the possible future applications of fuzzy reasoning. The conference I envisaged was eventually organized by a colleague, Dr. Edward Kirby, and myself and took place in Pitlochry, Scotland, during the week of 10–14 July 1995. This conference was actually the sixth in a series of Mathematical Chemistry Conferences organized under the auspices of the International Society for Mathematical Chemistry. The conference, which was the first to discuss fuzzy logic applied in the chemical domain, brought together a wide variety of experts from over a dozen different countries. Included among the participants were mathematicians, physicists, and a great variety of chemists ranging from pure theoreticians to biochemists. The title of the conference was: Are the Concepts of Chemistry All Fuzzy? As an appropriate theme, a quotation from the noted Dutch physicist Hendrik Kramers[5] was adopted:

> In the world of human thought generally and in physical science in particular, the most important and most fruitful concepts are those to which it is impossible to attach a well-defined meaning.

Most of the chapters in this book are refereed and substantially expanded versions of lectures delivered at this conference on the days that were devoted to the discussion of fuzzy logic and its chemical applications. The first two of these lectures were foundational: one, delivered by myself, covered the treatment of uncertainty in the sciences generally, and the other, delivered by Professor Klir, was on the basic notions of fuzzy set theory and fuzzy logic. These appear in this volume as Chapters 1 and 2, respectively. The next three chapters are elaborations of lectures given on the role of fuzzy reasoning in the description of physicochemical concepts. Chapter 3 by Professor Mislow is concerned with the concept of chirality, Chapter 4 by Dr. Amann discusses quantum-theoretical concepts, and Chapter 5 by Professor Mezey takes a long look at the concepts of molecular structure and molecular shape. The remaining four chapters cover a variety of recent applications of fuzzy logic in the chemical sciences. In Chapter 6, Professor Brickmann considers how linguistic variables may be applied in molecular recognition problems. Chapters 7 and 8, written respectively by Drs. Xu and Bangov, introduce the basic ideas of molecular fuzzy clustering techniques, whereas Chapter 9 by Professor Dumitrescu focuses more specifically on hierarchical clustering techniques as applied to the broad area of analytical chemistry.

All of these authors have done a tremendous job, and I take this opportunity to thank them for their excellent contributions to our book. The founder of fuzzy logic, Professor Lotfi Zadeh, also generously consented to write the Foreword to our book, and he too is thanked for his very welcome contribution. Also deserving of sincere thanks are a number of other persons who in several different ways have helped to ensure that this book saw the light of day. These certainly include Edward Kirby and his wife, Jean, who played an absolutely pivotal role in helping me to get the original conference organized. The first seven of the nine contributed chapters herein were initially presented as lectures at the Pitlochry conference. I am also deeply appreciative of the considerable assistance I received from my secretary, Sherri Page, who had the task of preparing much of the final manuscript, and from David Packer of Academic Press, who ensured that the final manuscript was smoothly transformed into this volume. It is our hope that the contents will afford our readers an illuminating and stimulating introduction to fuzzy logic in chemistry.

Dennis H. Rouvray

REFERENCES

1. T. S. Kuhn, *The Structure of Scientific Revolutions*. Univ. of Chicago Press, Chicago, 1962.
2. B. Kosko, *Fuzzy Thinking: The New Science of Fuzzy Logic*. Hyperion, New York, 1993.
3. D. McNeill and P. Freiberger, *Fuzzy Logic*. Simon and Schuster, New York, 1993.
4. C. von Altrock, *Fuzzy Logic*, Vol. 1. Oldenbourg, Munich, 1993.
5. H. A. Kramers, *in H. A. Kramers: Between Tradition and Revolution* (M. Dresden, ed.), p. 539. Springer-Verlag, New York, 1987.

1

The Treatment of Uncertainty in the Physical Sciences

DENNIS H. ROUVRAY
Department of Chemistry
University of Georgia
Athens, Georgia 30602

> *In science, we find all grades of certainty short of the highest.*
>
> Bertrand Russell
> *Our Knowledge of the External World (1914)*

I. GENERAL INTRODUCTION

An intense curiosity about the natural world we inhabit is one of the most enduring of human attributes. A passionate desire to know, to understand, and to interpret what is happening around us is an experience common to us all. Indeed, it is this insatiable thirst for knowledge that ultimately primes the wellspring of our many flourishing arts and sciences. Yet, we may ask, is mere curiosity enough to give us reliable knowledge? A number of related questions also come to mind. What kinds of knowledge is it possible for us to discover? Of what can we be absolutely certain? Is our impassioned quest for knowledge no more than a frustrating exercise in futility? These issues we propose to explore here in a scientific context. To whet the appetite, we begin by charting a course through the somewhat turbulent waters of what has been regarded as certain in the past. Many of the supposed certainties of earlier ages are now gone forever. Mention might be made of loss of belief in the absolute truths of religious dogma that so dominated thinking in the Middle Ages, our abandonment of the notion of a mechanical universe operating like clockwork that provided the backdrop to the Age of Enlightenment, and, more recently, doubts that we now harbor on the optimism of our Victorian forebears, who confidently

Fuzzy Logic in Chemistry

1

expected an unstoppable march in the progress of science and technology to solve all our ills. Over the centuries, the conceivable areas where we might look for certainty have been steadily whittled away, and this has resulted in substantial erosion of many of our belief systems. Uncertainty now characterizes much of our thinking about the world and it is no accident that modern science is preoccupied with themes such as chaos, quantum indeterminacy, fuzzy logic, and semantic analysis. Our current paradigms tend to embrace uncertainties rather than the alleged certainties of the past.

Broadly speaking, certainty comes in two major varieties.[1] The first of these, referred to as primitive certainty, derives directly from the human senses. It is the kind of certainty we experience when we perceive objects or events in our immediate environment. The second variety, known as derivative certainty, pertains to knowledge that is deduced or inferred from observations or occurrences that may not have been perceived directly. Let us focus, for the moment, on the first variety. The human quest for certainty of any kind is epitomized in the work of the French philosopher and mathematician René Descartes. He became convinced[2] that "it is much more custom and example that persuade us than any certain knowledge." In seeking for a certainty that was "so assured that all the most extravagant suppositions brought forward by the skeptics were incapable of shaking it", he eventually concluded[3] that the one thing he could be certain about was the fact that he was thinking. This realization he expressed[4] in the now famous statement, "I think, therefore I am." It was soon pointed out[5] that such certainty also pertains to all the other activities associated with living. Thus, it would be equally true to assert that "I eat, therefore I am" or even "I die, therefore I am." Similar sentiments have continued to be expressed into the present century. For instance, the philosopher Ludwig Wittgenstein argued[6] that "if you tried to doubt everything you would not get as far as doubting anything. The game of doubting itself presupposes certainty." The scientist Sir Arthur Eddington posed the question, "What is the ultimate truth about ourselves?" and decided[7] that although "various answers suggest themselves...there is one elementary inescapable answer. We are that which asks the question."

The second variety of certainty, so-called derivative certainty, will necessarily be less certain than the first variety because it is based on a process of deduction or inference. Typically, to demonstrate the validity of a derivative certainty, a foundational statement has to be posited, the truth of which is taken to be self-evident. This statement is then used as the starting point in a chain of logical reasoning that eventually leads to the derivative certainty. Such a method of arriving at certainties is clearly open to several objections. The foundational statement itself may be suspect, in which case one or more additional statements will be required to support it. There is also a risk of becoming encumbered in an infinite regress of statements if such a process were to be continued indefinitely. One of the statements would ultimately have to be deemed certain and its truth self-evident. Another problem is that the reasoning process may be regarded as objectionable. This process, which commonly employs inductive logic, involves reaching some new conclusion that is not expressed by the statement(s) one starts with. The inference may be from a number of particular instances to some general statement or law, or may employ analogy to move from one description of a system to some other description of it. Such reasoning and its many pitfalls have been closely examined by Hesse[8] and Rouvray.[9] Finally, it must be mentioned that all inferential reasoning rests on the assumption of the consistency and uniformity of nature, that is to say, that what was valid in the past will be valid in the future. Such an assumption is purely an article of faith that cannot be proven. Thus, it is well to bear in mind[10] that "unless some things are certain, nothing can be even probable" and that "there are a great many statements the truth of which we rightly do not doubt, and it is perfectly correct to say that they are certain. We should not be bullied by the skeptic into renouncing an expression for which we have a legitimate use."

Before leaving the subject of certainty and embarking upon a discussion of the role of uncertainty in the sciences, a few comments about thought experiments would seem to be in order. Such experiments, sometimes still referred to by the original German word *Gedankenexperimente*, involve no physical experimentation at all; they are conducted solely at the intellectual level. The performance of an experiment of this kind is not at all unusual in the sciences, for, as Casti[11] has pointed out, the starting point of any science is not typically a set of observations made in a laboratory, but rather the development of some hypothesis that is subsequently to be tested. Pure thought experiments can certainly yield remarkable insights into the nature of the real world. In the seventeenth century, Galileo,[12] for instance, was able to make the important discovery that all objects fall to earth with the same velocity regardless of their mass. This thought experiment was of course later confirmed by actual physical experimentation. In the present century, two of the most famous thought experiments were performed in the 1930s. These are now referred to as the Schrödinger cat paradox and the Einstein–Podolsky–Rosen (EPR) experiment. Both have been used to argue that our current conceptions of quantum theory are erroneous. The former,[13] based on the allowed superposition of ground and excited states of atoms, demonstrates how bizarre such a situation becomes if it is applied to the macroworld: a cat, for instance, would have to be simultaneously alive and dead until a determination had been made of its actual status. In the EPR experiment,[14] two particles are imagined to have collided. Subsequent measurement of either the position or momentum of one of these particles will automatically yield information on the position or momentum of the other particle, even though the latter particle could not know which of the two quantities the experimenter had chosen to measure. Thought experiments are not restricted to arcane topics in physics; they are of general applicability in the sciences. It has been suggested[15] that, because we can envision the existence of alternative universes, our world need not be the way it actually is. If this were true, reasoning alone could not reveal the laws of nature; some experimentation would always be necessary. Einstein,[16] however, felt that "our experience hitherto justifies us in believing that nature is the realization of the simplest conceivable mathematical ideas.... . In a certain sense, therefore, it [is] true that pure thought can grasp reality, as the ancients dreamed."

II. THE QUEST FOR CERTAINTY

One direct outcome of the continual striving of humankind to achieve some measure of certainty about the natural world has been the emergence of an abundance of intellectual support systems. Many of these, having been nurtured over some 3000 years, now appear to us to be of such towering stature that they are rightly regarded as permanent monuments to our civilization. We refer here especially to the systems of logic, mathematics, philosophy, and science that have their origins primarily in ancient Greece. As each of these systems bears elegant testimony to the prowess of human intellectual endeavor, it might seem reasonable to anticipate that such systems could furnish us with at least some certainties about the real world. It is with a view to determining just how much certain knowledge each of these systems can deliver that we now undertake a brief review of each of them in turn.

Logic is a discipline that concerns itself with the provision of valid general rules that may be applied in any reasoning process. Logic is thus basically prescriptive in nature in that it decrees how reasoning is to be carried out. In fact, logic provides us with a syntax for our reasoning. Any entity in the physical world, e.g., an object, a person, a concept, or an idea, can be subjected to the process of reasoning. This process is initiated by giving the entity a name and then operating on it by applying a series of prescribed steps known as the rules of logic. Collectively, these rules constitute a formal language, for instance, the language of Aristotelian logic. The rules eventually lead to some new conclusion that is theoretically valid new information. The only problem here is that conclusions are valid only when the entities being manipulated are precise. However, the entities will not be precise unless they are logical or mathematical concepts because the entities defined in everyday or even scientific language always have some degree of vagueness associated with them.[17] This means that, in general, logical propositions will be satisfied not only for one single fact, but rather for a range of facts, which implies that the notion of validity will itself be vague.[18] Logic can thus not be relied upon to yield absolutely certain results. For scientific discourse, logic may be viewed as a necessary but not a sufficient procedure to ensure the validity of the conclusions reached. In the words of Oldroyd,[19] "logical certainty in science has always been somewhat too strict a requirement...what we regard as logically acceptable is...a joint product of the way things behave in the world and the way we see the world." Is there any other possible approach that could improve on this state of affairs? The most that can be hoped for is the development of some sort of broad-based general logic that will satisfy virtually everybody. On the chances of establishing such a logic, Berry[20] wrote that any "rational reconstruction of...mathematics, philosophy and science will include some logic or other...[but] it must be a grand logic, for only such can satisfy their diverse needs. So conceived, no one of the three can be true unless the embedded logic is true... because it describes the world as the world in cold fact is."

Turning now to mathematics, we may anticipate that in essence mathematics will differ little from logic. Both use sets of prescribed rules to reach new conclusions and the languages of logic and mathematics embrace similar rules. Much of what has been said about logic thus applies equally to mathematics. Of greater import here, however, is the fact that neither logic nor mathematics can be relied upon to yield absolutely certain knowledge. There are two major reasons why mathematics fails to deliver certainty about the real world. The first is that mathematics takes as its universe of discourse objects that are abstracted and idealized versions of reality. Thus, mathematics pertains to objects such as points of zero size, lines of zero thickness, circles that are perfect, and so on. Such objects, of course, are not found in nature; they exist only in the imagination. In reality, for instance, all circles have a rough edge and so need to be represented as fractal objects rather than perfect geometrical ones.[21] Although reasoning about abstractions leads us to precise results, the latter are of restricted value in practice because they do not accurately characterize the real world. The situation here has been summed up in the words of Einstein, who wrote[22] that "as far as the propositions of mathematics refer to reality, they are not certain; and as far as they are certain, they do not refer to reality." The second reason for uncertainty about mathematical results is that the axioms of mathematics cannot be derived from the axioms of logic. In his now famous incompleteness theorem, Gödel demonstrated[23] that any consistent system of logic that includes all the statements of arithmetic will always be incomplete because it will contain statements that cannot be proved to be true or false within the rules of the system. Thus, mathematical proofs can never be totally trusted and, ultimately, mathematical truths are arrived at via human intuition. As Casti[24] has eloquently put it, "intuition [is] the rock-bottom basis for every school of mathematical epistemology." It has even been suggested[25] that Gödel's theorem shows that artificial intelligence systems will always be inferior to the human mind because even the most sophisticated computer cannot formulate an incompleteness theorem about its own system.

Philosophy is even less likely to result in certitudes than either logic or mathematics. This is because its inherent nature requires it to pose questions rather than answer them. Philosophers are continually putting forward propositions that are then subjected to scrutiny by other philosophers. The function of philosophy may thus be interpreted as the critical examination of all views about our world and ourselves in an attempt to ascertain whether the views are meaningful, and, if so, what that meaning is. It is hardly surprising that general agreement on anything is seldom, if ever, reached. Philosophy exists in a state of perpetual flux with all of its propositions subject to continual attack. In surveying the philosophical arguments posited up to his time, Descartes concluded that none could be regarded as certain; the same could, of course, be said of all the arguments up to the present time! The philosophy of science affords us no better anchorage. From roughly the 1920s until the 1960s, the philosophy of science was in the thrall of logical positivism, a doctrine that attempts to weld modern logic to the notion of empiricism. Empiricism holds that all knowledge is unreliable except that which derives from direct sensual experience. Logical positivism thus maintains that a solid foundation for science can be constructed on the union of logic and empiricism. Since the 1960s, however, the grip of logical positivism has progressively weakened, with the result that there is currently no one predominant philosophy of science.[26] Realism,[27] a somewhat less strident view, which asserts that reliable, nonsubjective knowledge of our world can be gained from scientific endeavor, has experienced something of a comeback. This development has effectively brought us full circle to where we were before the arrival of logical positivism on the scene. It is ironic that most practicing scientists were then and still are today de facto realists.

The realist view of science that is now so pervasive began its rise to prominence at the end of the Middle Ages when science and philosophy parted ways. Increasingly, the scientific community assumed that science had nothing to do with philosophy—often referred to as metaphysics—and that scientific theories alone were capable of providing an accurate portrayal of the physical universe. Although it must be admitted that by adopting such a stance much progress has been made, there is an inherent danger in making this assumption. In particular, there exists a serious risk that scientific theories will be deemed to offer not only a necessary description of nature, but also a sufficient one.[28] That is to say, the concepts elaborated within any scientific theory may be regarded as part of objective reality. The dangers associated with such thinking are all too well known: we need mention only the concepts of phlogiston, vital force, or the ether that were key concepts in some of the scientific theories of yesteryear. A second major danger is that posed by the now powerful movement toward the supposedly scientific philosophy of reductionism. This is found in many different guises and may involve the reduction of wholes to parts, of mind to brain, or of all the sciences to physics. One of its most outspoken proponents has even proclaimed[29] that "[s]cientists, with their implicit trust in reductionism, are privileged to be at the summit of knowledge, and to see further into truth than any of their contemporaries." Fortunately, wiser counsels are also heard within the scientific arena. Thus, Dyson[30] has commented that the "progress of science requires the growth of understanding in both directions, downward from the whole to the parts and upward from the parts to the whole. A reductionist philosophy, arbitrarily proclaiming that the growth of understanding must go only in one direction, makes no scientific sense."

III. PROBABILISTIC PANACEAS

Having thus explored various avenues that could conceivably lead us to certainty about the physical world, we are obliged to recognize that none can take us all the way to our ultimate goal. It would appear that the avenue of science offers the closest approach, provided we avoid a number of pitfalls such as too strong a faith in realism or reductionism. Great caution is clearly called for in moving ahead—a caution that all too often is lamentably lacking in practice. When it comes to science, it seems even philosophers are prone to throw caution to the winds. Recall that Ayer maintained[10] that "it is perfectly correct to say that [many scientific statements] are certain." The prevailing attitude toward science is thus somewhat paradoxical: although there is no secure logical route to arrive at certainty, many scientific pronouncements are greeted with universal acceptance. One might almost say that we are determined to accept the infallibility of science against all the evidence. The tendency to believe in science became especially marked in the nineteenth century, and it was during that century that no effort was spared to produce reliable apparatus that could be used to perform accurate experiments.[31] Thus, whereas Lavoisier stated[32] in 1789 that we are "never to search for truth but by the natural road of experiment and observation," by 1856 it was possible for Clerk Maxwell to thunder[33] that "[n]othing...can escape from the ordeal of the measuring rod and the balance...and all laws must be expressed with reference to exact quantities, so that we have...an absolute security against vagueness and ambiguity." As ever greater emphasis began to be placed on numerical results, the trend to clothe all scientific results in mathematical language became established. It has been asserted,[34] how-ever, that such a trend implies an inherent weakness in and a public distrust of the disciplines involved. If this is the case, it would help to explain the growing reliance that has been placed on probabilistic reason-ing, especially in the present century.

In fact, one of the oldest approaches to uncertainty has been the assignment of probabilities to uncertain events. Interest in probability theory has its origins in the study of games of chance and was first placed on a firm mathematical footing by the work of Laplace toward the end of the eighteenth century.[35] Our contemporary approach to probability theory was largely laid down in 1933 following the publication of a now famous monograph by Kolmogorov.[36] Probability theory rests on the simple idea that events in the past provide a reliable guide to events in the future. This may be expressed in symbols as follows. Suppose that some event (such as the tossing of a coin) is repeated a total of n times and that a specified outcome (the coin comes up heads) is observed r times. It is argued that, for all future events of the same type (coin tossing), the probability P of observing the specified outcome (heads) is

$$P = \lim_{n \to \infty} \frac{r}{r + (n - r)} = \lim_{n \to \infty} \frac{r}{n}. \tag{1}$$

Upon this bedrock idea, a highly complex and sophisticated statistics has been constructed that can be used to compute not only the frequency of occurrence of most events, but also many kinds of averages such as the expectation values of ensembles of particles. Does this fundamental idea provide a sound basis for all these computations? The philosopher Ayer insists[37] that probability theory cannot yield any certainty about future events and that it cannot even indicate what is likely to happen, yet, bearing this in mind, he goes on to state his personal belief that the future will probably resemble the past and so render probabilistic interpretations of our world valid. His presumption would appear to be borne out in practice, especially where large numbers of events or large populations of entities are concerned.

To understand why this is so, let us consider briefly systems containing a large number of atoms, typically a number greater than the Avogadro number (ca. 6.02×10^{23}). With such a large number, we may be reasonably sure that the probability of events occurring will be described by Eq. (1). Having said this, however, it is necessary to point out that Eq. (1) applies only to random occurrences, which is to say that all the events under consideration should be random ones and all the entities (particles) involved should exhibit only random behavior. Whether any of the phe-nomena of nature are strictly random is still a matter of some dispute at the present time.[38] Phenomena may appear to be random only because our knowledge of the underlying physics is lacking. In spite of all these reservations, however, it does seem possible to derive a fair number of scientific laws that are based ultimately on the statistics of large ensembles of particles. Consider, for instance, the radioactive decay of the atoms in elements, a process that is widely believed to be random in nature. The law of radioactive decay derived on the basis of random atomic decay takes the form

$$N = N_0 \exp(-\lambda t), \tag{2}$$

where N_0 is the initial number of radioactive atoms at time $t = 0$, N is the number of atoms remaining at some later time t, and λ is a constant known as the radioactive decay constant. From this law one can obtain reproducible values of the half-lives of isotopes for all of the known elements[39]; the half-life of an isotope is calculated from λ by use of the expression $(\ln 2/\lambda)$, which equals approximately $0.693/\lambda$.

Statistical analysis of the behavior of large ensembles of particles generally falls within the province of statistical mechanics or statistical thermodynamics. Analysis of this kind is capable of providing some powerful insights into the inner workings of the physical world. For instance, it can be employed to demonstrate that well-known gas laws such as those of Boyle or Charles are statistical in nature. It can even be exploited to yield information about ensembles that is not obtainable by application of the macromolecular laws of thermodynamics.[40] The analysis is typically carried out on ensembles that are regarded as isolated from their environment. All the possible configurations of the particles in terms of some parameter such as their energy are theoretically determined. It is then assumed that each of their configurations is equally likely to occur because there is no reason for preferring one configuration over any other. The resultant statistical organization of the ensemble is deemed to have the same significance as the state of the system in thermodynamics.[41] However, because the microstructure of the system is now supposedly known, we can draw a relationship between the statistical organization of the ensemble and the macroscopic entropy S of the system, namely,

$$S = k \log_e W, \tag{3}$$

where k is the Boltzmann constant $(1.38 \times 10^{-23} J K^{-1})$ and W is the number of possible configurations of the ensemble. This equation affords a key linkage between the two approaches and makes it feasible to calculate a wide range of properties such as the heat capacities for gaseous, liquid, and solid-state systems. In spite of many computational successes, it has been pointed out,[42] however, that the probability distributions assumed to hold for the particles in ensembles are of an entirely fictitious nature, although they allegedly "will not mislead us too much even should we take them to be real in simple cases."

During the latter half of the twentieth century, the limitations of classical statistical arguments have become increasingly clear, especially in the quantum domain, where they sometimes lead to awkward paradoxes.[9] The search for alternatives has rediscovered the method of Bayes, the foundations of which were laid as long ago as 1746. Over the past two decades the supposed advantages of this approach have been hotly debated.[43] Basically, Bayes' theorem affords a means of updating the probability that some event will occur when new data pertaining to that event come to hand. In symbols, the theorem may be expressed as

$$P(Y|XZ) = \frac{P(Y|Z)P(X|YZ)}{P(X|Z)}, \qquad (4)$$

where the symbol $P(A|B)$ signifies the probability that statement A is true assuming that B is true. Here X represents some new set of data that becomes available, Y is a statement or hypothesis, and Z is the initial data set. It is of course necessary to assume that $P(X|Z)$ is not zero. The new probability that statement Y will be true is thus given by the product of the initial probability of its truth and the probability that the new data are true (assuming the truth of statement Y and the data Z) divided by the probability that the new data are true (assuming only the truth of the data Z). However, as mentioned earlier, probabilities cannot offer any certainty about the course of future events, and critics of the Bayesian approach maintain that the probabilities used are purely subjective.[44] It would seem though that this approach might have an edge in elucidating quantum paradoxes such as the Schrödinger cat paradox.[13] The author of a reformulated version of quantum mechanics based on Bayesian statistics claims[45] to have explained not only the cat paradox, but also wave–particle duality and the EPR paradox. These difficulties, which he now alleges have been overcome, arose in the first place from the fundamental error of regarding the wave function and the state of the system as identical. On this and related matters, the jury is still out.

IV. QUANTUM INDETERMINACY

As the subject of quantum theory has now been broached, it is perhaps opportune to say something about quantum uncertainty at this point. The discovery toward the end of the nineteenth century that the laws of physics were in many cases based on statistics came as a considerable surprise. However, this surprise was nothing compared to the profound sense of shock experienced by the scientific community as a whole when the propositions of quantum theory first became widely known.[46] In fact, the reverberations of this shock are still keenly felt today, and the ongoing debate about what it all means has remained lively and controversial. Feynman[47] even went as far as to admonish us not to keep asking: "But how can it be like that?" because he claimed this would only lead us into "a blind alley from which nobody has yet escaped. Nobody knows how it can be like that." Perhaps the deepest shock came from the realization that physicists and others would henceforth be obliged to discuss fundamentally unobservable entitles in their modeling of nature. Quantum theory had revealed that it was inherently impossible to form any kind of picture of the elementary constituents of matter. Such notions flew in the face of some three centuries of patient endeavor which had sought to elucidate and clarify physical reality as far as possible. Starting from the work of Descartes, there had been a powerful movement to give vague metaphysical notions a precise mathematical interpretation. The underlying ethos of science was now being roundly challenged.

The proponents of quantum theory openly sought to reverse the course of science by encouraging metaphysical speculation about the reality of the physical world.[48] Heisenberg, for instance, declared[49] that for atoms "an interpretation of the Rydberg formula in terms of the circular and elliptical orbits of classical geometry makes no physical sense whatsoever" and that the concept of electronic orbits should thus be abandoned and replaced with something more appropriate. The eventual replacement concept was that of atomic orbitals, which effectively left us with a vague and ill-defined picture of the atom. It is interesting and somewhat ironic that, whereas Russell and other philosophers had pointed out that the logic in our minds was uncertain, the scientist Heisenberg would demonstrate to us that even the atoms in our brain are uncertain. Heisenberg's work led him inexorably to the radical conclusion that causality is no longer valid in the quantum realm. This must be the case he maintained, because it is impossible to know all the factors needed to characterize a system with complete accuracy.[50] Not surprisingly, the idea of relinquishing the age-old concept of causality provoked fierce opposition, the most notable and stubborn opponent being Albert Einstein. The current consensus is that Einstein was probably wrong in his view, although Dirac[51] has cautioned that "it might turn out that Einstein will prove to be right, because the present form of quantum mechanics should not be considered as the final form... ."

The heavy superstructure of modern quantum mechanics rests largely upon a set of mathematical relationships published in 1927 by Heisenberg.[50] These relationships are now usually referred to collectively as the uncertainty principle. Heisenberg showed that in any quantum-mechanical system, pairs of dynamical variables for particles can be simultaneously and sharply defined only if their operators commute. This means only if their operators H and K satisfy the equation

$$HK - KH = 0. \tag{5}$$

In cases where Eq. (5) is not satisfied, some uncertainty will always be associated with the variables. If the uncertainty of the first variable is denoted as ΔV_1 and that of the second variable as ΔV_2, the product of these two uncertainties obeys a relationship of the form

$$\Delta V_1 \cdot \Delta V_2 \geq h/2\pi, \tag{6}$$

where h is Planck's constant (6.626×10^{-34} Js). All of the pairs here must have the dimension of action, i.e., energy $\times$ time, when multiplied together. The most frequently cited pair of variables that satisfies Eq. (6) is the position and momentum, i.e., mass $\times$ velocity, of a quantum particle. For this particular pair, Eq. (6) becomes

$$\Delta x \cdot \Delta p_x \geq h/2\pi, \tag{7}$$

where x represents the position of the particle and p_x represents its momentum in the x direction. Another such pair of variables is, of course, energy and time. Further analysis of this principle is to be found in an article by Lieb.[52]

The interpretation of the uncertainty principle has remained a highly contentious issue. Its earliest interpreter, the physicist Niels Bohr, adopted a positivistic approach, i.e., an approach that focused on the relationships that exist between measurable quantities. Bohr envisaged noncommuting operators as characterizing so-called complementary variables. By this latter term, Bohr meant variables that are (1) always taken together as pairs since both are needed for a complete description of the experimental results, and (2) mutually exclusive in the sense that any sharpening of the value of one of the variables automatically leads to a greater uncertainty of the other variable in the pair. The wave and particle description of quantum entities may be seen as one example of complementary descriptions of matter. Bohr maintained that, because quantum mechanics is inherently statistical in nature, it is not possible to know reality as it actually is. The most that we can know is what appears to us when we become part of some experiment performed on reality. In the light of this interpretation, even the foregoing statement of the uncertainty principle needs to be revised. In that statement, reference was made to particles, although it is not strictly valid to imply that what we are observing experimentally is particulate in character. Thus, a better formulation of the principle would be: There exists no predictive law that makes reference to the simultaneous position and momentum of any quantum entity. Although Bohr's view still prevails as the orthodox interpretation of quantum mechanics, it is fair to point out that it has been subjected to intense criticism since it was first put forward.[53, 54] General agreement on the meaning of quantum ideas has yet to be reached.[55]

V. CHAOTIC PHENOMENA

In addition to the occurrence of random events and the inherent uncertainties in quantum mechanics, both of which are interpreted in terms of probability theory, there is another kind of uncertainty that can arise in fully determined systems. This rather surprising variety of uncertainty is now generally referred to as chaos; more strictly it should be described as deterministic chaos. In the case of probabilistic theories, chance occurrences are used to generate some sort of determinism for the system under study, whereas in the case of chaos theory, it is the deterministic nature of the system itself that is responsible for the uncertainty. Until the end of the nineteenth century it was generally believed that fully determined systems would always exhibit predictable behavior and that they would not be capable of displaying random behavior. Not surprisingly, therefore, any instances of such behavior tended to be ignored and go unnoticed. The first person to take seriously the possibility of chaotic phenomena in deterministic physical systems was the French mathematician Poincaré. In a thorough analysis[56] of the so-called three-body problem, which involved an investigation of the trajectories of three gravitationally interacting heavenly bodies, he arrived at an epoch-making discovery. Even though his initial equations were fully defined and the variables he employed (such as the positions and times pertaining to the bodies) were properly specified, he found that there was no way of reliably predicting the behavior of these bodies far into the future. Poincaré had demonstrated that there was indeed scope for complexity and unpredictability in supposedly deterministic Newtonian systems.

Poincaré had in fact uncovered the existence of chaos, a phenomenon so out of keeping with the tenor of his time that it was largely ignored for the next half century. The modeling of physical reality at the time was heavily based on the use of differential equations and in this respect Poincaré followed the trend. The use of such equations was no accident because it was commonly assumed that the smooth and continuous solutions obtained from these equations accurately reflected nature, which was deemed to be continuous and interpretable in terms of Euclidean geometry. The solutions obtained from such equations were collectively described as "classical attractors," by which was meant that the solutions could be characterized either as stable point equilibria, stable limit cycles, or stable tori in phase space.[57] A whole host of problems, ranging from the motion of a simple pendulum to the movements of the planets around the Sun, were considered to have been solved by means of these equations. Eventually, however, it became increasingly apparent that a variety of problems could not be satisfactorily resolved in this comparatively straightforward way. One example was the three-body problem tackled by Poincaré; another was that of the occurrence of turbulence in fluids with a high

Reynolds number. Such nonlinear problems tended to remain unresolved because there was no mathematical apparatus appropriate for them; each of them seemed to be a law unto itself.

The classical attractors and stable solutions supposedly characteristic of Newtonian systems were eventually replaced in chaotic systems by so-called *strange attractors* and unstable solutions. The concept of the strange attractor was first put forward[58] in 1971. From a geometrical standpoint, this attractor may be viewed as a continuum of points. However, the arrangement of the points differs from that in smooth curves or tori and assumes the form of a fragmented or, more accurately, a fractal pattern. Such patterns represent motion in the system that is either marginally or completely unstable. This means that virtually identical initial sets of conditions for the system will produce very different system dynamics in terms of the patterns traced out over a significant period of time. The instability arises from the fact that the differential equations used to model physical systems contain within them inherent sensitivities that become apparent when the input variables take on certain values. The equations are thus ill-conditioned in the sense that, for certain values of the input variables, solving such equations in terms of algebraic series will lead to divergent rather than convergent values of the output variables with respect to time. Thus, depending on the initial conditions, the dynamics of physical systems can be either stable or unstable. The extreme sensitivity to initial conditions that characterizes chaotic behavior implies that even tiny differences in the starting conditions will lead to totally different outcomes in the time evolution of the system. This in turn renders it impossible, in general, to predict events in the physical world with certainty.

Whatever quantitative mathematical model we ultimately adopt to describe a physical system, there will always inhere within that system a component of irreducible uncertainty. Consider, for instance, systems modeled in terms of Newtonian mechanics. We now know that such systems are far richer mathematically than was originally thought because they all have novel, chaotic solutions that have long remained unsuspected. Indeed, it has even been remarked[59] that mechanical systems are hardly the place to look for clockwork-like stability since all of them will exhibit chaotic behavior under appropriate conditions. This latter type of behavior is usually ascribed to the presence of random factors in the system and so it is generally considered that it can be dealt with by means of statistical analysis.[57] Examples of chaotic behavior are ubiquitous and are to be found in systems ranging in size from those of astronomic proportions to those of atomic dimensions. Detailed studies on the dynamics of the solar system have revealed,[60–62] for instance, that this system is in a chaotic state with the separation doubling every 3.5 million years between the paths of two solar systems starting from slightly different conditions.[62]

Such calculations also show that, fortunately for us, the Earth will probably stay at roughly the same distance from the sun for the next 100 million years, although this cannot be established with certainty! As one worker[63] in this field observed, "[c]haos appears to be a common feature of planetary systems...[so] it's not surprising that our real solar system is chaotic. It would be surprising if it weren't."

Chaos is not only a common feature in the astronomical realm; examples of it abound in all kinds of physical systems. In systems typically studied by physicists, one encounters, for instance, the erratic motion of pendulums subjected to forced oscillation or the unpredictable movements in fluids in the turbulent state.[64] Such systems are described as dissipative systems, that is, systems in which mechanical energy is gradually dissipated as heat. Turbulence is the highly irregular flow patterns observed in fluids that have a random, rotational component in their motion. This kind of flow cannot be said to be fully understood at present; indeed, it represents one of the most challenging problems in contemporary physics. Turbulent convection in the atmosphere is largely responsible for the weather patterns on our planet. However, these are so complex that they make weather forecasting impossible for more than a few days ahead.[65] In the world of chemistry, many examples of chaotic behavior have been discovered in systems that exhibit nonlinearity and feedback,[66] one notable example being that of the chemical oscillations in reactions such as the Belousov–Zhabotinsky reaction. Similar reactions also occur in the biological domain,[67] examples being the oscillating reactions associated with glycolysis or the heartbeat—both of which can exhibit chaos. Indeed, life itself might be viewed as an island of stability in a sea of chaos. It is thus hardly surprising that our world has been characterized[68] as "a world of instabilities and fluctuations, which are ultimately responsible for the amazing variety and richness of the forms and structures we see in nature."

VI. THE APPROACH TO UNCERTAINTY

The ubiquitous presence of uncertainty that we have encountered in preceding sections has given rise to the more mellow view that the goal of acquiring absolutely certain knowledge from any of the sciences is an unattainable one. Gone are the days when we might have expected to attain certitude about our world through the diligent application of scientific method. The fact is that there are just too many inherent uncertainties to make this feasible. The reluctant acceptance of this new view of reality has dramatically impacted our thinking in the present century and has in turn prompted the search for ways and means of coming to terms with the new situation. There have been numerous attempts to understand the origin of the problem. We might even say that it has been one of the foremost problems to be confronted in the twentieth century.[69] Even the question of whether it is the world or our description of it that is uncertain has not been finally resolved. As long ago as 1923, Russell averred[17] that "[a]part from representation…there can be no such thing as vagueness or precision; things are what they are, and there is an end of it." The currently emerging consensus would appear to support Russell. It has recently been claimed[70] that "[v]agueness is indeed one manifestation of the fact that our classifications are not fixed by natural boundaries…. The cause of our ignorance is conceptual."

Russell[17] was also of the opinion that (1) all language is vague (and this includes scientific language), (2) there exists a hierarchy of vagueness, (3) vagueness will necessarily invalidate the methods of classical logic, and (4) vagueness is not the same thing as generality. We start by saying something about the first two of these propositions. The notion that all language is vague follows directly from the classification problem alluded to previously. Starting with the philosopher Kant, it has been repeatedly pointed out[71] that the categories we base our thinking on are socially mediated and cannot be transcended to afford a view of the world as it really is. The second of the propositions of Russell suggests that it should be possible to classify vagueness itself. Such classifications can in fact be traced back all the way to Socrates, who attempted to define the various kinds of certain knowledge we can acquire. Here, however, we shall focus only on very modern work. Smets[72] suggested that there are three varieties of ignorance that could be subsumed under the three broad categories of incompleteness, imprecision, and uncertainty. The first of these refers to data that are lacking, the second to data that are not known to the precision required, and the third to data that are subjective, i.e., determined by human opinion. Smithson,[73] on the other hand, believed that there were only two basic kinds of ignorance, namely, error and irrelevance, although these could be subdivided into a host of other categories as shown in Fig. 1. Here error refers to a cognitive state of ignorance, whereas irrelevance pertains to the act of ignoring data. For applications in the field of artificial intelligence studies, Krause and Clark[74] proposed that uncertainty be divided up as shown in Fig. 2. The two basic kinds of uncertainty are now unary (pertaining to single propositions) and set-theoretic (pertaining to sets of propositions). Clearly, the various typologies that have been put forward are intended to serve differing purposes. There seems to be no overall agreement on how uncertainty should be analyzed.

It is this fact perhaps more than any other that would appear to account for the astonishingly uniform and consistent approach to uncertainty over the past three centuries.[75] Following the studies of the mathematicians Cardan in the seventeenth century and Laplace and Pascal in

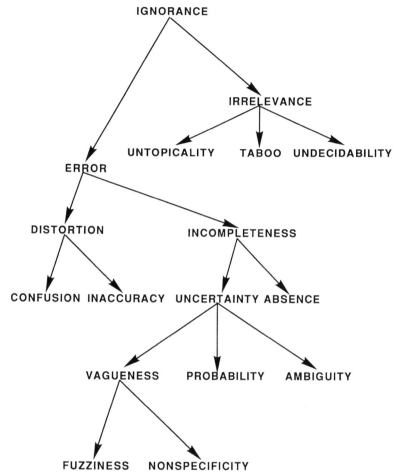

FIGURE 1 An attempted logical typology of uncertainty put forward by Smithson.[73] He proposed starting from the notion of ignorance and divided it into the notions of error and irrelevance, both of which he further subdivided into several other notions.

the eighteenth century,[35] there has existed until comparatively recent times a broad consensus that all problems associated in any way with uncertainty should be tackled by means of probabilistic analysis. Numerous examples could be cited. The nineteenth century witnessed the development of statistical mechanics and statistical thermodynamics, both of which were based on the probabilistic behavior of large ensembles of particles. In the present century, the radioactive decay of atoms has been characterized with the calculus of probabilities as has the position assumed by an electron, or indeed any subatomic species, within an atom or molecule. More recently, the growth of interest in the manifestations of

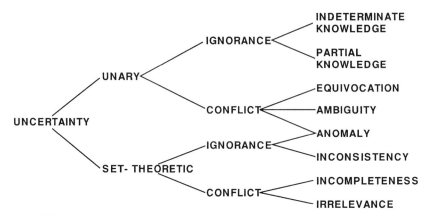

FIGURE 2   An attempted classification of uncertainty in the field of artificial intelligence studies put forward by Krause and Clark.[74] This work is c entered on the notion of uncertainty, unlike that of Smithson, who started from ignorance. Uncertainty is divided basically into unary and set-theoretic aspects, i.e., into individual propositions or to sets of propositions.

chaos has again led to reliance on probability theory. Thus, the motion exhibited by systems in a chaotic state has been interpreted as random motion, although it should be mentioned here that some authors, e.g., Ruhla,[76] have somewhat artificially attempted to differentiate between chaotic and random behavior. The distinction they draw is that chaos arises in systems governed by only a few variables, whereas random behavior is determined by many variables. Such a distinction is, of course, purely a semantic one since in the end both types of system are described in the same language of probability. One is reminded here of the perceptive remark of the American psychologist Maslow, who declared[77] that "it is tempting, if the only tool you have is a hammer, to treat everything as if it were a nail."

There has been considerable resistance even to the thought of introducing new, nonprobabilistic ways of characterizing uncertainty. Indeed, some proponents of traditional probability theory have begun to defend their fiefdom with such vehemence that we may now be said to be entering an era of turf wars in this domain. Thus, Kandel[78] has claimed that probability is "an objective characteristic; the conclusions of probability can, in general, be tested by experience." Lindley[79] went even further and maintained that "probability is the only sensible description of uncertainty and is adequate for all problems involving uncertainty." This line of reasoning eventually led him to issue "a challenge ... that anything that can be done by [other] methods can be better done with probability." In spite of such provocative pronouncements, however, the last word on the subject has not been spoken. Indeed, the search for nonprobabilistic ways to represent reality is rapidly forging ahead and the shortcomings of the traditional methods are being increasingly pointed out. Obviously, there is much life left yet in this controversy and much heated debate on the subject is to be expected over the next several decades. It will doubtless take that long before the matter is satisfactorily resolved. In the words of Howson,[80] "[i]t would be foolhardy to predict that philosophical probability has entered a final stable phase.... It would also probably be incorrect to pretend that there is likely in the near future to be any settled consensus as to which interpretations of probability make viable and useful theories and which are dead ends."

VII. MULTIVALUED LOGICS

Let us now explore some of the ideas put forward in the twentieth century that were to create so much controversy. Even before the beginning of this century, doubts had begun to be expressed about the validity of much of our scientific logic and thought processes. There was, for instance, the question of the objectivity of the concept of probability. As so neatly later summed up by Pauli,[81] "we here encounter the theme of the *one* expectation to be rendered objective and the *many* events. A closer analysis of this state of affairs is not easy; in particular, the transition from the logicomathematical formulation to experience raises profound epistemological problems." Another issue was that of vagueness and the means of coping with vague statements. As mentioned in the previous section, Russell[17] had pointed out that vagueness must perforce invalidate the methods of classical logic. Every statement that is framed in practice will be vague to some extent because it uses words that are themselves vague. According to Russell,[17] therefore, for any statement we might formulate there will not be "one definite fact necessary and sufficient for its truth, but a certain region of possible facts, any one of which would make it true." Moreover, he contended, we are not even able to assign a definite boundary to this region. Such arguments left logicians and philosophers with little alternative but to reexamine carefully the foundations of traditional methods of arriving at logical conclusions.

The origins of this process of reexamination can be traced back at least as far as the writings of the logician George Boole in 1854. In his now famous book on the laws of logic,[82] Boole used as his starting point three of the fundamental axioms used by Aristotle.[83] These axioms upon which Aristotelian logic rests were generally thought to be so self-evident that they required no proof. The three axioms, all now referred to as laws, are (1) the law of contradiction, which states that A cannot be both B and not B; (2) the law of the excluded middle, which states that A must be either B or not B; and (3) the law of identity, which states that A will always be A. In his book,[82] Boole sought to "exhibit the real nature of the ancient doctrine and to remove one or two prevailing misapprehensions respecting its extent and sufficiency." There is thus a suggestion that Aristotelian logic may not be sufficient in all cases. Boole's criticism was, however, very mild and Boole's work involved essentially an elaboration of Aristotelian logic. However, since Boole's time, increasingly strident criticisms of Aristotelian logic have been heard. Thus, the American philosopher Peirce claimed[84] that the vague "might be defined as that to which the principle of contradiction does not apply." British philosophers, such as Ayer[85] and Russell,[17] either rejected or had serious doubts concerning the law of the excluded middle. For a review of other objections up to the 1930s, the reader is referred to the work of Kattsoff.[86]

The growing opposition to Aristotelian logic and, in particular, to the law of the excluded middle, was paralleled by a movement to replace this law with some generally acceptable substitute.[87] Following several pioneering endeavors by early workers to establish a three-valued logic,[88] the foundations of this new kind of logic were finally laid in the 1920s independently by Post[89] and Łukasiewicz.[90-92] The underlying idea was that statements were not necessarily true or false, but[92] "must have a third value different from 0, or the false, and from 1, or the true." This third value was assigned a numerical value of one half by Łukasiewicz and was described as "the possible value." It was not long before Łukasiewicz realized that there was no need to stop at three truth values and indeed he generalized his work to infinite-valued systems.[92] In passing, however, we should mention that Post[89] too had presented a system based on a multivalued logic. The adoption of many truth values lying between the extremes of true and false had the effect of generating a sliding scale of degrees of truthfulness for each statement that could be made. It was a significant breakthrough, for, in the words of Łukasiewicz,[93] "[l]ogic changes from its very foundations if we assume that in addition to truth and falsehood there is also some third logical value or several such values." The first suggestion that multivalued logics might play a role in the physical sciences was made by Zwicky,[94] who insisted that "[f]ormulations of scientific truth intrinsically must be many-valued." A possible application in unifying quantum mechanics and relativity theory was immediately pointed out by Bell;[95] other applications have since been cited by Skala.[96]

With the advent of multivalued logics, the scene was largely set for the emergence of an offshoot field of endeavor that is now commonly described as fuzzy logic or fuzzy reasoning. With the benefit of hindsight, it seems almost inevitable that a discipline of this kind would have to be developed. What is far less obvious, however, is that fuzzy logic, after getting off to a somewhat hesitant start, should have blossomed so dramatically, especially over the past two decades. It seems that the time was ripe, for, by the second half of the twentieth century, several different strands of thought had intermeshed to form a powerful intellectual underpinning for the basic ideas in fuzzy logic. We have already mentioned that all language was seen to be vague and imprecise and that a number of philosophers such as Peirce and Russell had sought to understand the role of language in the characterization of reality. Moreover, remarkable breakthroughs arising from the development of quantum theory had demonstrated that uncertainty was an inherent part of the world that had to be taken into account. Perhaps the most important strand providing support for fuzzy logic, however, was the growth of multivalued logics. The latter had evolved because of (1) a widespread recognition that Aristotelian logic was inappropriate for most real-world situations; (2) a steady movement away from Aristotelian logic and its gradual replacement with a variety of multivalued logics; and (3) a new willingness to contemplate the use of alternate, i.e., nonprobabilistic, ways of characterizing uncertainty. The prevailing climate of thought was thus prepared for the appearance of fuzzy logic.

VIII. POSSIBILISTIC PARADIGMS

Even before fuzzy logic made its formal entry on the scene in 1965, a number of workers had anticipated its arrival. In a key paper on the subject of vagueness, published in 1937, the philosopher Black pointed out[98] that "all symbols whose application involves the recognition of sensible qualities are vague" and went on to suggest that "it is possible to define the notion of a ... consistency function, corresponding to each vague symbol and thus to classify, or even, theoretically, to measure, degrees of vagueness." The consistency function was a simple ratio of the numbers of observers who did and did not assign a given object to some class, assuming the numbers were sufficiently large to yield a statistically significant ratio. Black's investigation of the applicability of logic to vague symbols led him to infer that "the ordinary rules for the logical transformation of sets of statements produce conclusions whose degree of vagueness is the same order as those of the premises." The feasibility of constructing "a calculus which will provide a more adequate explication of classes in their scientific use than is afforded by the conventional calculus of classes" was taken up by Kaplan and Schott[99] in 1951. Their approach was interesting because it differed from standard set-theoretical treatments in that it regarded "every individual as an instance to some degree or other of the class, rather than taking some individuals to be instances without qualification and the rest not at all." The degree of membership was indicated by "some real number between 0 and 1." The contributions of several other workers, including those of Hempel and Quine, who also proposed using degrees of membership for classes, have been discussed by Engel.[100]

TABLE I Some of the Key Historical Developments in Our Understanding of Uncertainty That Led to the Emergence of Fuzzy Set Theory and Fuzzy Logic

Author(s)	Date(s)	Outline of contribution(s)
Boole	1854	Suggestion that Aristotelian logic may not be sufficient for all purposes
MacColl	1896	Development of a three-dimensional logic
Peirce	1902–1905	Discussion of vagueness in science and proposal of trichotomic mathematics
Vasilev	1910–1913	Development of a three-valued logic
Łukasiewicz	1917–1920	Development of a well-founded three-valued logic
Post	1921	Independent development of a three-valued logic
Russell	1923	Publication of a key paper on the definition of vagueness
Łukasiewicz	1930	Development of a multivalued logic
Zwicky	1933–1934	Idea of flexibility of scientific truth; use of multi-valued logics in science
Black	1937	Suggestion that degrees of vagueness could be measured by a consistency function
Kaplan & Schott	1951	Development of a calculus of classes with membership in the range 0 to 1
Zadeh	1965	Introduction of the concept of the fuzzy set
Zadeh	1975–1978	Development of possibility theory; fuzzy sets act as elastic constraints on system parameters

It is thus evident that by the end of the first half of this century a number of ideas key to the formation of fuzzy logic were in place. A number of these ideas are listed chronologically in Table. I. The seminal work on the subject, however, did not appear until 1965. In this work Zadeh[101] introduced the concept of the fuzzy set and defined it simply as "a class of objects with a continuum of grades of membership." Zadeh felt[101] that the fuzzy set would be useful for the characterization of classes of objects that were not sets in a strict mathematical sense and that it could provide "a natural way of dealing with problems in which the source of imprecision is the absence of sharply defined criteria of class membership rather than the presence of random variables." He suggested that fuzzy sets be used in essentially the same way as ordinary sets—now generally described as crisp sets—had been, although fuzzy sets[101] would likely have "a much wider scope of applicability, particularly in the fields of pattern classification and information processing." The fuzzy set is clearly a generalization of the crisp set in that each member of a fuzzy set belongs to the set only to some extent, i.e., each member has a grade of membership in the set. This grade takes the form of a membership function for each member and is indicated by a real number on the line segment $[0, 1]$.

We shall not discuss the properties or mathematics of fuzzy sets here, as this topic constitutes the subject matter of Chapter 2. What we will mention, however, is that fuzzy set theory, unlike most other attempts to establish a new approach to uncertainty, gradually gained in acceptance and eventually went on to become a stellar performer.

Broadly speaking, there were two distinct phases in the early development of fuzzy set theory, both of which were initiated by Zadeh. The first phase, which started in 1965, focused on fuzzy classes of objects and the role they had in describing real-world situations. This phase was concerned with applications of fuzzy sets to areas such as fuzzy clustering, fuzzy classification, and interpolative reasoning in the fuzzy control of systems and devices. The second phase began around 1975 and was driven by the idea that fuzzy sets could be regarded as elastic constraints on the possible parameters or states that systems could assume.[102] This line of thinking gave rise to a flourishing field of research known as possibility theory.[103, 104] The key ideas in possibility theory are that fuzzy knowledge of a system imposes an elastic constraint on the system that can be expressed in terms of a possibilistic distribution π, which involves mapping the system onto the line segment $[0, 1]$. Such distributions are then used to describe systems by possibilistic statements, e.g., a mapping to zero would mean a completely impossible situation for the system in question. Thus, fuzzy sets play the same role in possibility theory that measure theory plays in probability theory. Since both approaches characterize uncertainty numerically by making mappings onto the line segment $[0, 1]$, it may be asked wherein they differ. Their fundamental difference has been highlighted by Kosko,[105] who considered objects that either satisfy or do not satisfy Aristotle's law of the excluded middle (discussed in Section VII). For an object A and its opposite A_{opp}, the relationship

$$A \cap A_{\mathrm{opp}} = \varnothing \qquad (8)$$

will be strictly valid when this law of Aristotle holds, but not otherwise. Probability theory deals with objects that satisfy Eq. (8), whereas possibility theory treats objects that do not.

As pointed out by Kosko,[105] probability is concerned with event occurrence, namely, the uncertainty that some event will occur, whereas possibility theory is concerned with event ambiguity, that is, the extent to which some event occurs. The difference between them has been brought out in discussions on the nature of quantum particles such as electrons. Indeed, it might even be claimed that quantum theory was the first area of application of fuzzy reasoning in the physical sciences. Arguments on how particles were best to be described erupted even before the advent of fuzzy logic.[94, 106] The traditional approach originally expounded by Born[107, 108] envisaged the description of particles in terms of a probabilistic distribution in space. For a particle represented by the wave function ψ, the probability of the particle being located in the volume of space dV is $\psi\psi^* dV$; integrated over all space, the probability is of course unity. From a possibilistic perspective, this probability would actually measure the degree to which a deterministic electron occurs in the volume dV. The numerical values involved are the same; only the interpretation of their significance differs. The idea of interpreting the basic notions of quantum theory in terms of possibilistic rather than probabilistic descriptors was first adumbrated in the work of Prugovečki.[109] He pointed out the need for unsharp probability measures in discussing the simultaneous determination of particle variables that satisfy a Heisenberg-type relationship. His measure, which he termed a complex probability measure, was almost identical to fuzzy measures that were later put forward. The advantages of adopting such measures in quantum theory have since been confirmed by a variety of different workers.[110-112] Some recent applications of fuzzy reasoning in the quantum-theoretical domain are presented in Chapter 4.

Applications of fuzzy reasoning to areas other than quantum theory began to be made in the 1970s. The first physicochemical concept to be fuzzified was that of chirality. Along with the more specific concepts of homochirality and heterochirality, chirality was shown[113] to be "a primitive fuzzy concept." Mislow and Bickart[113] pointed out that, although chirality is sharply defined when applied to the idealized shapes and figures encountered in Euclidean geometry, when applied to real objects such as molecules the concept assumes a fuzzy character. This is because the position of any real-world system along the chiral–nonchiral continuum will always be a function of the time scale of measurement and the resolving power of the instrument used for the measurement. Accordingly, it is permissible to speak of degrees of chirality and to compare molecular systems that may be more or less chiral than other systems. A full account of this and related work is found in Chapter 3. The concepts of molecular structure and molecular symmetry are also fuzzy in nature. This was established by Maruani and Mezey,[114-117] who replaced the idealized geometry used to describe the nuclear framework of molecular species by a more realistic set of fuzzy nuclear geometries. Just as fuzzification of chirality did not necessarily destroy the rank ordering of molecular species, so the fuzzy nuclear geometries retained some of the group-theoretical relations that hold for the idealized molecular geometry case. This fact made it possible to define fuzzy measures for both molecular structure and molecular symmetry. The relevant details are presented in Chapter 5 along with a discussion of fuzzy measures for molecular shape.

IX. CONCLUDING REMARKS

We began our discourse on uncertainty in the physical sciences with the observation that complete certainty about anything in our world is an unattainable goal. This means, of course, that all our statements, theories, and laws pertaining to physical reality will be uncertain to a degree. In spite of this fact, however, scientific pronouncements are generally accepted as valid—especially by those not directly involved with the area of science concerned! Uncertainty was formally admitted into science over 300 years ago when studies began to be made on the outcomes of games of chance. Since that time, probability arguments have remained with us and have been employed in the description and characterization of myriad scientific systems. The great durability of the concept of probability is clearly due to its serving us so well in so many different contexts. As probabilistic notions were gradually elaborated into a more fully fledged statistical theory, the latter was increasingly employed to model supposedly random phenomena. Thus, in the nineteenth century the radioactive decay of atoms and the behavior of ensembles of gas molecules confined to a box were modeled in this way, and in the present century the position of quantum particles in space has been given a probabilistic interpretation. Even the manifestations of chaotic behavior in physical systems have been explained in terms of probabilistic arguments. Probability reigned supreme as the only meaningful way to understand and characterize uncertainty until the beginning of the twentieth century. It might almost be said[118] that here was an instance of our becoming "so dazzled by the concepts we invent that we can see the world only through their glare."

Around the turn of the century, however, several totally unexpected developments in the physical sciences occurred that had the effect of calling into question the hitherto unchallenged supremacy of probabilistic reasoning. Three such developments were (1) the founding of quantum mechanics, which showed that uncertainty is inherent in our world even at the subatomic level, (2) the discovery that even deterministic systems could exhibit chaotic behavior, and (3) the realization that traditional Aristotelian logic had its limitations when applied to real-world systems. The latter development was particularly significant from our standpoint because it ushered in first three-valued logic and eventually many-valued logics. The world was no longer being viewed in strictly black and white terms, all the shades of gray were now being included. In this new kind of intellectual environment it was only a matter of time until fuzzy logic would make its appearance on the scene. Fuzzy logic, which may be viewed as an infinite-valued logic, generalized Aristotelian logic just as fuzzy sets had generalized crisp sets. More to the point, it soon began to demonstrate its efficacy in describing and predicting the behavior of the real world. Starting with quantum-theoretical concepts, fuzzy reasoning quickly manifested its suitability for the description of physicochemical concepts. Moreover, with its now numerous successful applications within the physicochemical realm, several examples of which are explored elsewhere in this volume, possibilistic reasoning may be said to have established itself as the principal challenger to probabilistic reasoning in the description of our uncertain world.

REFERENCES

1. B. Russell, *Our Knowledge of the External World*, p. 75. Allen & Unwin, London, 1914.
2. E. S. Haldane and G. R. T. Ross (transl.), *The Philosophical Works of Descartes*, Vol. 1, p. 91. Cambridge Univ. Press, Cambridge, UK, 1911.
3. E. S. Haldane and G. R. T. Ross (transl.), *The Philosophical Works of Descartes*, Vol. 1, p. 101. Cambridge Univ. Press, Cambridge, UK, 1911.
4. E. S. Haldane and G. R. T. Ross (transl.), *The Philosophical Works of Descartes*, Vol. 1, p. 101. Cambridge Univ. Press, Cambridge, UK, 1911. What Descartes actually wrote was in French: *Je pense, donc je suis*. This can also be translated as: I am thinking, therefore I am.
5. P. Gassendi (1641), cited in E. S. Haldane and G. R. T. Ross (transl.), *The Philosophical Works of Descartes*, Vol. 2, p. 137. Cambridge Univ. Press, Cambridge, UK, 1911.
6. L. Wittgenstein, *On Certainty*, para. 115. Harper, New York, 1969.
7. A. Eddington, *New Pathways in Science*, p. 311. Cambridge Univ. Press, Cambridge, UK, 1935.
8. M. B. Hesse, *Models and Analogies in Science*, p. 101. Univ. of Notre Dame Press, Notre Dame, IN, 1966.
9. D. H. Rouvray, *J. Chem. Inf. Comput. Sci.* **34**, 446 (1994).
10. A. J. Ayer, *The Problem of Knowledge*, pp. 41, 73. Macmillan & Co., London, 1965.
11. J. L. Casti, *Searching for Certainty*, p. 31. Morrow, New York, 1991.
12. W. A. Wallace, *Prelude to Galileo*, p. 145. Reidel, Dordrecht, 1981.
13. E. Schrödinger, *Naturwissenschaften* **23**, 807, 823, 844 (1935).
14. A. Einstein, B. Podolsky, and N. Rosen, *Phys. Rev.* **47**, 777 (1935).
15. P. Davies, *New Sci.* **146**, 26 (1995).
16. A. Einstein, *The World as I See It*, p. 36. Covici Friede, New York, 1934.
17. B. Russell, *Australasian J. Philos. Psychol.* **1**, 84 (1923).
18. T. Williamson, *Vagueness*, p. 270. Routledge, London, 1994.
19. D. Oldroyd, *The Arch of Knowledge*, p. 369. Methuen, New York, 1986.
20. G. Berry, *Synthese* **19**, 215 (1968).
21. L. Nottale, *Fractal Space–Time and Microphysics*, Chap. 3. World Scientific, Singapore, 1993.
22. A. Einstein, *Geometrie und Erfahrung*, p. 3. Springer, Berlin. What Einstein actually wrote was in German: *Insofern sich die Sätze der Mathematik auf die Wirklichkeit beziehen sind sie nicht sicher, und insofern sie sicher sind, beziehen sie sich nicht auf die Wirklichkeit.*
23. K. Gödel, *On Formally Undecidable Propositions* (B. Meltzer, transl.), p. 37. Basic Books, New York, 1962.
24. J. L. Casti, *Searching for Certainty*, p. 401. Morrow, New York, 1991.
25. J. R. Lucas, *Philosophy* **36**, 112 (1961).
26. J. A. Kourany (ed.), *Scientific Knowledge*, p. ix. Wadsworth, Belmont, CA, 1987.
27. B. van Fraassen, *in Scientific Knowledge* (J. A. Kourany, ed.), p. 343. 1987.
28. H. J. Priestley, *Australasian J. Philos. Psychol.* **1**, 208 (1923).
29. P. W. Atkins, *in Nature's Imagination* (J. Cornwell, ed.), p. 122. Oxford Univ. Press, London, 1995.
30. F. Dyson, *in Nature's Imagination* (J. Cornwell, ed.), p. 1. Oxford Univ. Press, London, 1995.
31. M. N. Wise (ed.), *The Values of Precision*. Princeton Univ. Press, Princeton, NJ, 1995.

32. A. Lavoisier, *Elements of Chemistry* (R. Kerr, transl), p. xviii, Dover, New York, 1965. The original, *Traité élémentaire de Chimie*, was published in Paris in 1789.

33. J. C. Maxwell, in *The Scientific Letters and Papers of James Clerk Maxwell* (P. M. Harman, ed.), Vol. 1, p. 425. Cambridge Univ. Press, Cambridge, UK, 1990.

34. T. M. Porter, *Trust in Numbers*, p. 200. Princeton Univ. Press, Princeton, NJ, 1995.

35. J. Hoffmann-Jørgensen, *Probability with a View toward Statistics*, Vol. 1, Chap. 1. Chapman and Hall, New York, 1994.

36. A. N. Kolmogorov, *Foundations of the Theory of Probability* (N. Morrison, transl.), Chelsea, New York, 1950. The original, *Grundbegriffe der Wahrscheinlichkeitsrechnung*, was published in Berlin in 1933.

37. A. J. Ayer, *The Problem of Knowledge*, p. 78. Macmillan & Co., London, 1965.

38. C. S. Sharma, in *Modelling Under Uncertainty* (S. B. Jones and D. G. S. Davies, eds.), p. 151. Institute of Physics, Bristol, UK, 1986.

39. C. M. Lederer and V. S. Shirley (eds.), *Table of Isotopes*, 7th ed. Wiley-Interscience, New York, 1978.

40. L. A. Girifalco, *Statistical Physics of Materials*, Chap. 1, p. 4. Wiley-Interscience, New York. 1973.

41. M. H. Everdell, *Statistical Mechanics and its Chemical Applications*, Chap. 1, p. 21. Academic Press, London, 1975.

42. N. Cartwright, *How the Laws of Physics Lie*, p. 154. Clarendon Press, Oxford, 1983.

43. P. Urbach, *British J. Philos. Sci.* **43**, 311 (1992).

44. C. Juhl, *Philos. Sci.* **60**, 302 (1993).

45. S. Youssef, *Modern Phys. Lett. A* **6**, 225 (1991).

46. G. Gamow, *Thirty Years that Shook Physics: The Story of Quantum Theory*. Doubleday, New York, 1966.

47. R. Feynman, *The Character of Physical Law*, p. 129. MIT Press, Cambridge, MA, 1967.

48. J. Honner, *The Description of Nature*, p. 3. Clarendon, Oxford, 1987.

49. Quoted by J. Gerber, *Arch. Hist. Exact Sci.* **5**, 349 (1969). What Heisenberg actually said was in German: *Es ist wirklich meine Überzeugung, dass eine Interpretation der Rydberg-Formel im Sinne von Kreis und Ellipsenbahnen in klassischer Geometrie nicht den geringsten physikalischen Sinn hat ... .*

50. W. Heisenberg, *Z. Phys.* **43**, 172 (1927).

51. P. A. M. Dirac, in *Directions in Physics* (H. Hora and J. R. Shepanski, eds.), p. 10. Wiley, New York, 1975.

52. E. Lieb, in *Encyclopedia of Physics* (R. Learner and G. Trigg, eds.), p. 1078. Addison-Wesley, Reading, MA, 1981.

53. M. Bunge, *British J. Philos. Sci.* **6**, 141 (1955).

54. S. Y. Auyang, *How is Quantum Field Theory Possible?* Oxford Univ. Press, London, 1995.

55. W. C. Price and S. S. Chissick (eds.), *The Uncertainty Principle and Foundations of Quantum Mechanics: A Fifty Years' Survey*. Wiley, New York, 1977.

56. H. Poincaré, *Acta Math.* **13**, 1 (1890).

57. J. L. McCauley, *Chaos, Dynamics and Fractals: An Algorithmic Approach to Deterministic Chaos*. Cambridge Univ. Press, Cambridge, UK, 1993.

58. D. Ruelle and F. Takens, *Commun. Math. Phys.* **20**, 167 (1971).

59. M. Wilson, *Philos. Sci.* **56**, 502 (1989).

60. J. Moser, *Math. Intell.* **1**, 65 (1978).

61. I. Peterson, *Newton's Clock: Chaos in the Solar System*, Chap. 7. Freeman, New York, 1993.

62. J. A. Laskar, *Nature* **338**, 237 (1989).

63. G. D. Quinlan and S. Tremaine, *Astronom. J.* **100**, 1694 (1990).

64. C. Ruhla, *The Physics of Chance: From Blaise Pascal to Niels Bohr* (G. Barton, transl.), Chap. 6. Oxford Univ. Press, London, 1992.

65. D. Ruelle, *Math. Intell.* **2**, 126 (1980).
66. S. K. Scott, *Chemical Chaos*. Clarendon, Oxford, 1991.
67. R. J. Field and L. Györgyi (eds.), *Chaos in Chemistry and Biochemistry*. World Scientific, Singapore, 1993.
68. G. Nicolis and I. Prigogine, *Exploring Complexity*: *An Introduction*, p. ix. Freeman, New York, 1989.
69. M. Smithson, *Ignorance and Uncertainty*: *Emerging Paradigms*, p. viii. Springer-Verlag, New York, 1989.
70. T. Williamson, *Vagueness*, p. 269. Routledge, London, 1994.
71. D. Oldroyd, *The Arch of Knowledge*, p. 347. Methuen, New York, 1986.
72. P. Smets, *Inform. Sci.* **57**, 135 (1991).
73. M. Smithson, *Ignorance and Uncertainty*: *Emerging Paradigms*, Springer-Verlag, New York, 1989.
74. P. Krause and D. Clark, *Representing Uncertain Knowledge*: *An Artificial Intelligence Approach*, p. 7. Kluwer, Dordrecht, The Netherlands, 1993.
75. M. Kline, *Mathematics in Western Culture*, Chap. 23. Oxford Univ. Press, London, 1953.
76. C. Ruhla, *The Physics of Chance*: *From Blaise Pascal to Niels Bohr* (G. Barton transl.), p. 127. Oxford Univ. Press, London, 1992.
77. A. H. Maslow, *The Psychology of Science*: *A Reconnaissance*, p. 15. Harper & Row, New York, 1966.
78. A. Kandel, *Fuzzy Mathematical Techniques with Applications*, p. 74. Academic Press, San Diego, 1986.
79. D. V. Lindley, *Statist. Sci.* **2**, 17 (1987).
80. C. Howson, *British J. Philos. Sci* **46**, 1 (1995).
81. W. Pauli, *Dialectica* **8**, 112 (1954).
82. G. Boole, *An Investigation of the Laws of Thought*, Chap. 15, p. 226. Walton and Maberley, London, 1854.
83. R. McKeon (ed.), *The Basic Works of Aristotle*, p. 87. Random House, New York, 1941.
84. C. Hartshorn and P. Weis (eds.), *Collected Papers of Charles Sanders Peirce*, Vol. 5, Sect. 5.505. Harvard Univ. Press, Cambridge, MA, 1935.
85. A. J. Ayer, *The Origins of Pragmatism*, p. 115. Freeman, New York, 1968.
86. L. Kattsoff, *Philos. Sci.* **1**, 149 (1934).
87. N. Rescher, *Topics in Philosophical Logic*, Chap. 6, p. 54. Reidel, Dordrecht, 1968.
88. P. Rutz, *Zweiwertige und Mehrwertige Logik*, p. 23. Ehrenwirth, Munich, 1973.
89. E. Post, *Amer. J. Math.* **43**, 163 (1921).
90. J. Łukasiewicz, *Ruch Filozoficzny* **5**, 169 (1920).
91. J. Łukasiewicz and A. Tarski, *Soc. Sci. Lett. Varsovie, Classe III* **23**, 30 (1930).
92. J. Łukasiewicz, *C.R. Soc. Sci. Lett. Varsovie, Classe III* **23**, 51 (1930).
93. J. Łukasiewicz, *in Selected Works* (L. Borkowski, ed.), p. 246. North-Holland, Amsterdam, 1970.
94. F. Zwicky, *Phys. Rev.* **43**, 1031 (1933).
95. E. T. Bell, *Phys. Rev.* **43**, 1033 (1933).
96. H. J. Skala, *Fuzzy Sets Syst.* **1**, 129 (1978).
97. C. S. Peirce, *in Dictionary of Philosophy and Psychology* (J. M. Baldwin, ed.), p. 748. Macmillan Co., New York, 1905.
98. M. Black, *Philos. Sci.* **4**, 427 (1937).
99. A. Kaplan and H. F. Schott, *Methodos* **3**, 165 (1951).
100. R. E. Engel, *J. Philos.* **86**, 23 (1989).
101. L. A. Zadeh, *Inform. Control* **8**, 338 (1965).
102. L. A. Zadeh, *Fuzzy Sets Syst.* **1**, 3 (1978).
103. D. Dubois and H. Prade, *Possibility Theory*. Plenum, New York, 1988.

104. G. de Cooman, D. Ruan, and E. E. Kerre, *Foundations and Applications of Possibility Theory*. World Scientific, Singapore, 1995.
105. B. Kosko, *Internat. J. Gen. Systems* **17**, 211 (1990).
106. H. Margenau, *Philos. Sci.* **1**, 118 (1934).
107. M. Born, *Z. Phys.* **37**, 863 (1926); **38**, 803 (1926).
108. M. Beller, *Stud. Hist. Philos. Sci.* **21**, 563 (1990).
109. E. Prugovečki, *Canad. J. Phys.* **45**, 2173 (1967).
110. R. Giuntini, *Found. Phys.* **20**, 701 (1990).
111. G. Cattaneo and F. Laudisa, *Found. Phys.* **24**, 631 (1994).
112. A. J. van der Wal, cited in G. de Cooman, D. Ruan, and E. E. Kerre, *Foundations and Applications of Possibility Theory*, p. 234. World Scientific, Singapore, 1995.
113. K. Mislow and P. Bickart, *Israel J. Chem.* **15**, 1 (1977).
114. J. Maruani and P. G. Mezey, *C.R. Acad. Sci. Paris Ser. II* **305**, 1051 (1987); **306**, 1141 (1987).
115. P. G. Mezey and J. Maruani, *Mol. Phys.* **69**, 97 (1990).
116. J. Maruani and P. G. Mezey, *J. Chim. Phys.* **87**, 1025 (1990).
117. P. G. Mezey and J. Maruani, *Int. J. Quantum Chem.* **45**, 117 (1993).
118. G. Johnson, *Fire in the Mind*, p. 112. Knopf, New York, 1995.

2

From Classical Mathematics to Fuzzy Mathematics: Emergence of a New Paradigm for Theoretical Science

GEORGE J. KLIR

Center for Intelligent Systems
Binghamton University, SUNY
Binghamton, New York 13902

I. INTRODUCTION

The purpose of this chapter is to examine a major paradigmatic change that is currently taking place in science—a change that concerns the role of uncertainty in science. This change is manifested by a transition from the traditional attitude toward uncertainty in science, according to which uncertainty is undesirable and the ideal is to eliminate it, to an alternative attitude, according to which uncertainty is fundamental and its avoidance is counterproductive.

The traditional attitude toward uncertainty is well characterized by the following statement made by William Thomson[34] (known later as Lord Kelvin) in 1883:

> In physical science a first essential step in the direction of learning any subject is to find principles of numerical reckoning and practicable methods for measuring some quality connected with it. I often say that when you can measure what you are speaking about and express it in numbers, you know something about it; but when you cannot measure it, when you cannot express it in numbers, your knowledge is of a meager and unsatisfactory kind: it may be the beginning of knowledge but you have scarcely, in your thoughts, advanced to the state of science, whatever the matter may be.

This statement captures well the spirit of science in the nineteenth century: scientific knowledge should be expressed in precise numerical terms; imprecision and other types of uncertainty do not belong to science.

When uncertainty was viewed as unscientific, there was little motivation to seriously study it. This explains why uncertainty has traditionally been neglected by science.[30] It was only with the emergence of statistical mechanics at the beginning of this century[11] that uncertainty became recognized as useful, or even essential, in certain scientific inquiries. However, this recognition was strongly qualified: uncertainty was conceived solely in terms of probability theory. It took more than a half century to liberate uncertainty from its narrow confines of probability theory and to study its various other (nonprobabilistic) manifestations as well as their utility in science and technology.

The focus in this chapter is on the mathematics pertaining to fuzzy set theory and its role in science. However, other approaches to uncertainty are also briefly introduced.

II. TYPES OF UNCERTAINTY

The classical mathematical theories by which certain types of uncertainty can be expressed are classical set theory and probability theory. In terms of set theory, uncertainty is expressed by any given set of possible alternatives in situations where only one of the alternatives may actually happen. For example, when an interval of values of a variable is predicted by a given mathematical model, the set of values in the interval represents a *predictive uncertainty*; when an unsettled historical question allows a set of possible answers rather than a unique one, the set represents a *retrodictive uncertainty*; when medical diagnosis of a patient results in a set of possible diseases rather than a single disease, the set represents a *diagnostic uncertainty*; when design requirements are specified in terms of sets of alternatives, the sets represent a *prescriptive uncertainty*.

Uncertainty expressed in terms of sets of alternatives results from the nonspecificity inherent in each set. Large sets result in less specific predictions (retrodictions, prescriptions, etc.) than their smaller counterparts. One area of mathematics that deals with this kind of uncertainty is interval analysis.[23]

The second classical theory of uncertainty—*probability theory*—expresses uncertainty in terms of a classical measure on subsets of a given universal set of alternatives.[9, 13] The measure is a function that, according to the situation, assigns a number in the unit interval $[0, 1]$ to each subset of the universal set. This number, called the *probability* of the subset, expresses the likelihood that the unknown unique alternative (the correct prediction, the actual disease of the patient, etc.) is in the subset.

Due to the additivity of classical measures,[13] the probability of each subset is uniquely determined from the probabilities assigned to the smallest nonempty subsets of the universal set, each consisting of exactly one alternative. These subsets are usually called *singletons*. Since singletons are mutually exclusive, any nonzero probabilities assigned to two or more of them conflict with one another. Uncertainty expressed by probability theory is thus based on the conflict among the likelihood claims assigned to singletons.

Classical set theory and probability theory have recently been challenged as inadequate to capture the full scope of uncertainty. The challenge originates from two substantially broader frameworks for dealing with uncertainty, fuzzy set theory[18] and fuzzy measure theory.[36] While fuzzy set theory is a generalization of classical set theory, fuzzy measure theory is a generalization of classical measure theory and, hence, it is also a generalization of probability theory.

As is well known, fuzzy set theory was introduced by Zadeh[37] in 1965. Its objects—fuzzy sets—are sets whose boundaries are not required to be precise; that is, membership in a fuzzy set is not necessarily a matter of affirmation or denial, but it is, in general, a matter of degree. The degrees of membership of elements of a designated universal set in a given set are usually characterized by numbers in the unit interval $[0, 1]$, but various other characterizations are also employed.

From the standpoint of science, as it is still predominantly understood, the idea of a fuzzy set is extremely radical. When accepted, one has to give up classical bivalent logic, generally presumed to be the principal pillar of science. Instead, we obtain a logic in which propositions are not required to be either true or false, but may be true or false to different degrees. As a consequence, some laws of bivalent logic do not hold any more, such as the law of the excluded middle or the law of contradiction. At first sight, this seems to be at odds with the very purpose of science. However, this is not the case. As argued in Section V, scientific methodology is significantly enhanced when the truth of scientific propositions is allowed to be expressed in terms of degrees.

The second broad framework for dealing with uncertainty—fuzzy measure theory—was founded by Sugeno[32] in 1974, even though some basic ideas of fuzzy measures had already been recognized by Choquet[3] in 1953. Fuzzy measure theory is an outgrowth of classical measure theory,[13] which is obtained by replacing the additivity requirement of classical measures with the weaker requirements of monotonicity (with respect to set inclusion) and continuity (or semicontinuity) of fuzzy measures.

Given a universal set X and a nonempty family $\mathscr{C}$ of subsets of X (usually with some algebraic structure), a *fuzzy measure* on $\langle X, \mathscr{C} \rangle$, which is suitable for characterizing uncertainty, is a function

$$g: \mathscr{C} \to [0,1]$$

that satisfies the following requirements:

(g1) $g(\varnothing) = 0$ and $g(X) = 1$ (boundary requirements);

(g2) For all $A, B \in \mathscr{C}$, if $A \subseteq B$, then $g(A) \le g(B)$ (monotonicity);

(g3) For any increasing sequence $A_1 \subset A_2 \subset \cdots$ in $\mathscr{C}$, if $\bigcup_{i=1}^{\infty} A_i \in \mathscr{C}$, then

$$\lim_{i \to \infty} g(A_i) = g\left(\bigcup_{i=1}^{\infty} A_i \right)$$

(continuity from below);

(g4) For any decreasing sequence $A_1 \supset A_2 \supset \cdots$ in $\mathscr{C}$, if $\bigcap_{i=1}^{\infty} A_i \in \mathscr{C}$, then

$$\lim_{i \to \infty} g(A_i) = g\left(\bigcap_{i=1}^{\infty} A_i \right)$$

(continuity from above).

When X is a finite set, requirements (g3) and (g4) are trivially satisfied. Functions g, for which only one of these requirements is satisfied, are called *semicontinuous fuzzy measures*. For any pair $A, B \in \mathscr{C}$ such that $A \cap B = \varnothing$, fuzzy measures are capable of capturing any of the following situations:

(a) $g(A \cup B) > g(A) + g(B)$, which expresses a cooperative action or synergy between A and B;

(b) $g(A \cup B) = g(A) + g(B)$, which expresses the fact that A and B are noninteractive;

(c) $g(A \cup B) < g(A) + g(B)$, which expresses some sort of incompatibility between A and B.

Probability theory is capable of capturing only situation (b), and it is thus severely restrictive.

III. FUZZY SETS AND FUZZY LOGIC: AN OVERVIEW

In each particular application of classical set theory as well as fuzzy set theory, all the sets of concern (classical or fuzzy) are subsets of a fixed set, which consists of all objects relevant to the applications. This set is called a *universal set* and it is always denoted in this chapter by X. To distinguish classical (nonfuzzy) sets from fuzzy sets, the former are referred to as *crisp sets*.

In classical set theory, a common way of defining an arbitrary crisp set A on a given universal set X is to distinguish elements of X that belong to A from those that do not belong to A by the characteristic function of A. This function assigns the number 1 to each element of X that is a member of A; and it assigns the number 0 to the remaining elements of X. It is important to realize that the numbers 1 and 0 are employed here only as convenient symbols and have no numerical significance.

Fuzzy sets are defined on any given universal set by functions analogous to characteristic functions of crisp sets. These functions are called *membership functions*. To define a fuzzy set A on a given universal set X, the membership function of A assigns to each element x of X a number in the unit interval $[0,1]$. This number is viewed as the degree of membership of x in A.

Contrary to the symbolic role of numbers 1 and 0 in characteristic functions of crisp sets, numbers assigned to relevant objects by membership functions of fuzzy sets have a numerical significance. This significance is preserved when crisp sets are viewed (from the standpoint of fuzzy set theory) as special fuzzy sets.

Two distinct notations are most commonly employed in the literature to denote membership functions. In one of them, the membership function of a fuzzy set A is denoted by μ_A and its form is

$$\mu_A \colon X \to [0,1].$$

For each $x \in X$, the value $\mu_A(x)$ is the degree of membership of x in A. In the second notation, the membership function is denoted by A and has, of course, the same form:

$$A \colon X \to [0,1].$$

Again, $A(x)$ is the degree of membership of x in A.

According to the first notation, the symbol of the fuzzy set is distinguished from the symbol of its membership function. According to the second notation, this distinction is not made, but no ambiguity results from this double use of the same symbol since each fuzzy set is completely and uniquely defined by one particular membership function. The second notation is adopted in this chapter; it is simpler and, by and large, more popular in the current literature on fuzzy sets.[18]

An important property of fuzzy sets is their capability to express gradual transitions from membership to nonmembership. This expressive capability has great utility. For example, it allows us to capture, at least in a crude way, the meanings of expressions in natural language, most of which are inherently vague. Crisp sets are hopelessly inadequate for this purpose. However, it is important to realize that meanings of expressions in natural language are strongly dependent on the context within which the expressions are used. For example, the linguistic expression "high temperature" has very different meanings when applied to a patient, a nuclear reactor, or a weather forecast at some place and time. This means that membership functions must be constructed in the context of each particular application.

As an example, several possible membership functions that are reasonable for defining the set of real numbers that are "close to 3" are shown in Fig. 1. Which of these functions captures best the concept "close to 3" depends on the context within which the concept is applied. It turns out, however, that most current applications of fuzzy set theory are not overly sensitive to changes in shapes of the membership functions employed. Since triangular shapes (function A in Fig. 1) and trapezoidal shapes (Fig.

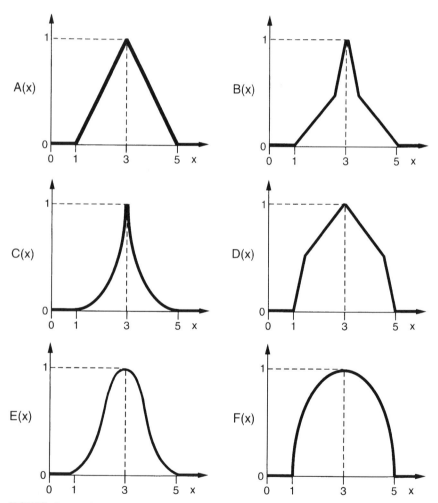

FIGURE 1   Possible membership functions of fuzzy sets of real numbers that are close to 3.

2) are easy to construct and handle, they are predominant in current applications of fuzzy set theory.

The problem of constructing membership functions in the contexts of various applications is not a problem of fuzzy set theory per se. It is a problem of knowledge acquisition, which is a subject of a relatively new field referred to as knowledge engineering.

In some applications, membership functions can easily be determined by clearly recognized prototypes and suitable similarity functions. In other applications, neural networks can be employed for constructing membership functions by learning from data exemplifying the linguistic expression involved. Many other methods for constructing membership functions are described in the literature,[18] but this topic is beyond the scope of this chapter.

A. Basic Concepts of Fuzzy Sets

Given a fuzzy set A defined on X, the largest value of the membership function A for all $x \in X$ is called the *height* of A and it is usually denoted by $h(A)$. When $h(A) = 1$, A is called a *normal fuzzy set*; otherwise, it is called a *subnormal fuzzy set*.

Among the most important concepts associated with fuzzy sets are the concepts of an α-*cut* and a *strong* α-*cut*. Given a fuzzy set A defined on X and a number α in the unit interval $[0, 1]$, the α-cut of A, denoted by $^\alpha A$, is the crisp set that consists of all elements of A whose membership degrees in A are greater than or equal to α; that is,

$$^\alpha A = \{x \mid A(x) \geq \alpha\}.$$

The strong α-cut $^{\alpha+}A$ has a similar meaning but the condition "greater than or equal to" is replaced with the condition "greater than." Hence,

$$^{\alpha+}A = \{x \mid A(x) > \alpha\}.$$

The α-cut ^1A (for $\alpha = 1$) is usually called the *core* of A, while the strong α-cut ^{0+}A (for $\alpha = 0$) is called the *support* of A.

The introduced concepts are illustrated by the trapezoidal membership function A in Fig. 2. This function is defined on the closed interval $[a, b]$ of real numbers (i.e., $X = [a, b]$), which may represent the range of values of a physical variable.

Given any arbitrary fuzzy set A and any two numbers $\alpha_1, \alpha_2 \in [0, 1]$ such that $\alpha_1 < \alpha_2$, it is obvious, by the given definitions, that the set inclusions

$$^{\alpha_2}A \subseteq {}^{\alpha_1}A \quad \text{and} \quad {}^{\alpha_2+}A \subseteq {}^{\alpha_1+}A$$

always hold. This means that α-cuts and strong α-cuts of any fuzzy set form two distinct families of nested crisp sets. It is well established[18] that either of these families uniquely represents the fuzzy set A via the

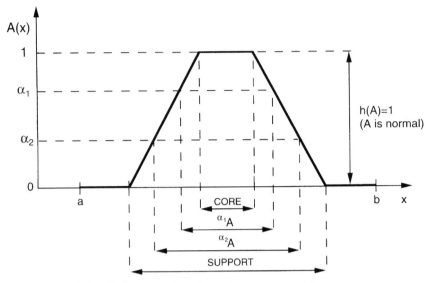

FIGURE 2   Illustration of some basic concepts of fuzzy sets.

formulas

$$A(x) = \sup_{\alpha \in [0,1]} \alpha \cdot {}^{\alpha}A(x)$$

$$= \sup_{\alpha \in [0,1]} \alpha \cdot {}^{\alpha+}A(x)$$

for all $x \in X$. Either of these representations allows us to extend various properties of crisp sets and operations on crisp sets to their fuzzy counterparts. In each extension, a given classical (crisp) property or operation is required to be valid for each crisp set involved in the representation. For example, the classical property of convexity of crisp sets can be extended to fuzzy sets by the requirement that all α-cuts (or strong α-cuts) of a convex fuzzy set be convex in the classical sense.

Any property or operation extended from classical set theory into the domain of fuzzy set theory that is preserved in all α-cuts is called a *cutworthy property or operation*; if it is preserved in all strong α-cuts, it is called a *strong cutworthy property or operation*. It is important to realize that only some properties and operations involving fuzzy sets are cutworthy or strong cutworthy. They are of special significance since they bridge fuzzy set theory with classical set theory. They are like reference points from which other fuzzy properties or operations deviate to various degrees.

Contrary to their classical counterparts, operations on fuzzy sets are not unique. This is a natural consequence of the fact—well-established by numerous psychological experiments—that logical connectives (not, and, or, etc.) in linguistic expressions have different meanings when applied by human beings to different vague concepts in different contexts. To be able to capture the different meanings, we need to characterize the operations associated with the logical connectives as broadly as possible.

Fuzzy complements, intersections, and unions have been characterized and studied on axiomatic grounds. Efficient procedures are now available by which various classes of functions can be generated, each of which covers the whole recognized semantic range of the respective operation. In addition, averaging operations for fuzzy sets, which have no counterparts for crisp sets, have also been investigated in this way. This rather theoretical subject, which is beyond the scope of this overview, is thoroughly covered in ref. 18.

In this chapter, it is sufficient to introduce only the most common fuzzy operations, which are the only fuzzy operations that are cutworthy (and also strongly cutworthy). These operations, usually called the *standard fuzzy operations*, are defined for all $x \in X$ by the following formulas:

$$\overline{A}(x) = 1 - A(x) \quad \text{(standard fuzzy complement)},$$

$$(A \cap B)(x) = \min[A(x), B(x)] \quad \text{(standard fuzzy intersection)},$$

$$(A \cup B)(x) = \max[A(x), B(x)] \quad \text{(standard fuzzy union)}.$$

Given two fuzzy sets A, B defined on the same universal set X, A is said to be a subset of B if and only if

$$A(x) \leq B(x)$$

for all $x \in X$. The usual notation $A \subseteq B$ is used to signify the subsethood relation. The set of all fuzzy subsets of X is called the *fuzzy power set* of X and is denoted by $\mathcal{F}(X)$.

Any fuzzy power set with the subsethood relation is a lattice, in which the standard fuzzy intersection and union play the roles of the meet and the join, respectively. The lattice is distributive and complemented under the standard fuzzy complement. Contrary to the Boolean lattice, which is associated with classical power sets, it does not satisfy the law of the excluded middle and the law of contradiction. Such a lattice is usually called a *DeMorgan lattice*.

B. Fuzzy Numbers and Fuzzy Arithmetic

Fuzzy sets that are defined on the set $\mathbb{R}$ of real numbers (i.e., $X = \mathbb{R}$) have special significance in fuzzy set theory. They can be interpreted as fuzzy numbers provided they satisfy the following requirements:

(a) They are normal fuzzy sets
(b) Their supports are bounded
(c) Their α-cuts are closed intervals for every $\alpha \in (0, 1]$

Fuzzy sets on $\mathbb{R}$ that satisfy these requirements capture various linguistic expressions, describing approximate numbers or intervals, such as "numbers that are close to a given real number" or "numbers that are around a given interval of real numbers." Moreover, we can define meaningful arithmetical operations on these fuzzy sets via the α-cut representation. At each α-cut, these operations are the standard arithmetical operations on closed intervals[23]:

$$[a,b] + [d,e] = [a+d, b+e],$$
$$[a,b] - [d,e] = [a-e, b-d],$$
$$[a,b] \cdot [d,e] = [\min(ad, ae, bd, be), \max(ad, ae, bd, be)],$$
$$[a,b]:[d,e] = [\min(a:d, a:e, b:d, b:e), \max(a:d, a:e, b:d, b:e)],$$

where the division is not defined when $0 \in [d,e]$. Operations on fuzzy numbers are thus defined as cutworthy in terms of the corresponding operations on closed intervals of real numbers.

Fuzzy arithmetic enables us to evaluate algebraic expressions in which values of variables are fuzzy numbers[21]. It also enables us to deal with algebraic equations in which coefficients and unknowns are fuzzy numbers. Furthermore, fuzzy arithmetic is a basis for developing fuzzy calculus and, eventually, fuzzy mathematical analysis[4,5,10]. Although a lot of work has already been done along these lines, enormous research effort is still needed to fully develop these mathematical areas.

C. Fuzzy Systems

In general, a *fuzzy system* is any system whose variables (or, at least, some of them) range over states that are fuzzy numbers rather than real numbers. These fuzzy numbers may represent linguistic terms such as "very small," "medium," and so on, as interpreted in a particular context. If they do, the variables are called *linguistic variables*.

Each linguistic variable is defined in terms of a base variable, whose values are real numbers within a specific range. A base variable is a variable in the usual sense, as exemplified by any physical variable (e.g., temperature, pressure, electric current, magnetic flux, etc.) as well as any other numerical variable (e.g., interest rate, blood count, age, performance, etc.). In a linguistic variable, linguistic terms representing approximate values of a base variable, relevant to a particular application, are captured by approximate fuzzy numbers. That is, each linguistic variable consists of the following elements:

- A *name*, which should capture the meaning of the base variable involved
- A *base variable* with its *range of values* (a closed interval of real numbers)

- A *set of linguistic terms* that refer to values of the base variable
- A *semantic rule*, which assigns to each linguistic term its meaning—an appropriate fuzzy number defined on the range of the base variable

An example of a linguistic variable is shown in Fig. 3. Its name is "performance," which captures the meaning of the associated base variable—a variable that expresses the performance (in percentage) of a goal-oriented entity (a person, machine, organization, method, etc.) in some context by real numbers in the interval $[0, 100]$. Linguistic values (states) of the linguistic variable are "very small," "small," "medium," "large," and "very large." Each of these linguistic terms is assigned one of the trapezoidal-shaped fuzzy numbers by a semantic rule, as shown in Fig. 3.

D. Fuzzy Relations

When fuzzy sets are defined on universal sets that are Cartesian products of two or more sets, they are called fuzzy relations. For any Cartesian product of n sets, the relations are called n-dimensional. From the standpoint of fuzzy relations, ordinary fuzzy sets may be viewed as degenerate, one-dimensional relations. All concepts and operations applicable to fuzzy sets are applicable to fuzzy relations as well. However, fuzzy relations involve additional concepts and operations that emerge from their multidimensionality.

Concepts that are important for general n-dimensional fuzzy relations include projections to lower-dimensional spaces, cylindric extensions of projections, and cylindric closures. These concepts are simple generalizations of their classical counterparts,[18] and it is not essential to cover them here. It is more important to introduce some key concepts regarding fuzzy binary relations, which have a broad applicability.

Membership functions of any fuzzy binary relation R on $X \times Y$ have the form

$$R\colon X \times Y \to [0, 1].$$

For each pair $\langle x, y \rangle \in X \times Y$, the membership degree $R(x, y)$ indicates how strongly element x is related to element y according to R. The inverse of R, denoted by R^{-1}, is a relation on $Y \times X$ defined by

$$R^{-1}(y, x) = R(x, y)$$

for all pairs $\langle y, x \rangle \in Y \times X$. Clearly

$$(R^{-1})^{-1} = R$$

for any fuzzy binary relation.

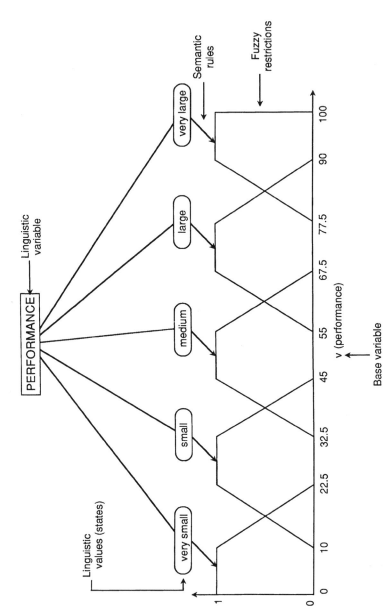

FIGURE 3 Example of linguistic variable.

Given two fuzzy binary relationships R and Q defined on $X \times Y$ and $Y \times Z$, respectively, their *standard composition* $R \circ Q$ is defined by the formula

$$(R \circ Q)(x, z) = \max_{y \in Y} \min[R(x, y), Q(y, z)]$$

for all pairs $\langle x, z \rangle \in X \times Z$. This composition is associative and it also satisfies the property

$$(R \circ Q)^{-1} = Q^{-1} \circ R^{-1}.$$

For fuzzy relations on $X \times X$, the following cutworthy properties are recognized:

- *Reflexivity*: $R(x, x) =$ for all $x \in X$
- *Symmetry*: $R(x, y) = R(y, z)$ for all $x, y \in X$
- *Antisymmetry*: $R(x, y) > 0$ and $R(y, x) > 0$ imply that $x = y$ for all $x, y \in X$
- *Transitivity*: $R(x, z) \geq \max_{y \in Y} \min[R(x, y), R(y, z)]$ for all $x, z \in X$.

Employing these definitions, we can characterize important classes of fuzzy relations in the same way as their crisp counterparts. *Fuzzy equivalence relations* are reflexive, symmetric, and transitive; *fuzzy compatibility relations* are reflexive and symmetric; *fuzzy partial orderings* are reflexive, antisymmetric, transitive, etc. Each of these relations is cutworthy; that is, each α-cut of a fuzzy relation of a particular type is a crisp relation of the same type.

Consider fuzzy binary relations P, Q, and R, defined on sets $X \times Y$, $Y \times Z$, and $X \times Z$, respectively, for which

$$P \circ Q = R.$$

This means that a set of equations of the form

$$\max_{y \in Y} \min[P(x, y), Q(y, z)] = R(x, z)$$

is satisfied for all $x \in X$ and $z \in Z$. These equations are called *fuzzy relation equations*.

The problem of solving fuzzy relation equations arises whenever two of the relations are given and the third is to be determined. When P and Q are given, the problem of determining R is trivial. When R and Q (or P) are given, to determine P (or Q) is considerably more difficult, but it is very important in many applications. Efficient methods for solving this problem have been developed,[18] but their coverage is beyond the scope of this overview.

E. Fuzzy Logic

The term "fuzzy logic" is often interpreted in two ways. In a broad interpretation, fuzzy logic is viewed as a system of concepts, principles, and methods for dealing with modes of reasoning that are approximate rather than exact. In a narrow interpretation, it is viewed as a generalization of the various many-valued logics. This narrow interpretation of fuzzy logic is not of interest in this chapter.

In its broad interpretation, fuzzy logic is based upon fuzzy set theory. It utilizes all resources developed within fuzzy set theory for formulating various forms of sound approximate reasoning. For this purpose, it is necessary to establish a connection between degrees of membership in fuzzy sets and degrees of truth of fuzzy propositions. Such a connection can be made only under the assumption that they both refer to the same objects. Let X denote the set of these common objects. Then, given a fuzzy set A defined on X, the membership degree $A(x)$ for any $x \in X$ may be interpreted as the degree of truth of the fuzzy proposition "x is a member of A." Conversely, given an arbitrary proposition "x is F," where $x \in X$ and F is a fuzzy (vague) property (such as small, large, medium, normal, etc.), its degree of truth for each $x \in X$ may be interpreted as the membership degree $A(x)$ by which a fuzzy set characterized by the property F is defined. Under this correspondence, operations of negation, conjuction, and disjunction of fuzzy propositions are defined in exactly the same way as the operations of fuzzy set complementation, intersection, and union, respectively.

Fuzzy logic deals with propositions expressed in natural language. The linguistic expressions involved may contain fuzzy linguistic terms of any of the following types:

- *Fuzzy predicates*: tall, young, expensive, near, far,...
- *Fuzzy truth values*: true, false, fairly true, very true,...
- *Fuzzy quantifiers*: many, few, almost all,...
- *Fuzzy probabilities*: likely, highly unlikely, very likely,...
- *Fuzzy modifiers* (linguistic hedges): very, quite, extremely,...

All of these linguistic terms except fuzzy modifiers are represented in each context by appropriate fuzzy sets. Fuzzy predicates are represented by fuzzy sets defined on universal sets of elements to which the predicates apply. Fuzzy truth values and fuzzy probabilities are represented by fuzzy sets defined on the unit interval $[0, 1]$. Fuzzy quantifiers are either absolute or relative; they are represented by appropriate fuzzy numbers defined either on the set of natural numbers or on the interval $[0, 1]$. Fuzzy modifiers are operations by which fuzzy sets representing the various other linguistic terms are appropriately modified to capture the meaning of the modified linguistic terms.

In a crude way, it is useful to distinguish the following four types of fuzzy propositions, each of which may, in addition, be quantified by an appropriate fuzzy quantifier.

1. *Unconditional and unqualified propositions* are expressed by the canonical form

$$p: x \text{ is } F,"$$

where F is a fuzzy set representation of a fuzzy predicate relevant to x and p is a symbol of the proposition. The degree of truth of p, $T(p)$, is in this case the same as the degee of membership of x in F; that is,

$$T(p) = A(x).$$

2. *Unconditional and qualified propositions* have the canonical form

$$p: "x \text{ is } F \text{ is } S,$$

where S is a fuzzy set representing either a fuzzy truth value or a fuzzy probability, which makes the proposition either truth-qualified or probability-qualified, respectively. For example, the proposition "The rate of inflation is very low is fairly true" is a truth-qualified proposition; on the other hand, the proposition "The temperature (at a given place, time, etc.) is around 75°F is very likely" is probability-qualified.

3. *Conditional and unqualified propositions* have the canonical form

$$p: "\text{If } x \text{ is } F, \text{ then } y \text{ is } G,"$$

where F and G are fuzzy sets that represent fuzzy predicates relevant to x and y, respectively. These propositions may also be viewed as propositions of the form

$$"(x, y) \text{ is } R,"$$

where R is a fuzzy relation that is determined by the choice of the operation of fuzzy implication. Similarly to fuzzy conjunction and disjunction, fuzzy implication is not unique and depends on the context in which the proposition is used.

4. *Conditional and qualified propositions* have the canonical form

$$p: "\text{If } x \text{ is } F, \text{ then } y \text{ is } G \text{ is } S,"$$

which is basically a combination of the previous forms.

Reasoning based on fuzzy propositions of the four types, possibly quantified by various fuzzy quantifiers, is usually referred to as *approximate reasoning*. Although approximate reasoning is currently a subject of intensive research, its basic principles are already well established. For example, the most common inference rules of classical logic, such as *modus ponens*, *modus tollens*, and *hypothetical syllogism*, have already been generalized within the framework of fuzzy logic. In general, approximate reasoning draws upon the methodological apparatus of fuzzy set theory, such as operations on fuzzy sets, manipulations of fuzzy relations, and fuzzy arithmetic.[18]

F. Fuzzy Logic and Possibility Theory

Consider an arbitrary proposition in the canonical form "*x* is *P*," where *x* is an object from some universal set *X* and *P* is a predicate relevant to the object. To qualify for a treatment by classical logic, the proposition must be devoid of any uncertainty; that is, it must be possible to determine whether the proposition is true or false. Any proposition that does not satisfy this requirement, due to some inherent uncertainty in it, is thus not admissible in classical logic. This is overly restrictive since uncertainty-free propositions are rather rare in human affairs.

The source of uncertainty in any particular proposition of the form "*x* is *P*" is either the predicate *P* or the object *x*. In the former case, the proposition is uncertain because the definition of *P* is vague. The vagueness of *P* makes it impossible to determine, in general, whether the proposition is true or false. The proposition is thus rejected by classical logic. The only way to make it acceptable is to depart from classical logic by allowing the truth of the proposition to be a matter of degree. This course has been taken by fuzzy logic, where the degree of truth of the proposition expresses the extent to which the object is compatible with the predicate.

The second type of uncertainty in propositions of the given type results from information deficiency regarding the object *x*. While the predicate *P* is in this case defined precisely, information about *x* is insufficient to determine whether or not *x* satisfies *P*. The proposition is in this case either true or false, but its actual truth status cannot be determined. However, it is useful to assign a number in the unit interval $[0, 1]$ to the proposition to express the degree of evidence that the proposition is true. Assigning degrees of evidence to relevant propositions is a topic dealt with in measure theory.

The two types of uncertainty in propositions are very different. The first emerges from vagueness inherent in language and results in degrees of truth; the second emerges from information deficiency and results in degrees of evidence. While the second type of uncertainty is strongly dependent on available information regarding the object involved, the first is totally independent of this information.

Propositions may also contain uncertainties of both types. To deal with information contained in such propositions, a measure-theoretic counterpart of fuzzy set theory was introduced by Zadeh under the name *possibility theory*.[40] The following are basic notions of the theory.

Let $\mathscr{X}$ denote a variable that takes values in a universal set X and let the equation $\mathscr{X} = x$, where $x \in X$, be used for describing the fact that the value of $\mathscr{X}$ is x. Consider now that a fuzzy proposition of the form "$\mathscr{X}$ is F" is given, where F is a fuzzy set on X. Viewing F as a characterization of an elastic constraint on possible values of $\mathscr{X}$, it is reasonable to interpret the membership degree $F(x)$ for each $x \in X$ as the degree of possibility that $\mathscr{X} = x$. According to this interpretation, given a fuzzy set F on X and the proposition "$\mathscr{X}$ is F," the degree of possibility $r_F(x)$ of $\mathscr{X} = x$ for each $x \in X$ is defined as numerically equal to the degree of membership $F(x)$. Formally,

$$r_F(x) = F(x)$$

for all $x \in X$. Function r_F is called a *possibility distribution function.*

In this interpretation, the possibility degree that the value of $\mathscr{X}$ is in any given crisp set A is equal to the greatest possibility degree for all $x \in A$. That is, given a possibility distribution function r_F, the associated possibility measure Pos_F is defined for all crisp subsets A of X via the formula

$$\text{Pos}_F(A) = \sup_{x \in A} r_F(x),$$

where the supremum may be replaced with the maximum if it exists. An extension of this definition for fuzzy sets $A \in \mathscr{F}(X)$ is given by the formula

$$\text{Pos}_F(A) = \sup_{x \in X} \min[A(x), r_F(x)],$$

which subsumes the previous formula as a special case.

From these basic properties of possibility measures, the full calculus of possibility theory, analogous to the calculus of probability theory, has been developed. Its primary role is to deal with incomplete information expressed in terms of fuzzy propositions.[6, 7, 18] Due to limited space, it is not possible to cover here details of this calculus.

IV. SCIENTIFIC PARADIGMS AND PARADIGM SHIFTS

As mentioned in Section I, it is my contention that science is currently undergoing a major paradigm shift regarding attitudes toward uncertainty, and that this shift is reflected by the emergence of fuzzy mathematics. Before examining this paradigm shift, it is appropriate to briefly explain what scientific paradigms and paradigm shifts are.

The concept of a scientific paradigm was introduced in a highly influential book by Kuhn.[19] In general, a *scientific paradigm* is a set of concepts, presuppositions, beliefs, habits of mind, theories, standards, principles, and methods that are shared and accepted by the scientific community in a given field. According to Kuhn, work in any given field of science is normally pursued under a particular paradigm that is generally accepted in the field at that time. The paradigm, based upon past scientific achievements in the field, defines implicitly the legitimate problems and methods, as well as specific rules and standards for scientific practice. Moreover, education in the field is based upon the paradigm. Each period during which science is practiced under a particular paradigm is called a period of *normal science*.

As long as no serious difficulties are encountered in dealing with the problems of interest, the accepted paradigm is unchallenged. However, when it becomes increasingly difficult or even impossible to deal with certain problems (due to persistent paradoxes, anomalies, etc.), the time is ripe for the emergence of a new paradigm. As a rule, the emerging paradigm is not immediately accepted by the scientific community. The time period between the emergence of a new paradigm and its general acceptance is called a *paradigm shift*. Although the duration of paradigm shifts varies considerably from case to case, several characteristic stages seem to be involved in virtually every paradigm shift. These stages, which are also visible in the ongoing paradigm shift regarding uncertainty, are discussed in Section VI.

A crisis of one paradigm, followed by a paradigm shift, and resulting eventually in the acceptance of a new paradigm is called a *scientific revolution*. Some of the best known scientific revolutions are associated with the names of Copernicus, Newton, Lavoisier, Darwin, Maxwell, and Einstein. They are well characterized by Kuhn in his book.[19] Kuhn's characterization of progress in science as a sequence of periods of normal science that are connected by relevant paradigm shifts is being increasingly accepted as a realistic outlook.

V. FROM CLASSICAL SETS TO FUZZY SETS: A GRAND PARADIGM SHIFT

My purpose in this section is to explain why the various novel theories of uncertainty are important for science and why it is reasonable to view their use in science as a new scientific paradigm. I discuss several roles of these uncertainty theories in science, in particular the roles played by fuzzy set theory and fuzzy logic.

The first role of uncertainty to be discussed here involves the relationship between mathematical models and the associated phenomena of the real world. To make my arguments understandable, it is essential that I first express my views regarding this relationship.

As a system scientist, I subscribe to the constructivist view of reality and knowledge.[12, 15] According to this view, distinctions, classes, separations, systems, etc., are all artificial; they are not made by nature, but constructed by the perceptual and mental capabilities of the human mind. That is, systems do not exist in the real world, ready made to be discovered by us. Instead, we construct them by making appropriate distinctions, be they made in the experiential world by our perceptual capabilities, enhanced by measuring instruments, or conceived in the world of ideas by our mental capabilities.

Mathematical modeling attempts to connect constructions made in the experiential world with those made in the world of ideas. Since we cannot perceive with infinite detail, we cannot measure with infinite resolution, and we cannot compute in the experiential world with infinite precision, there exists a gap between any system constructed in terms of experiential distinctions and a mathematical system that attempts to model it. While this gap cannot be fully avoided, it can be at least partially bridged by choosing appropriate mathematical constructions for our modeling purposes.

The need to bridge the gap between a mathematical model and experience is well characterized in a penetrating study by the American philosopher Max Black[2]:

> It is a paradox, whose importance familiarity fails to diminish, that the most highly developed and useful scientific theories are ostensibly expressed in terms of objects never encountered in experience. The line traced by a draftsman, no matter how accurate, is seen beneath the microscope as a kind of corrugated trench, far removed from the ideal line of pure geometry. And the "point-planet" of astronomy, the "perfect gas" or thermodynamics, or the "pure species" of genetics are equally remote from exact realization. Indeed the unintelligibility at the atomic or subatomic level of the notion of a rigidly demarcated boundary shows that such objects not merely are not but could not be encountered. While the mathematician constructs a theory in terms of "perfect" objects, the experimental scientist observes objects of which the properties demanded by theory are and can, in the very nature of measurement, be only approximately true. Mathematical deduction is not useful to the physicist if interpreted rigorously. It is necessary to know that its validity is unaltered when the premise and conclusion are only "approximately true". But the indeterminacy thus introduced, it is necessary to add in criticism, will invalidate the deduction unless the permissible limits of variation are specified. To do so, however, replaces the original mathematical deduction by a more complicated mathematical theory in respect of whose interpretation the same problem arises, and whose exact nature is in any case unknown. This lack of exact correlation between a scientific theory and its empirical interpretation can be blamed either upon the world or upon the theory. We can regard the shape of an orange or a tennis ball as imperfect copies of an ideal form of which perfect knowledge is to be had in pure geometry, or we can regard the geometry of spheres as a simplified and imperfect version of the spatial relations between the members of a certain class of physical objects. On either view there remains a gap between scientific theory and its application which ought to be, but is not, bridged. To say that all language (symbolism, or thought) is vague is a favorite method for evading

the problems involved and lack of analysis has the disadvantage of tempting even the most eminent thinkers into the appearance of absurdity. We shall not assume that "laws" of logic or mathematics prescribe modes of existence to which intelligible discourse must necessarily conform. It will be argued, on the contrary, that deviations from the logical or mathematical standards of precision are all pervasive in symbolism; that to label them as subjective aberrations sets an impassable gulf between formal laws and experience and leaves the usefulness of the formal sciences an insoluble mystery.

It is easy to see that the concepts of vagueness and precision mentioned in this quote are closely related to the concepts of fuzziness and crispness, which distinguish fuzzy sets from crisp sets. Observe that Black associates these concepts with language. This is compatible with the constructivist viewpoint, according to which vagueness and precision are not inherent in the real world, but emerge from language in which we describe our constructions (classes, systems, etc.). In particular, vagueness of a symbol (a linguistic term) in a given language results from the existence of objects for which it is intrinsically impossible to decide whether the symbol does or does not apply to them according to the linguistic habits of some speech community using the language. That is, fuzziness of fuzzy sets and fuzzy propositions is a language-based uncertainty, which is fundamentally different from uncertainty based on information deficiency.[16]

A general conclusion reached by Black[2] is that dealing with vague propositions requires a logic that does not obey some of the laws of classical logic, notably the law of the excluded middle and the law of contradiction. The same conclusion was arrived at by Bertrand Russell[25] in 1923, as seen from the following quote:

> Vagueness and precision alike are characteristics which can only belong to a representation, of which language is an example. They have to do with the relation between a representation and that which it represents. Apart from representation, whether cognitive or mechanical, there can be no such thing as vagueness or precision; things are what they are, and there is an end of it.... The law of excluded middle is true when precise symbols are employed, but it is not true when symbols are vague, as, in fact, all symbols are.... The notions of "true" and "false" can only have a precise meaning when the symbols employed—words, perceptions, images, or what not—are themselves precise. Since propositions containing nonlogical words are the substructure on which logical propositions are built, it follows that logical propositions also, so far as we can know them, become vague through the vagueness of "truth" and "falsehood." We can see an ideal of precision, to which we can approximate indefinitely; but we cannot obtain this ideal. Logical words, like the rest, when used by human beings, share the vagueness of all other words.... All traditional logic habitually assumes that precise symbols are being employed. It is therefore not applicable to this terrestrial life, but only to an imagined celestial existence.

The first paragraph in this quote shows that Russell, similarly to Black, views vagueness and precision as characteristics of a representation rather than characteristics of the things being represented; that is, his view is compatible with the constructivist view. Notice Russell's remark in the second paragraph that all symbols are vague. He elaborates on this point a little more:

> Let us consider the various ways in which common words are vague, and let us begin with such a word as "red." It is perfectly obvious, since colours form a continuum, that there are shades of colour concerning which we shall be in doubt whether to call them red or not, not because we are ignorant of the meaning of the word "red," but because it is a word the extent of whose application is essentially doubtful....All words describing sensible qualities have the same kind of vagueness which belongs to the word "red." This vagueness exists also, though in a lesser degree, in the quantitative words which science has tried hardest to make precise, such as a metre or a second. I am not going to invoke Einstein for the purpose of making these words vague. The meter, for example, is defined as the distance between two marks on a certain rod in Paris, when that rod is at a certain temperature. Now the marks are not points, but patches of a finite size, so that the distance between them is not a precise conception. Moreover, temperature cannot be measured with more than a certain degree of accuracy, and the temperature of a rod is never quite uniform. For all these reasons the conception of a metre is lacking in precision. The same applies to a second. The second is defined by relation to the rotation of the earth, but the earth is not a rigid body, and two parts of the earth's surface do not take exactly the same time to rotate; moreover all observations have a margin of error. There are some occurrences of which we can say that they take less than a second to happen, and others of which we can say that they take more, but between the two there will be a number of occurrences of which we believe that they do not all last equally long, but of none of which we can say whether they last more or less than a second. Therefore, when we say an occurrence lasts a second, all that it is worthwhile to mean is that no possible accuracy of observation will show whether it lasts more or less than a second.

Other distinguished scholars expressed similar views to those of Russell and Black regarding the inadequacy of classical logic for representing aspects of the experimental world. For example, Pierre Duhem, a distinguished French physicist and philosopher, made the following observation in his 1906 book[8]

> It is impossible to describe a practical fact without attenuating by the use of the word "approximately" or "nearly"; on the other hand, all the elements constituting the theoretical fact are defined with rigorous exactness. A practical fact is not translated therefore by a single theoretical fact but by a kind of bundle including an infinity of different theoretical facts....A mathematical deduction is of no use to the physicist so long as it is limited to asserting that a given *rigorously* true proposition has for its consequence the *rigorous* accuracy of some such other proposition. To be useful to the physicist, it must still be proved that the second proposition remains *approximately* exact when the first is only *approximately* true. And even that does not suffice. The range of these two approximations must be delimited; it is necessary to fix the limits of error which can be made in the result when the degree of precision of the methods of measuring the data is known. Such are the rigorous conditions that we are bound to impose on mathematical deduction if we wish this absolutely precise language to be able to translate without betraying the physicist's idiom, for the terms of this latter idiom are and always will be vague and inexact like the perceptions which they are to express. On these conditions, but only on these conditions, shall we have a mathematical representation of the *approximate*. But let us not be deceived about it; this "mathematics of

approximation" is not a simpler and cruder form of mathematics. On the contrary, it is a more thorough and more refined form of mathematics, requiring the solution of problems at times enormously difficult, sometimes even transcending the methods at the disposal of algebra today.

The gap between mathematical theories and their applications, which is the theme of all the foregoing quotes, arises from one common source: the discrepancy between the precision required by classical set theory (and its associated logic) and the inherent resolution limits of our perceptual capabilities as well as measuring instruments. Consider, for example, measurements of a physical quantity taken by a particular instrument. Due to the finite resolution of the instrument employed, appropriate quantization of the measurement is inevitable. Assume, for example, that the considered range of the quantity is $[0, 1]$ and that the measuring instrument allows us to measure with the accuracy of one decimal digit. Then, measurements are values taken from the collection of values 0, 0.1, 0.2, ..., 0.9, 1, which stand for the intervals $[0, 0.05)$, $[0.05, 0.15), ..., [0.85, 0.95), [0.95, 1]$. This example of the usual quantization is illustrated in Fig. 4a.

Although the usual quantization is mathematically convenient, it completely ignores uncertainties induced by unavoidable measurement errors around the boundaries between the individual intervals. This is highly unrealistic. Quantization forced by the limited resolution of the measuring instrument involved can be made more realistic by replacing the crisp intervals with fuzzy intervals or fuzzy numbers. This is illustrated for our example in Fig. 4b. Fuzzy sets are in this example fuzzy numbers expressed by the shown triangular membership functions, which express the linguistic

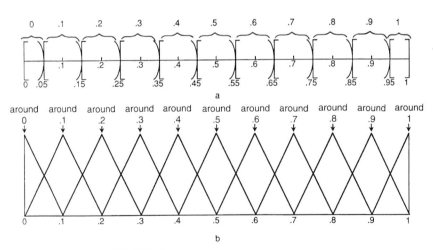

FIGURE 4 Quantization versus granulation.

labels "around 0," "around 0.1," "around 0.2," etc. Instead of triangular membership functions, other types of functions (trapezoidal, bell-shaped, etc.) may be used to utilize available knowledge regarding the various characteristics of measurement errors in a given context. Fuzzy quantization, as exemplified by Fig. 4b, is often referred to as *granulation*.

Contrary to quantization, granulation allows us to capture relevant measurement uncertainties in experimental data. This implies that data based upon appropriate granulation are more accurate than the corresponding data based on the usual quantization.

As is well known, fuzzy sets are holistic concepts, each representing a potentially infinite family of nested crisp sets (α-cuts), each defined for a particular number (α) in the unit interval.[18] When we operate on fuzzy sets, we operate in fact simultaneously on all crisp sets in the associated families. This is quite powerful, both conceptually and computationally.

Considering the holistic nature of fuzzy sets, it is not surprising that they were envisioned by proponents of holism early in this century. To illustrate this connection of holism with fuzzy sets, I chose the following short excerpt from a book by Smuts[31] (published in 1926), which is perhaps the most comprehensive book on holism ever published.:

> The science of the nineteenth century was like its philosophy, its morals and its civilization in general, distinguished by a certain hardness, primness and precise limitation and demarcation of ideas. *Vagueness*, indefinite and blurred outlines, anything savoring of mysticism, was abhorrent to that great age of limited exactitude. The rigid categories of physics were applied to the indefinite and hazy phenomena of life and mind. Concepts were in logic as well as in science narrowed down to their most luminous points, and the rest of their contents treated as non-existent. Situations were not envisaged as a whole of clear and vague obscure elements alike, but were analyzed merely into their clear, outstanding, luminous points. A "cause," for instance, was not taken as a whole situation which at a certain stage insensibly passes into another situation, called the effect. No, the most outstanding feature in the first situation was isolated and abstracted and treated as the cause of the most outstanding and striking feature of the next situation, which was called the effect. Everything between this cause and this effect was blotted out, and the two sharp ideas or rather situations of cause and effect were made to confront each other in every case of causation like two opposing forces. This logical precision immediately had the effect of making it impossible to understand how the one passed into the other in actual causation. We have to return to the fluidity and plasticity of nature and experience in order to find the concepts of reality. When we do this we find that round every luminous point in experience there is a gradual shading off into haziness and obscurity. A "concept" is not merely its clear luminous center, but embraces a surrounding sphere of meaning or influence of smaller and smaller dimensions, in which the luminosity tails off and grows fainter until it disappears.

Let me examine now the nature of mathematical models that involve probability. Due to the additivity requirement of probability theory, elementary events are required to be pairwise disjoint and the probability of each is required to be expressed by a real number. The requirement of disjoint events makes good sense mathematically. Unfortunately, it becomes problematic when we leave the world of mathematics. Let me outline an argument advanced by Viertl[35] to show the inadequacy of probability theory at the experimental level. The argument is based on the fact that there is no method of measurement that is free from error.

Consider, for example, two disjoint events, A and B, defined in terms of adjoining intervals of real numbers, as shown in Figure 5a. Due to the effect of measurement errors, observations at the sharp boundary between events A and B are totally meaningless and should be completely discounted. Moreover, observations in the close neighborhood of this boundary (within the reach of measurement errors) are not fully credible and should be discounted according to some discount-rate function, as illustrated in Fig. 5a. When the same measurements are taken for the union $A \cup B$, as shown in Fig. 5b, the discount-rate function is not applicable. Hence, the same observations produce more evidence for the single event $A \cup B$ than for the two disjoint events A and B. The evidential support for $A \cup B$ is thus not equal to the sum of the evidential supports for A and B. That is, the additivity requirement of probability measures is violated; the correct measure is in this case superadditive. Alternatively, an appropriate granulation can be used to define probabilities on fuzzy events.[38]

Let me turn now to the second role of uncertainty in science and technology—its role in managing complexity and controlling computational cost. This role can perhaps be best explained in the context of systems modeling.

In constructing a system as a model of some aspects of reality or as a model of some desirable manmade object, we always attempt to maximize its usefulness. This aim is closely connected with the relationship among three key characteristics of every systems model: complexity, credibility,

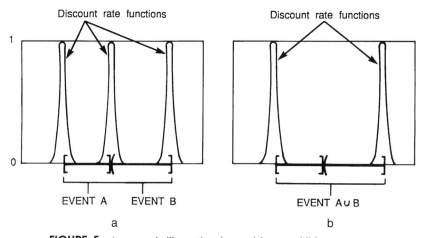

FIGURE 5   An example illustrating the need for nonadditive measures.

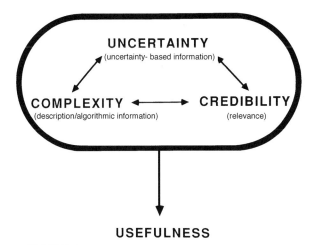

FIGURE 6   The role of uncertainty in systems modeling.

and uncertainty (Fig. 6). This relationship, which is a subject of current study in systems science, is not as yet fully understood. We know, however, that uncertainty has a pivotal role in any efforts to maximize the usefulness of systems models. While uncertainty associated with the purpose of the model (such as prediction, retrodiction, explanation, or prescription) is not desirable for its own sake, it becomes very valuable when considered in connection with the other two characteristics of systems models. A slight increase in the relevant uncertainty may often significantly reduce the complexity of the model and, at the same, increase its credibility (i.e., relevance to the purpose for which the model was constructed). Uncertainty is thus an important commodity in the modeling business, which can be traded for gains in the other essential characteristics of systems models. To utilize uncertainty in this way, we need to understand it as broadly as possible. It is this important positive role of uncertainty, I believe, which is primarily responsible for the rapidly growing interest, during the last three decades or so, in studying uncertainty in all its manifestations.

The explained connection of uncertainty with complexity and relevance in systems models is well captured in the following concise statement by Zadeh:[39]

> As the complexity of a system increases, our ability to make precise and yet significant statements about its behavior diminishes until a threshold is reached beyond which precision and significance (or relevance) become almost exclusive characteristics.

The poet Saki expressed (in his *Aphorisms*) the same idea in a simple, poetic way: "A little inaccuracy sometimes saves tons of explanations."

The new uncertainty theories have also considerably greater expressive power and, consequently, they can effectively deal with a broader class of problems. An important feature of fuzzy set theory is its capability to capture the vagueness of linguistic terms in statements of natural language. This, in turn, provides us with a greater capability of modeling human common-sense reasoning, decision making, and other aspects of human cognition. When employing this capability in machine design, the resulting machines are human friendlier.

VI. STAGES IN THE PARADIGM SHIFT

Although each paradigm shift has many unique special characteristics, all paradigms seem to involve several, surprisingly similar common stages. The purpose of this section is to express my personal observations about these stages in the ongoing paradigm shift regarding the emergence of broad uncertainty theories. Each stage is distinguished from other stages by some characteristic position taken by opponents of the new paradigm.

When a new paradigm is suggested, it is initially ignored by the scientific community. From the various quotes in Section V, it is clear that the need for a departure from classical bivalent logic was expressed by some authors during the first half of this century (Russell, Duhem, Smuts, Black). However, all suggestions advanced by these authors were totally ignored at that time.

An important turning point in the paradigm shift was the publication of the seminal paper by Zadeh.[37] The significance of this paper was that it contained a specific proposal, which challenged classical set theory as well as classical bivalent logic. It also challenged the seemingly unique connection between uncertainty and probability theory, which had previously been taken for granted: it showed that uncertainty represented by fuzzy sets is of a different type than probabilistic uncertainty and, hence, it is beyond the scope of probability theory to capture it.

The specificity of Zadeh's proposal and the fact that it was oriented to science and technology (contrary to it predecessors, which were oriented to philosophy) and published in a respectable scientific journal fathered a strong negative reaction from some influential scientists and mathematicians within a few years of its publication. This is fairly typical of the second stage of a paradigm shift. The following are two representative samples of this initial, highly hostile and emotional reaction:[41]

> No doubt Professor Zadeh's enthusiasm for fuzziness has been reinforced by the prevailing political climate in the U.S.—one of unprecedented permissiveness. "Fuzzification" is a kind of scientific permissiveness; it tends to result in socially appealing slogans unaccompanied by the discipline of hard scientific work and patient observation. *(Rudolf E. Kalman, 1972)*

> Fuzzy theory is wrong, wrong, and pernicious. I cannot think of any problem that could not be solved better by ordinary logic.... The danger of fuzzy theory is

that it will encourage the sort of imprecise thinking that has brought us so much trouble. *(William Kahan, 1975)*

In spite of the predominantly negative initial reaction to the proposed fuzzy set theory and, later, fuzzy measure theory, some scientists and mathematicians recognized the significance of these theories and committed some of their research work to their development. Through this work, the new theories were steadily advancing and it was increasingly more difficult for their opponents to dismiss them without resorting to some more serious arguments. This led to the third stage in the paradigm shift, characterized by organized debates between supporters and opponents of the new paradigm.

Several debates, focusing primarily on the relationship between probability theory and the various novel uncertainty theories, are documented in the literature.[1, 14, 22, 27] A frequently raised issue in these debates was the lack of solid mathematical foundations of the new theories. The most extreme claims advanced by opponents of the new uncertainty theories were expressed by Lindley[20] (Italics added):

> *The only* satisfactory description of uncertainty is probability. By this I mean that *every* uncertainty statement *must* be in the form of a probability; that several uncertainties *must* be combined using the rules of probability; and that the calculus of probabilities is adequate to handle *all* situations involving uncertainty. Probability is *the only* sensible description of uncertainty and is adequate for *all* problems involving uncertainty. *All* other methods are inadequate.... *Anything* that can be done with fuzzy logic, belief functions, upper and lower probabilities, or *any* other alternative to probability, can *better* be done with probability.

We can see in these debates once again that most opponents of fuzzy set theory from the area of probability theory attempt to compare probabilities with degrees of truth (or degrees of membership). However, these are not comparable. As explained in Section III in the context of possibility theory, degrees of truth result from linguistic uncertainty, while probabilities result from information deficiency. These two types of uncertainty may be combined, but their comparison is meaningless.

It is significant that some probabilists have been supportive of fuzzy set theory and recognized its complementarity to probability theory. One of them is Kapur, a well-known contributor to classical (probability-based) information theory. The following excerpt from a published interview[26] expresses his views regarding fuzzy set theory:

> In mathematics, earlier, algebra and topology were fighting for the soul of mathematics. Ultimately both are co-existing and are enriching each other. Similarly today there is a struggle between probability theory and fuzzy set theory to capture the soul of uncertainty. I am sure ultimately both will co-exist and enrich each other. Already the debate has led to a deeper understanding of what we mean by uncertainty....I believe that uncertainty is too deep a concept to be captured by probability theory alone. Probability theory has had a long history, while fuzzy set theory is relatively of recent origin. Let it grow to its full strength.

Another probabilist endorsing fuzzy set theory is Viertl, who has been active in developing reasoning methods that combine probability theory with fuzzy logic.[35]

With the continuing debate, during which foundations of the new theories considerably advanced, arguments claiming the lack of solid foundations became less convincing. As a result, many opponents have gradually become more conciliatory. The focus of the debate has shifted from questions regarding foundations to questions regarding applicability.

The lack of proven practical applications of the new uncertainty theories during their early stages of development was a favorite criticism of their opponents. At the very beginning, they were able to embarrass advocates of the theories by simple questions such as: "Can you show us at least one practical application of these new theories?" When some practical applications of the new theories were eventually developed, the opponents made their questions more and more demanding. When evidence of practical applicability of fuzzy set theory and, to a lesser degree, the other novel uncertainty theories became overwhelming (in the early 1990s, primarily in Japan), qualitatively new questions were posed by the opponents: "Can you show us at least one problem which can be solved with the help of these new theories, but which cannot be solved without them?"

Although the last question is still open, there exist now examples of problems that have been solved with the help of fuzzy set theory, while all previous attempts to solve them by other known methods had failed. Perhaps the most visible example is the fuzzy control system (designed by Michio Sugeno and his team) by which a pilotless helicopter is controlled by simple vocal commands, a system that has already been implemented and successfully tested.[33] Another, more esoteric example is a successful fuzzy control system for controlling a triple inverted pendulum.[42]

Responses of opponents of fuzzy set theory to these examples of very successful applications of fuzzy set theory have been of two kinds. In the first kind of response, the examples are accepted as legitimate applications of fuzzy set theory, but it is maintained that some traditional methodology (classical control theory, Bayesian methodology, classical logic, etc.) would solve the problems even better. An example of this kind of response is the following excerpt from a personal letter I received from Anthony Garrett, one of my professional acquaintances and a devoted Bayesian, after I informed him about the fuzzy helicopter control:

> I quite accept that you saw a successful fuzzy system [fuzzy helicopter control] in Japan. What I am saying is that, unless it is isomorphic to Bayesian logic circuitry, there will be circumstances in which it reaches the wrong decision and the Bayesian circuitry gets it right .... I'm afraid I continue to believe that there are no areas which fuzzy logic can treat, but which Bayesian can't.

While this kind of response does not challenge fuzzy set theory in any way, it challenges the Bayesian methodology to substantiate the expressed claims.

Responses of the second kind (to the evidence of highly successful applications of the novel uncertainty theories) are substantially different. They acknowledge that the applications concerned are successful, but question the role of the new theories in these applications. A highly visible example of a response of this kind (addressing applications of fuzzy logic) is a paper by Charles Elkan, which is entitled "The Paradoxical Success of Fuzzy Logic." I deem it important to examine some aspects of this paper more carefully.

Elkan's paper was published in two versions. Version 1 is included in the *Proceedings of the Eleventh National Conference on Artificial Intelligence* (MIT Press, 1993); version 2 was published in ref. 27, together with comments by 22 respondents.

In both versions of his paper, Elkan argues that the apparent success of fuzzy logic in many practical applications is paradoxical since fuzzy logic collapses "under many ostensibly reasonable logical equivalencies" into the classical, two-valued logic. To demonstrate this collapse of fuzzy logic, he uses one definition and one theorem. Using the theorem, he argues that fuzzy logic must encounter all the technical difficulties of classical logic in expert systems and other applications. To understand the essence of Elkan's criticism, we have to examine his definition and theorem.

In both versions of the paper, Elkan uses the same notation: A, B denote assertions; $t(A), t(B) \in [0,1]$ denote the degrees of truth in A, B, respectively; and $\wedge$, $\vee$, and $\neg$ denote logical connectives of conjunction, disjunction, and negation, respectively. The two versions also use the same definition, which intends to be a definition of a system of fuzzy logic (observe that $t(A), t(B) \in [0,1]$ and the operations employed are the standard fuzzy connectives):

DEFINITION 1. Let A and B be arbitrary assertions. Then

$$t(A \wedge B) = \min\{t(A), t(B)\},$$

$$t(A \vee B) = \max\{t(A), t(B)\},$$

$$t(\neg A) = 1 - t(A),$$

$$t(A) = t(B) \quad \text{if } A \text{ and } B \text{ are logically equivalent}.$$

Beyond this definition, the two versions of the paper differ.

In version 1, the definition is qualified by the statement: "In the last case of this definition, let logically equivalent mean equivalent according to the rules of classical two-valued propositional calculus." This qualification means, in effect, that Definition 1 defines the system of classical, two-valued logic. The definition is employed in the following theorem:

THEOREM 1. *Given the four postulates of Definition 1, for any two assertions A and B, either $t(B) = t(A)$ or $t(B) = 1 - t(A)$.*

Observe that the theorem is false if Definition 1 is applied without the qualification; with the qualification, it is true and its proof is trivial: by definition. Elkan, in fact, uses an elaborate and rather long proof, which is totally redundant.

In version 2, Definition 1 is left unqualified, but Theorem 1 is qualified:

THEOREM 1 (Qualified).   *Given the formal system of Definition 1, if*

$$\neg(A \wedge \neg B) \quad and \quad B \vee (\neg A \wedge \neg B)$$

are logically equivalent, then for any two assertions A and B, either $t(B) = t(A)$ *or* $t(B) = 1 - t(A)$.

Observe that Definition 1, if unqualified, defines a standard system of fuzzy logic. As well known (and easy to prove), this system satisfies all laws of classical logic except the law of the excluded middle [since, in general, $t(A \vee \neg A) \neq 1$ in this system] and the law of contradiction [since, in general, $t(A \wedge \neg A) \neq 0$ in this system]. When one of these laws is imposed on the system, the degrees of truth become constrained to 0 and 1: clearly if $t(A) \in [0, 1]$, then

$$\max[t(A), 1 - t(A)] = 1$$

can be satisfied only for either $t(A) = 0$ or $t(A) = 1$. Similarly,

$$\min[t(A), 1 - t(A)] = 0$$

can be satisfied only for either $t(A) = 0$ or $t(A) = 1$. Moreover, when one of the laws is imposed on the system, the other law is implied by the operations in Definition 1 and the assumption that $t(A) \in [0, 1]$. Requiring, for example, that $t(A \vee \neg A) = 1$ for any given assertion A implies that $t(A \wedge \neg A) = 0$:

$$\max[t(A), 1 - t(A)] = 1 \quad \Rightarrow \quad 1 - \max[t(A), 1 - t(A)] = 0$$
$$\Rightarrow \quad \min[t(A), 1 - t(A)] = 0.$$

Remark.   To accept the law of contradiction, but reject the law of the excluded middle, the intuitionists had to invent a special negation operator, defined by

$$t(\neg A) = \begin{cases} 1 & \text{when } t(A) = 0, \\ 0 & \text{otherwise.} \end{cases}$$

The system of intuitionist logic, based on this operator is, of course, fundamentally different from the system introduced by Definition 1.

The logical equivalence imposed in Theorem 1 on the system of Definition 1 constrains the system in the same way as the requirement that the system satisfy the law of excluded middle. To show this, we recognize that the system of Definition 1 satisfies commutativity, distributivity, and De Morgan's laws. Using these laws, the first expression Theorem 1, $\neg(A \wedge \neg B)$, may be replaced with $\neg A \vee B$ (by De Morgan's law), and this, in turn, may be replaced with $B \vee \neg A$ (by commutativity). Similarly, the second expression, $B \vee (\neg A \wedge \neg B)$, may be replaced with $(B \vee \neg A) \wedge (B \vee \neg B)$ (by distributivity). Now, the requirement that $B \vee \neg A$ and $(B \vee \neg A) \wedge (B \vee \neg B)$ be logically equivalent can be satisfied (under the operations of Definition 1) iff $t(B \vee \neg B) = 1$, i.e., iff we require the law of the excluded middle. As shown earlier, this implies the law of contradiction and, consequently, the usual system of classical two-valued logic. This means that Theorem 1 may as well be restated in a more natural way and its proof substantially simplified as follows:

THEOREM 1 (Alternative formulation). *Given the formal system of Definition 1, if $t(A) = t(B)$ is required when A and B are logically equivalent according to the rules of classical two-valued logic, then for any two assertions A and B, either $t(B) = t(A)$ or $t(B) = 1 - t(A)$.*

Proof. Trivial

Fuzzy logic must violate, by definition, some laws of classical logic or, alternatively, it must violate the truth functionally.[17] Hence, Elkan's theorem (however trivial) does not say anything about fuzzy logic at all. The technical limitations implied by the theorem do not apply to fuzzy logic, but to classical two-valued logic.

VII. CONCLUSIONS

The nature and quality of the most recent criticisms of the new uncertainty theories, as exemplified by the criticism of fuzzy logic offered by Elkan, is a strong indicator that the paradigm shift is now considerably advanced. However, it is substantially more advanced within the engineering community, where the significance of fuzzy logic is now generally recognized, than within the scientific community.

Examining the various disciplines of science, physics and chemistry seem to be thus far the disciplines least affected by the emergence of fuzzy logic. This may be explained by the fact that the use of classical mathematics has been quite successful in these disciplines and, consequently, there is no need to revise the various theories based on classical mathematics. Thus far, contributions describing the use of fuzzy logic in chemistry have been rare.[24, 28, 29] This is likely to change within the next decade or so. The expressive richness of fuzzy mathematics will undoubtedly lead to more realistic theories even in these "hardest" of scientific disciplines. In this sense, the *Sixth International Conference on Mathematical Chemistry* held in Pitlochry, Scotland, at which the prospective role of fuzzy logic in chemistry was for the first time critically examined, will likely be recognized as an important turning point.

ACKNOWLEDGMENT

Work on this chapter was supported in part by the Office of Naval Research (Grant N00014-94-1-0263).

REFERENCES

1. J. C. Bezdek, ed., *IEEE Trans. Fuzzy Systems* (Special issue) **2**, 1–45 (1994).
2. M. Black, *Philos. Sci.* **4**, 427–455 (1990). Reprinted in *Int. J. Gen. Systems* **17**, 107–128 (1990).
3. G. Choquet, *Ann. Inst. Fourier* **5**, 131–295 (1953–1954).
4. P. Diamond and P. Kloeden, *Metric Spaces of Fuzzy Sets: Theory and Application.* World Scientific, Singapore, 1994.
5. D. Dubois and H. Prade, *Fuzzy Sets and Systems* **8**, 1–17, 105–116, 225–233 (1982).
6. D. Dubois and H. Prade, *Possibility Theory.* Plenum, New York, 1988.
7. D. Dubois, J. Lang, and H. Prade, *Handbook of Logic in Artificial Intelligence and Logic Programming* (D. M. Gabbay et al., eds.), pp. 439–513. Clarendon, Oxford, UK, 1994.
8. P. Duhem, *The Aim and Structure of Physical Theory.* Princeton Univ. Press, Princeton, NJ, 1954. (Originally published in French in 1906.)
9. T. Fine, *Theories of Probability.* Academic Press, New York, 1973.
10. J. Fridrich, *Int. J. Gen. Systems* **22**, 381–389 (1994).
11. J. W. Gibbs, *Elementary Principles in Statistical Mechanics.* Yale Univ. Press, New Haven, 1902. Reprinted by Ox Bow Press, Woodbridge. CT, 1981.
12. E. v. Glasersfeld, *Radical Constructivism: A Way of Knowing and Learning.* The Farmer Press, London, 1995.
13. P. R. Halmos, *Measure Theory.* Van Nostrand, Princeton, NJ, 1950.
14. G. J. Klir, *Int. J. Gen. Systems* **15**, 347–378 (1989).
15. G. J. Klir, *Facets of Systems Science.* Plenum Press, New York, 1991.
16. G. J. Klir, in *Advances in Computers* (M. C. Yovits, ed.), pp. 255–232. Academic Press, San Diego, 1993.
17. G. J. Klir, in *Advances in Fuzzy Theory and Technology* (P. P. Wang, ed.), Vol. 2, pp. 3–47. Duke Univ. Press, Durham, NC, 1994.
18. G. J. Klir and B. Yuan, *Fuzzy Sets and Fuzzy Logic: Theory and Applications.* Prentice Hall, Englewood Cliffs, NJ, 1995.
19. T. S. Kuhn, *The Structure of Scientific Revolutions.* Univ. of Chicago Press, Chicago, 1962.
20. D. V. Lindley, *Statist. Sci.* **2**, 17–24 (1987).
21. M. Mares, *Computation Over Fuzzy Quantities.* CRC Press, Boca Raton, FL, 1994.
22. M. McLeish, ed., *Computer Intelligence* **4**, 57–142 (1988).
23. R. E. Moore, *Methods and Applications of Interval Analysis.* SIAM, Philadelphia, 1979.
24. M. Otto and R. R. Yager, *Rev. Int. Syst.* **6**, 465–481 (1992).
25. B. Russell, *Australian J. Psychol. Philos.* **1**, 84–92 (1923).
26. A. K. Seth, *Bull. Math. Assoc. India* **22**, 1–42 (1990).
27. L. Shastri, ed., *IEEE Expert* **9**, 2–49 (1994).

28. D. Singer, *Fuzzy Sets and Systems* **47**, 39–48 (1992).
29. D. Singer and P. G. Singer, *Int. J. Systems Sci.* **24**, 1363–1376 (1993).
30. M. Smithson, *Ignorance and Uncertainty: Emerging Paradigms*. Springer-Verlag, New York, 1989.
31. J. C. Smuts, *Holism and Evolution*. Macmillan, London, 1926.
32. M. Sugeno, "*Theory of Fuzzy Integrals and its Applications*," Ph.D. dissertation, Tokyo Institute of Technology, Tokyo, 1974.
33. M. Sugeno *et al.*, *Proceedings of International Fuzzy Engineering Symposium 1991*, pp. 1120–1121, 1991, Yokohama, Japan.
34. W. Thomson, *Popular Lectures and Addresses*. MacMillan, London, 1891.
35. R. Viertl, *Probability and Bayesian Statistics* (R. Viertl, ed.), pp. 471–475.
36. Z. Wang and G. J. Klir, *Fuzzy Measure Theory*. Plenum Press, New York, 1992.
37. L. A. Zadeh, *Inform. Control* **8**, 338–353 (1965).
38. L. A. Zadeh, *J. Math. Anal. Appl.* **23**, 421–427 (1968).
39. L. A. Zadeh, *IEEE Trans. on Syst. Man., Cybernetics* **1**, 28–44 (1973).
40. L. A. Zadeh, *Fuzzy Sets and Systems* **1**, 3–28 (1978).
41. L. A. Zadeh, *Int. J. Gen. Systems* **17**, 95–105 (1990).
42. J. Zhang, *et al.*, *Inform. Sci.* **72**, 271–284 (1993).

3

Fuzzy Restrictions and Inherent Uncertainties in Chirality Studies

KURT MISLOW

Department of Chemistry
Princeton University
Princeton, New Jersey 08544

I. INTRODUCTION

The definitions of "chiral" and the associated terms "homochiral" and "heterochiral" made their first appearance in a footnote of a lecture, entitled "The Molecular Tactics of a Crystal," that Sir William Thomson, who had been elevated to Lord Kelvin in 1892, delivered to the Oxford University Junior Scientific Club on May 16, 1893.[1] The famous footnote reads:

> I call any geometrical figure, or group of points, *chiral*, and say that it has chirality, if its image in a plane mirror, ideally realized, cannot be brought to coincide with itself. Two equal and similar right hands are homochirally similar. Equal and similar right and left hands are heterochirally similar or "allochirally" similar (but heterochirally is better). These are also called "enantiomorphs," after a usage introduced, I believe, by German writers. Any chiral object and its image in a plane mirror are heterochirally similar.

The term "chiral" was thus given a strictly geometrical meaning by Kelvin: an object is chiral if and only if it cannot be rendered congruent with its mirror image by proper rotations combined with translations in the object's space (a "direct" congruence). Alternatively, a chiral geometrical object is one whose point group does not contain any improper rotations—simple or rotary reflections.

From the context of the above quotation, "similar" may be taken to have the same meaning as "congruent;" according to Kelvin's formulation, therefore, the only allowed operations are isometries—transformations in which the distances that separate any two points within the object remain invariant. The terms "homochiral" and "heterochiral" then acquire the following meaning: two rigidly chiral objects are homochiral if and only if they can be rendered directly congruent, whereas two rigidly chiral objects are heterochiral if and only if they can be rendered congruent by an improper rotation combined with proper rotation-translations in the object's space (an "opposite" congruence).

According to Kelvin's definition, every geometrical object is either chiral or achiral, and no object is both. The term "chirality," as applied to geometrical objects, including models of molecules, is therefore clear-cut and unambiguous. Whether one calls a chemical system "chiral" or "achiral" is, however, based on experimental observations; the two-valued logic that is appropriate for geometrical objects then no longer applies and it becomes necessary to resort to fuzzy logic. Uncertainty, if not fuzziness, also attends any attempt to quantify the chirality of geometrical objects, let alone of chemical systems: the reason is that many different measures of chirality exist, and they all give different and sometimes conflicting results. Finally, attempts to partition sets of chiral objects into homochirality classes may, under certain circumstances, be doomed to failure, so that it becomes meaningless to speak of objects in such sets as being "right-handed" or "left-handed."

Uncertainty and inexactitude are inherent to science. As indicated in the preceding text and as described subsequently in further detail, the study of chirality and associated phenomena proves to be no exception.

II. MANIFESTATIONS OF CHIRALITY AND THE CHOICE OF MODELS

The symmetry of the model of a molecule or of a molecular ensemble depends on the conditions of the relevant physical (or chemical) measurement, and may vary for the same system according to time scale of observation and instrumental sensitivity. Whether the model of a chemical system is chiral or achiral may therefore depend on the conditions of observation. There is no ambiguity when chirality properties are observed: the hemihedrality of quartz crystals, the optical rotation of hexahelicene, and the enantiospecificity of hog-kidney acylase, for example, are all unmistakable manifestations of an underlying structural chirality. On the other hand, achirality is not so simply implied by the absence of such observations.

Consider, for example, a vessel that contains a statistically significant number of molecules, such as a mole of helium atoms at room temperature. With a high degree of probability bordering on certainty, this ensemble is asymmetric at any instant in time because, a priori, *any system is expected to be asymmetric unless constrained to be otherwise*, and no such constraints are enforced in the present example. Thus, there is no reasonable expectation that at any instant in time the ensemble of helium atoms in our container will be achiral or that at any later instant a mirror image configuration will occur: there are just too many degrees of freedom. In short, geometrical achirality is practically unattainable in a sample with a statistically significant number of molecules. Nevertheless, all feasible measurements on this system will give the appearance of achirality, due to time-averaged cancellations of randomly fluctuating, local chiral effects. We have called such systems *stochastically achiral.*[2]

This analysis may be extended to formally achiral molecules that are composed of four or more atoms. The motions in such polyatomic molecules are restricted by the restoring forces imposed by bonding, and stochastic achirality is here the result of internal vibrations. Thus, for example, molecular deformations in some vibrational states impart chirality to the methane molecule, but the sense of chirality averages to zero under the conditions of measurement. As this discussion makes clear, the conventional T_d symmetry of methane is a property solely of the model.

Similar considerations apply to systems at equilibrium in which the sample consists of molecules that are chiral and that undergo rapid enantiomerization on the time scale of observation, for example *cis*-1,2-difluorocyclohexane and ethylmethylpropylamine. The ground-state geometries of these molecules are asymmetric, but, under the dynamic conditions described in the foregoing text, an achiral model (with C_s symmetry for the two cited examples) is a proper representation of the system. In that sense the dynamic model differs in no significant respect from the model of a conventionally achiral system, such as methane. Of course, the probability is small that at any instant the enantiomeric mixture at equilibrium is exactly equimolar; the absence of observable chirality phenomena such as optical activity is the result of rapid cancellations of random statistical fluctuations of activity in the time domain of observation. In other words, although at any instant in time the mixture (with a high degree of probability) has an excess of one enantiomer or the other, under the conditions of measurement it *effectively* contains an equal number of enantiomeric molecules.

A. Cryptochirality

Consider a sample of racemic 2-butanol at room temperature, prepared by hydrogenation of 2-butanone under achiral conditions. In contrast to *cis*-1,2-difluorocyclohexane and ethylmethylpropylamine, molecules of 2-butanol show no evidence of enantiomerization on the leisurely time scale that is associated with optical rotation measurements or with the separation of enantiomers. Because the probability is exactly one half that any macroscopic sample has an odd number of molecules, there is an even chance that at least one chiral molecule in our sample remains uncompensated. That is, the racemic mixture is not *strictly* racemic. Even if the sample has an even number of molecules, the probability is extremely small that it is strictly racemic. According to Mills,[3] "when 10,000,000 dissymmetric [i.e., chiral] molecules are produced under conditions which favour neither enantiomorph, there is an even chance that the product will contain an excess of more than 0.021% of one enantiomorph or the other. It is practically impossible for the product to be absolutely optically inactive." Furthermore, although the percent enantiomeric excess decreases with increasing size of sample, the actual number of molecules in excess increases and the probability of obtaining a strictly racemic sample decreases. In light of these facts, it is safe to conclude that hydrogenation of 2-butanone under achiral conditions will yield 2-butanol enriched in one of the two enantiomers. That is, "total asymmetric synthesis," even in the absence of a chiral bias, is unavoidable on statistical grounds alone. Note, however, that the probability of any given experiment yielding an excess of, say, the "right-handed" enantiomer is exactly one half, so that overall parity is still preserved for a statistically significant set of experiments.

As the preceding discussion suggests, conditions may exist under which even the most powerful measuring device available will be incapable of detecting a significant difference between samples of different enantiomeric composition above the noise level of stochastic achirality. We call such a system *cryptochiral*, because the model demands an excess of one enantiomer over the other in the time domain of observation, while the chirality phenomenon to be observed falls below the threshold of the operational null and thus is undetectable.[2] Note that a cryptochiral substance is operationally indistinguishable from a stochastically achiral one because, at and below the operational null, enantiomorphous systems can be neither differentiated from each other nor distinguished from achiral ones.

The classic example of cryptochirality comes from triglyceride chemistry. The biogenous long-chain acid triglycerides of the type $CH_2(R)CH(R')CH_2(R')$ or $CH_2(R)CH(R')CH_2(R'')$ are optically inactive, even though they are presumed to be enantiomerically pure since they can be prepared from D-mannitol by a series of reactions that do not involve racemization. It is therefore clear that "the natural triglycerides, though they do not show a rotation, are not necessarily racemic, but might easily occur in either of the two enantiomorphic forms."[4] Symmetry arguments tell us that here, as in other examples of cryptochiral compounds where optical activity falls below the operational null,[5, 6] the effect on the plane of polarized light, though immeasurably small, is not identically zero. The absence of demonstrable activity merely reveals the inadequacy of the detection device.

Another example may help to clarify matters further. In the series of enantiomerically pure compounds $H(CH_2)_nCHOH(CH_2)_nD$, the observed chirality properties are expected to decrease in magnitude with an increase in n. Similarly, the degree of asymmetric induction in an asymmetric synthesis that utilizes this carbinol is expected to decrease in magnitude with an increase in n; in this case all experiments yield cryptochiral products even though the model still demands different levels of activity below the operational null for different values of n.

Let us return to the example of 2-butanol. On the basis of any conceivable experimental observation, a sample 2-butanol with an enantiomeric excess $(R - S)/(R + S)$ of, say, $(10^{23} - 1)/(10^{23} + 1)$ is seen to be chiral and operationally indistinguishable from one that is enantiomerically pure in the strictest sense. On the other hand, a cryptochiral sample with an enantiomeric excess of, say, $1/(10^{23} + 1)$ would appear to be achiral because any chirality property measured on it would fall way below the operational null. It might be argued that a sample with an enantiomeric excess even as minute as $1/(10^{23} + 1)$ cannot properly be considered achiral, since, if it were to be so regarded, one would then be forced into the position of having to admit to the existence of a value for the enantiomeric excess, somewhere between $1/(10^{23} + 1)$ ("achiral") and 1 ("chiral" and enantiopure), beyond which the mixture could no longer be described as achiral. This value would vary, depending on the particular chirality property, and would furthermore be tied to the operational null. By way of this argument—which is completely analogous to the classical paradox induced by the question of whether a man with only one more hair than a bald man is still bald—one is inexorably led to the conclusion that it is impractical, not to say unreasonable, to draw a sharp line between chiral and achiral molecular ensembles: in contrast to the crisp classification of geometrical objects into chiral and achiral ones, in the present case one deals with a fuzzy borderline distinction, and the qualifier "operationally" must be implicitly attached to "achiral" or "racemic" whenever one uses these terms with reference to observable properties of a macroscopic sample.

It thus appears that proper application of the terms "chiral" and "achiral" to real (chemical) systems requires "fuzzification." To place this concept in proper perspective, we need to take a look at the role of inexactitude in scientific communication.[2]

B. Fuzzy Set Theory and Chemistry

As chemists, we often employ inexact terms, such as fast/slow, strong/weak, concentrated/dilute, hot/cold, etc., yet we are confident that these words suffice to carry the desired message, unburdened by superfluous precision, within the context of the report. Although quantification of these terms by an appropriate measurement is always possible, it is often unnecessary, as in the specification of color. Indeed, it is generally recognized that terms such as the above play a vital role in the efficient communication of information when the message is conveyed in a natural (in contrast to machine) language, e.g., young/old, tall/short, or rich/poor. One cannot help agreeing with Black[7] that all empirical concepts are "loose" in the sense that they are not sharply bounded.

A formal classification theory for such terms, which are neither very exact nor very inexact, was introduced by Zadeh,[8] who, among others, saw the need for generalizing the theory of sets to admit the possibility of unsharp boundaries between classes of imprecisely defined objects. Whereas the characteristic function can only have a value of 0 or 1 (in harmony with the Aristotelian logic of the excluded middle), membership in a fuzzy set is a matter of degree. This degree (or grade) of membership (μ) may have any value in the continuum of the real interval [0, 1]; in the special case where the value of μ is restricted to 0 or 1, the degree of membership becomes the characteristic function.

For example, let us examine the proposition, "the solution is hot." If the universe of discourse is the domain of temperatures (T), one might choose to assign to a subset of temperatures, say 0–300°C, a range of μs such that $\mu = 0$ and 1 corresponds to 0 and 300, respectively. This constitutes a fuzzy restriction on the meaning of the predicate "hot." The meaning of the proposition "the solution is hot" is then elastically constrained by the value of μ assigned to the temperature of the solution; for example, for $T = 200$°C, μ might have a value of 0.8. Similarly, the truth value of the proposition is neither 0 nor 1, as it would have to be in two-valued logic, but something in between. In short, the principle of cutting, i.e., a partitioning into the equivalence classes "hot" and "not hot," is inapplicable because the categories are fuzzy.

The mapping of T from the subdomain of temperatures onto the interval [0, 1] is the membership or compatibility function $\mu(T)$. The form of this function is subject to some relatively nonarbitrary constraints: to be consistent with one's primitive notions of hotness, $\mu(T)$ should increase monotonically and smoothly with T, and it should be roughly sigmoidal in shape. However, since a whole family of curves will fit this description, the choice of numerical values for the parameters of $\mu(T)$ will to a considerable extent be an arbitrary one. Therefore, the threshold criteria, which are given by $\mu(T)$, are themselves fuzzy.

C. Chirality as a Primitive Fuzzy Concept

Just as hot, young, rich, tall, and hirsute may be expressed by compatibility functions μ(°C), μ(years), μ($), μ(cm), and μ(no. of hairs), so the fuzzy "chiral" may be expressed on the scalar of a numerically valued variable that quantitatively expresses some manifestation of chirality, e.g., optical rotation, chemical shift difference, or rate constant, with the appropriate domain understood. It is therefore permissible to speak of degrees of chirality and to compare molecules or molecular ensembles by the use of expressions such as "more" or "less" chiral or achiral. For example, in the series of enantiomerically pure compounds $H(CH_2)_nCHOH(CH_2)_nD$ discussed previously, one has a clear sense that samples with large values of n are somehow less chiral, or more achiral, than those with small values. As another example, consider the series of enantiomerically pure molecules $^{14}CH_3(CH_2)_nCH(CH_3)(CH_2)_nCH_3$. It has been asserted that the compound with $n = 5$ "would have zero rotation within experimental error on the most sensitive polarimeter now available."[10] Despite their cryptochirality, however, it is still possible to speak of molecules with $n = 4$ and $n = 6$ as being more chiral and less chiral, respectively, than the molecule with $n = 5$, both with reference to molecular structure and to chirality properties.

The ranking of chiral systems such as the foregoing by their structures and associated properties is intuitively obvious, and is arrived at independently of any theory that accounts for these properties. For this reason we are led to the conclusion that chirality in a real system is a primitive fuzzy concept.[2]

Fuzzification does not necessarily destroy rank ordering. Just as younger and hotter are unambiguous binary relations, even though young and hot are fuzzy concepts, so more or less chiral or achiral provides an unambiguous order of chirality for a given membership function. Linguistic hedges[9] may also be applied, as in expressions such as "much more chiral." With reference to a phenomenon observed just above the threshold of the operational null, one might speak of the system as "barely (or feebly) chiral." Although the terms "chiral" and "achiral," like "pure" and "perfect," are absolutes, there is no likelihood of confusion when these words are used in their fuzzy (real system) sense. After all, a chemist may say "purer" rather than "less impure" without danger of being misunderstood. "More perfect" is another locution hallowed by usage, and here too the meaning is unmistakable. Thus, although "chiral" and "achiral" are not formally fuzzy terms when used in the absolute (geometrical) sense, they become fuzzy in the context of chemical usage.

To conclude: it is natural to abandon two-valued logic in science because, when one deals with the phenomena of nature, one enters "a stage in logic in which we recognize the utility of imprecision".[11] The vagueness of "chiral" and "achiral," when used with reference to a chemical system, should therefore be regarded as a virtue rather than as a defect.

D. On Quantifying the Chirality of Geometrical Objects

We saw that it is natural to attach a quantitative meaning to the chirality of molecular structures and that it is possible to rank the chirality of closely related structures. There is no reason not to apply measures of chirality as well to geometrical objects, including geometrical models of molecules and molecular ensembles. We recently provided an overview of the scope and limitations of the many chirality measures that have been developed.[12, 13] Here we describe some of the uncertainties that are inherent in the application of such measures, and we illustrate them with reference to the two-dimensional chirality of the right triangle, one of the simplest of geometrical figures.

Consider two right triangles, A and B, with internal angles of $18.8°/71.2°/90°$ and $30°/60°/90°$, respectively, and their degrees of chirality χ as calculated by two different measures.[12] According to a measure based on geometric chirality products,[14] A ($\chi = 0.074$) is more chiral than B ($\chi = 0.057$), whereas, according to a measure based on symmetry coordinates,[15] B ($\chi = 0.500$) is more chiral than A ($\chi = 0.313$). As this example illustrates, the relative ranking of objects according to degree of chirality depends on the particular function that is chosen as the chirality measure.

Equally uncertain is the shape of the most chiral member in a given class of objects. Once again, we illustrate this problem with reference to right triangles. The answer to the question "What is the shape of the most chiral right triangle?" depends on the particular function that is chosen as the chirality measure: according to measures based on geometric chirality products,[14] on symmetry coordinates,[15] on common volumes,[16] and on Hausdorff distances,[12, 17] the smallest internal angles are $18.8°$, $30.0°$, $37.5°$, and $35.2°$, respectively.[12] Similar difficulties are encountered in response to the question "What is the shape of the most chiral triangle?" According to the Hausdorff measure, the internal angles of the most chiral triangle are $21.02°$, $43.69°$, and $115.29°$.[17b] A dramatically different answer is provided by the aforementioned three other measures. According to these measures, the extremal triangles that correspond to the supremum of their respective functions are only approached in the limit, and the most chiral triangles are therefore infinitely flat![12]

We have seen that different chirality measures give incommensurable values for the shape and the relative degree of chirality of a given object. In short, there is no such thing as a unique scale by which chirality can be measured. If a choice among the numerous chirality measures that have been discussed in the literature[13] is to have any relevance to the world of observables, it must be predicated on the measure's ability to correlate the calculated degree of chirality with some experimentally determined chirality property. That is, the measure must successfully model some pseudoscalar physical or chemical property. As will be discussed subsequently, this goal may prove to be elusive.

III. THE HOMOCHIRALITY PROBLEM

Any arbitrarily selected set C of chiral geometrical objects c_i can always be mapped one-to-one onto the set C' of the corresponding enantiomorphs c_i'. The set $C \cup C'$ is therefore a union of heterochiral pairs:

$$C \cup C' = \bigcup_{i=1} \{c_i, c_i'\}.$$

While a sense of chirality can be defined arbitrarily for the members of each individual heterochiral pair in such a set, so that one member within each pair is "right-handed" and the other "left-handed," it does not follow from this that the "right-handed" member of one pair bears a well-defined relationship to the "right-handed" member of another pair. In other words, it is not, in general, possible to partition the members of the set $C \cup C'$ into two homochirality classes, right-handed (**R**) or left-handed (**L**). This is the homochirality problem of the title.

A. Homochirality Classes

The existence of homochirality classes requires a uniform rule or procedure, such as the one provided by Kelvin's definition. This definition implies, however, that consideration be restricted to sets of isometrically congruent objects. Accordingly, a set $C \cup C'$ of geometrical objects can be partitioned into **R** and **L** classes only if members of $C \cup C'$ are pairwise either directly or oppositely congruent. Seen in this light, Kelvin's definition seems unduly restrictive. For example, if some standard helix were to be defined as right-handed, we could easily recognize right- and left-handedness in all sorts of helices, regardless of pitch and radius, and we should be able to partition all of them into **R** and **L** homochirality classes. Similarly, all human hands, though clearly never identical, can nevertheless be readily classified as "right" or "left," regardless of differences in size, shape, or color, because they all share an easily recognizable property appropriately enough called handedness. As a chemical example, hog-kidney acylase hydrolyzes only the natural enantiomers of N-acylamino acids, regardless of the structure of the R group in RCHNHAcCOOH. It would thus seem reasonable to regard N-acyl-L-amino acids as homochirally similar and as members of the same homochirality class.[2]

It is clear from these examples that even though a set of chiral objects may not be isometrically congruent, they may nevertheless share some characteristic feature that makes it possible to recognize a common sense of chirality. As an example from geometry, consider the set of triangles in Euclidean two-space (E^2) that are unlabeled, i.e., the set of triangles whose vertices are undifferentiated except for their positions in coordinate space. Scalene triangles, such as those shown in Fig. 1, are chiral in E^2, and the enantiomorphs are characterized by the orientation, clockwise (**R**) vs anticlockwise (**L**), of the sides arranged in the order largest (l) > medium (m) > smallest (s). Enantiomorphous triangles are represented as points in a pair of two-dimensional shape spaces that are separated by a one-dimen-

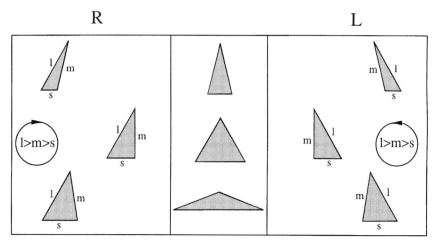

FIGURE 1 Unlabeled triangles in the plane. Enantiomorphs are characterized by the orientation, clockwise vs anticlockwise, of the sides arranged in the order largest (l) > medium (m) > smallest (s). **R** and **L** spaces are separated by a subspace of achiral triangles.

sional boundary; the points in this boundary space represent achiral triangles.[14] There is no constraint on the shape of the individual triangles. That is, the triangles need not be isometrically congruent. Nevertheless, the triangles within each shape space are all homochirally similar because they share the same chiral characteristic, **R** or **L**, given the order $l > m > s$. Going beyond Kelvin's definition thus allows us to partition all triangles into **R** and **L** homochirality classes, regardless of size and shape. The requirement for an acceptable division between right- and left-handed objects,[18] as applied to triangles, is satisfied by the following conditions: (1) every scalene triangle is either in an **R** or in an **L** space, and not on the boundary between them, (2) every triangle in **R** is matched by its enantiomorph in **L**, and (3) achiral triangles are not in either **R** or in **L**, but on the boundary between them. Note, however, that the existence of an achiral boundary is not a requirement in all cases for a separation into **R** and **L** classes. As an example, human hands naturally fall into disjoint sets that share no achiral boundary.

The triangles within each shape space of Fig. 1 can be rendered congruent by continuous deformation only along pathways that do not contain achiral points ("chiral pathways"). That is, the triangles are all chirally connected. This characteristic applies generally to the members of any homochirality class. On the other hand, the conversion of any given scalene triangle into its enantiomorph by continuous deformation in E^2 *inevitably* requires passage through an achiral (in E^2) isosceles or equilateral triangle, that is, along an achiral pathway. The triangles in the **R** and

L spaces of Fig. 1 constitute homochirality classes because enantiomorphous triangles cannot be chirally connected: passage between the two spaces is impossible without crossing an achiral boundary.

The homochirality problem arises whenever there exists a pathway of continuous deformation that connects two enantiomorphous objects but does *not* require passage through an achiral point. Under these circumstances, it becomes meaningless to speak of such objects as right-handed or left-handed. The case of the asymmetric tetrahedron is the simplest example.

The asymmetric tetrahedron with unlabeled vertices is the three-dimensional analog of the unlabeled scalene triangle. While conversion of such a tetrahedron to its enantiomorph by way of an achiral tetrahedron is certainly not excluded, as illustrated in Fig. 2 for interconversion by way of a C_s-symmetric intermediate, achiral pathways are easily circumvented because the set of achiral unlabeled tetrahedra in E^3, unlike the set of achiral unlabeled triangles in E^2, does not form a boundary between heterochiral sets.

An asymmetric tetrahedron can thus be converted to its enantiomorph by continuous deformation along a chiral pathway without ever having to pass through an achiral intermediate.[17a] One of an infinite number of such pathways is shown in Fig. 3. Enantiomorphous unlabeled tetrahedra are therefore chirally connected. As a consequence, it is impossible to isolate a geometrical characteristic, such as the $l > m > s$ sequence for the triangles in Fig. 1, that would allow partition of the set $C \cup C'$ of unlabeled tetrahedra into homochirality classes **R** and **L**: where, along the chiral pathway that connects two enantiomorphous tetrahedra, would one cross the boundary between **R** and **L**? In short, it is impossible to speak of chiral unlabeled tetrahedra in general as being either right-handed or left-handed.

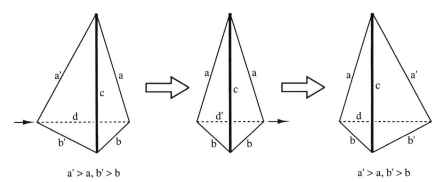

a' > a, b' > b   a' > a, b' > b

FIGURE 2 Conversion of an unlabeled asymmetric tetrahedron (left) into its mirror image (right) by continuous deformation (small arrows) of the geometric figure along an achiral pathway. The achiral intermediate (center) has C_s symmetry.

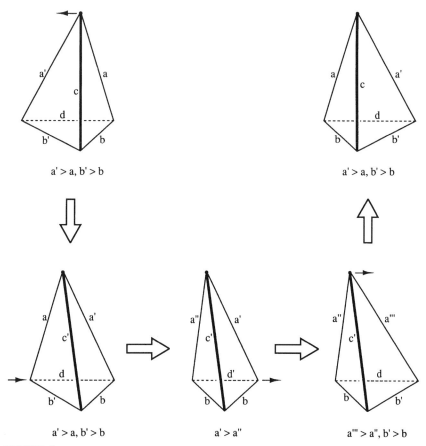

FIGURE 3 Conversion of an unlabeled asymmetric tetrahedron (top left) into its mirror image (top right) by continuous deformation (small arrows) of the geometric figure along a chiral pathway.

The unlabeled triangle is the simplex in E^2 (2-simplex) and the unlabeled tetrahedron is the simplex in E^3 (3-simplex); evidently, whether enantiomorphous n-simplexes can be partitioned into homochirality classes depends on the dimension of E^n. Recall that an n-simplex is a convex hull of $n + 1$ points that do not lie in any $(n - 1)$-dimensional subspace and that are linearly independent; that is, whenever one of the points is fixed, the n vectors that link it to the other n points form a basis for an n-dimensional Euclidean space E^n. An n-simplex may be visualized as an n-dimensional polytope (a geometrical figure in E^n bounded by lines, planes, or hyperplanes) that has $n + 1$ vertices, $n(n + 1)/2$ edges, and is bounded by $n + 1$ $(n - 1)$-dimensional subspaces. It has been shown that the homochirality problem for the simplex in E^3 is shared by all n-simplexes in E^n, $n > 2$.[19] Triangles are therefore unique in that they alone among all simplexes can be collected into homochirality classes, i.e., partitioned into heterochiral sets. In addition, since each n-simplex is the simplest geometrical construction in its E^n, it follows that no geometrical object in E^3 or higher dimensions can be assigned to homochiral **R** or **L** classes.

B. Ruch's Model

In the course of their systematic development of a general theory of chirality products, Ruch and co-workers[20] developed a model in which a set of n ligands is partitioned among the n sites of an achiral permutation skeleton. The permutations are transpositions of ligands on sites that are related by mirror planes of the skeleton. The skeletons are of two types[20d]: those in which every mirror plane contains $n - 2$ sites (category a) and those in which there are also mirror planes that contain fewer than $n - 2$ sites (category b). Skeletons in category a are exemplified by the regular tetrahedron (T_d), the trigonal pyramid (C_{3v}), the trigonal bipyramid (D_{3h}), and the tetragonal disphenoid (D_{2d}), while skeletons in category b are exemplified by the square pyramid (C_{4v}), the octahedron (O_h), and the cube (O_h).

Consider now the transposition of two ligands at the vertices of a regular tetrahedron. All six mirror planes of the regular tetrahedron contain $n - 2 = 2$ vertices, and the skeleton therefore belongs to category a. Each ligand is associated with a continuously varying, scalar parameter λ. As Ruch has pointed out,[20i] ligands may be symbolized by spheres of variable diameter λ that are centered at the vertices of an achiral polyhedron. If the transposition of the two ligands at sites a and b in Fig. 4 is then visualized as a continuous change in the diameters of the corresponding spheres, there inevitably comes a point at which the two diameters are the same and the model becomes achiral. The same is true for the transposition of any one of the six pairs of ligands on the tetrahedral frame. In general, the pathway of continuous deformation that connects two enantiomorphs in models belonging to category a necessarily requires passage through an achiral model. The requirement for an acceptable division between right- and left-handed objects is therefore satisfied for all skeletons in this category, and the models can be assigned to homochiral **R** and **L** classes.

In contrast to the tetrahedron, two of the four mirror planes of the square pyramid contain only one of five vertices and therefore fewer than $n - 2 = 3$ ligand sites. This skeleton therefore belongs to category b. If the transposition of the two ligands at sites a and b in Fig. 5 is also visualized as a continuous change in the diameter of the corresponding spheres, there is no point at which the model becomes achiral. The same is true of

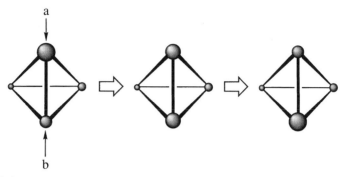

FIGURE 4 Conversion of an asymmetric tetrahedron (left) into its mirror image (right) along an achiral pathway. Four spheres with different but variable diameters are centered at the vertices of a regular tetrahedron. The spheres labeled a and b shrink and expand, respectively, until their diameters are switched. Along this path a point is reached at which the diameters are the same; this is the achiral intermediate (center) with C_s symmetry.

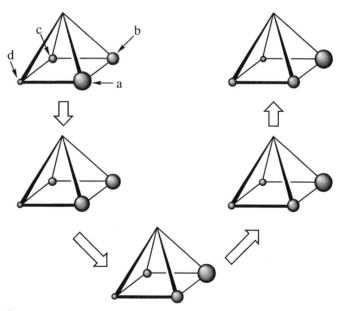

FIGURE 5 Conversion of an asymmetric square pyramid (top left) into its mirror image (top right) along a chiral pathway. Four spheres with different but variable diameters are centered at the vertices of the square base. The spheres labeled a and b shrink and expand, respectively, until their diameters are switched. This is followed by a similar switch of the diameters of spheres c and d.

the subsequent step, the transposition of ligands at sites c and d. Hence, the enantiomorphs are chirally connected. In general, the pathway of continuous deformation that connects two enantiomorphs in models belonging to category b need not entail passage through an achiral model. The requirement for an acceptable division between right- and left-handed objects is therefore not satisfied for skeletons in this category, and the models cannot be assigned to homochiral **R** and **L** classes.

C. Chiral and Achiral Enantiomerization Pathways

Enantiomerization of chiral tetraatomic molecules by unimolecular processes unavoidably requires the intermediacy of achiral structures, as, for example, in the racemization of NHDF by inversion at nitrogen or of H_2O_2 by rotation about the $O—O$ bond. Abstractly considered, it is easily seen that any chiral set of four differently labeled points $ABCD$ in E^3 cannot be converted to the enantiomorphous set without passage through a planar state, i.e., by an achiral pathway. Similarly, any chiral set of three differently labeled points ABC in E^2 cannot be converted to the enantiomorphous set without passage through a one-dimensional state in which all three vertices are collinear (Fig. 6). In general, once all their points are differently labeled, such "maximally labeled" simplexes in E^n require passage through a $(n-1)$-dimensional boundary space to reach the enantiomorphs, regardless of whether the arrays of the corresponding unlabeled points are chiral in E^n or not. The $(n-1)$-dimensional space

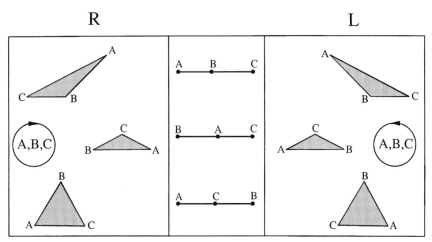

FIGURE 6 Labeled triangles in the plane. Enantiomorphs are characterized by the orientation, clockwise vs anticlockwise, of the vertices arranged in the order $A \to B \to C$. **R** and **L** spaces are separated by a subspace of one-dimensional (degenerate) triangles.

thus splits the n-dimensional space into two subsets that contain the enantiomorphs; the boundary space itself consists of points that represent objects that are chiral in E^{n-1} but achiral in E^n. Maximally labeled n-simplexes $(n > 2)$, in contrast to the unlabeled ones discussed in the preceding text, can therefore be partitioned into heterochiral sets, identified as **R** and **L** for the triangles in Fig. 6.

Enantiomorphs of maximally labeled sets that are more complex than simplexes are, however, chirally connected. In a chemical context, this means that any chiral molecule composed of five or more different atoms is in principle capable of conversion to its enantiomer by chiral as well as achiral pathways, *provided*, of course, that a feasible route exists on the potential energy surface.[21, 22] Consider, for example, rotation about the N—O bond and inversion at the nitrogen atom in the deuterated hydroxylamine derivative NHDOH. Figure 7 shows a graph whose vertices represent various conformations $(A-D)$ and their respective enantiomers (A^*-D^*), and whose edges represent inversions (i) at nitrogen and 90° rotations (r) about the N—O bond. Each of these conformations can be converted to its enantiomer either by an achiral or by a chiral pathway. For example, the single inversion that converts A into A^* represents an achiral pathway by way of a planar intermediate conformation (indicated by a solid circle at the center of the edge). Alternatively, the three-step itineraries $A \rightarrow D \rightarrow B^* \rightarrow A^*$ and $A \rightarrow B \rightarrow D^* \rightarrow A^*$ are the shortest chiral pathways for the enantiomerization of A. Similar alternatives exist for $C \rightarrow C^*$. The shortest achiral pathways for $B \rightarrow B^*$ are $B \rightarrow A \rightarrow A^* \rightarrow B^*$ or $B \rightarrow C \rightarrow C^* \rightarrow B^*$, while B and B^* are chirally connected by way of $B \rightarrow D^* \rightarrow C^* \rightarrow B^*$ or $B \rightarrow D^* \rightarrow A^* \rightarrow B^*$. Similar alternatives exist for $D \rightarrow D^*$.

The existence of chiral pathways in this molecule is made possible by the existence of the two independent degrees of freedom that govern internal motion, rotation, and inversion. As molecular complexity increases, the number of degrees of freedom also increases and, unless an achiral pathway is energetically much preferred, it becomes more and more likely that enantiomerization proceeds by a chiral pathway. For example, it is extremely improbable that reversal of helicity in a polymeric chain involves an achiral intermediate or transition state. There is a strong resemblance here to the stochastic achirality of ensembles of achiral molecules discussed previously.

A formal limit is reached when, due to structural constraints, all achiral pathways along the enantiomerization trajectory become energetically inaccessible under normal laboratory conditions. Chiral pathways then remain the only alternative. In 1954 it was pointed out that a compound of the type 4-[(R)-*sec*-butyl]-4'-[(S)-*sec*-butyl]-2,2',6,6'-tetramethylbiphenyl consists entirely of asymmetric molecules that undergo rapid enantiomerization, and that "conformational racemization, in the

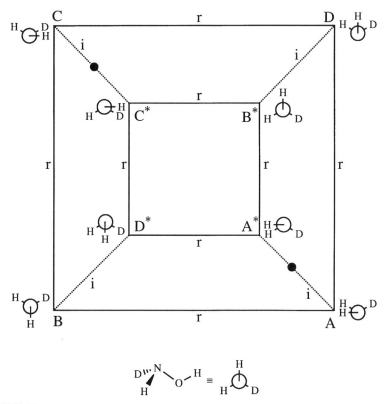

FIGURE 7 Graph of conformational interconversion paths for NHDOH. Edges represent 90° rotations (r) about the N—O bond or inversions (i) at nitrogen. Planar conformations are represented by solid circles.

case of this type of compound, cannot proceed via a symmetric [i.e., achiral] intermediate."[23a] The following year saw the actualization of a molecule of this type with the synthesis of $(1R)$-menthyl $(1S)$-menthyl $2,2',6,6'$-tetranitro-$4,4'$-diphenate[23b] (Fig. 8). The special property of this molecule can be understood with reference to the schematic sketch at the top of Fig. 8. If the four blocking groups (R) in the $2,2',6,6'$-positions are large enough, the biphenyl cannot become planar. Under these circumstances, the only achiral conformation of the biphenyl moiety, taken by itself, is D_{2d}, with the two σ planes intersecting at the central bond axis. At the same time, the only achiral conformation that is available to the two substituent groups -Cabc at the biphenyl $4,4'$-positions (taken together as a unit separate from the biphenyl) is either C_s, with a σ plane perpendicular to the central bond axis (as shown in Fig. 8), or C_i. Applying Curie's principle of superposition, either one of the two intersections

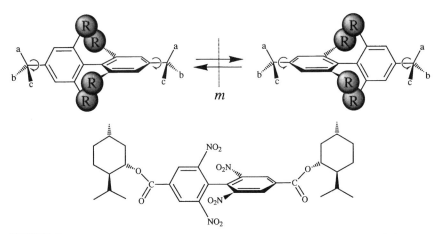

FIGURE 8 Top: Schematic drawings of two biphenyls with bulky substituents (R) in the $2,2',6,6'$ positions and enantiomorphous -Cabc groups in the $4,4'$ positions. The two mirror-image-related (m) biphenyls are interconverted by a 90° conrotatory twist about two single bonds, as indicated by the arrows. Bottom: ($1R$)-Menthyl ($1S$)-menthyl $2,2',6,6'$-tetranitro-$4,4'$-diphenate. Reproduced with permission from *Croat. Chem. Acta.* **69** 485 (1996).

$D_{2d} \cap C_i = C_1$ or $D_{2d} \cap C_s = C_1$ results in complete desymmetrization. In other words, the identity is the only element common to the point groups of D_{2d} and C_i, and the same is true of D_{2d} and C_s since the σ plane in C_s is perpendicular to both σ planes in D_{2d}. It follows that the molecule is asymmetric in all possible conformations and that the compound exists as a mixture of transient dl pairs. The d and l conformations can, however, interconvert by torsion around the bonds to -Cabc; for example, either one of the two enantiomorphous conformations in Fig. 8 can convert into the other by a twist of the biphenyl moiety (or, equivalently, of the two end groups) in either direction. We are therefore dealing with a "chemically achiral"[24] compound composed of asymmetric molecules whose enantiomeric conformations interconvert exclusively by chiral pathways.

Because it was the first of its kind, this prototype of "Euclidean rubber gloves"[25] secured a place in the textbook literature[26] and served to illustrate the use of symmetry and permutation groups in the description of nonrigid molecules.[24, 27, 28] Since that time, however, other compounds have been found that share with ($1R$)-menthyl ($1S$)-menthyl $2,2',6,6$-tetranitro-$4,4'$-diphenate the curious property of being composed of asymmetric molecules that interconvert exclusively by chiral pathways. Examples from our laboratory include asymmetric molecular propellers, whose enantiomerization by the two-ring flip mechanism involves an exclusively chiral pathway,[29] and certain bis(9-triptycyl)methane derivatives in which the two triptycyl groups behave as highly mobile (i.e., essentially frictionless) and tightly meshed (i.e., securely interlocked) bevel gears.[30] The cogs in one rotor of these molecular gears are benzene rings that fit snugly into the V-shaped notches formed by two rings in the other rotor, and the system undergoes dynamic gearing (correlated disrotation) with a low barrier, comparable in magnitude to that in ethane. Gear slippage is energetically disallowed under normal conditions of observation, i.e., on the laboratory time scale and at ambient temperatures. Throughout the course of this gearing motion, a molecule such as bis(9-triptycyl)carbinol (Fig. 9) remains asymmetric for the following reason: the only achiral symmetry of gear-meshed bis(9-triptycyl)methane is C_s, and while the symmetry of the C—CH(OH)—C fragment is also C_s, the two σ-planes are mutually perpendicular. As in the case of $(1R)$-menthyl $(1S)$-menthyl $2, 2', 6, 6'$-tetranitro-$4, 4'$-diphenate, bis(9-triptycyl)carbinol is therefore asymmetric in all possible conformations and the compound exists as a mixture of transient dl pairs. These d and l conformations can interconvert by torsion around the bonds to the central carbinol carbon; for example, either one of the enantiomorphous conformations in Fig. 9 can convert into the other by a correlated 60° disrotation of both triptycyl rotors. Bis(9-triptycyl)carbinol is thus revealed as yet another example of a chemically achiral compound composed of asymmetric molecules that interconvert exclusively by chiral pathways.

Among other molecules in this class are symmetrically substituted cyclic trans olefins of the type depicted in Fig. 10. All conformations of these molecules are asymmetric; the reason, as in the examples already discussed, is that the plane of the trans olefin unit remains at all times perpendicular to the average plane of the $-(CH_2)_n CHR(CH_2)_n-$ fragment. Given a sufficiently large ring size $(n > 3)$, internal rotation about the single bonds of the olefinic moiety interconverts all conformational enantiomers. An example of this type is the meso diepoxide derived from $trans, trans, trans$-$1, 5, 9$-cyclododecatriene.[31]

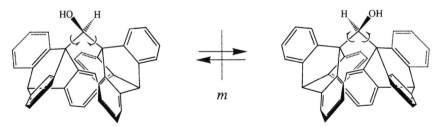

FIGURE 9 Two mirror-image-related (m) conformations of bis(9-triptycyl)carbinol. The carbinols are interconverted by a 60° disrotatory twist about the two C—CH(OH) bonds, as indicated by the arrows.

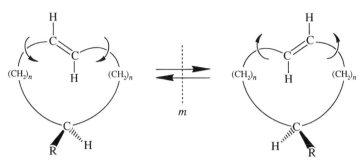

FIGURE 10 Two mirror-image-related (*m*) conformations of a symmetrically substituted cyclic trans olefin. The enantiomorphs are interconverted by internal rotation of the olefinic moiety about two C—C single bonds, as indicated by the arrows.

The absence of an achiral boundary along the conformational enantiomerization trajectory of chemically achiral compounds such as (1*R*)-menthyl (1*S*)-menthyl 2,2′,6,6′-tetranitro-4,4′-diphenate precludes partitioning of the molecular conformations into **R** and **L** classes. However, there is no sharp dividing line, in principle, between chemically achiral compounds and chiral molecules whose enantiomers are connected exclusively by reaction sequences that do not involve any achiral intermediates or transition structures. The pseudorotation of a phosphorane of the type Pabcde by the Berry mechanism[32] is an example of such an enantiomerization (Fig. 11). In this intramolecular rearrangement, the trigonal–bipyramidal educt is converted to the enantiomeric product in a sequence of steps that involve trigonal–bipyramidal intermediates connected by square–pyramidal transition structures. Educt, product, intermediates, and transition structures are all asymmetric, and the structures along the pseudorotation coordinate are thus all asymmetric. The complete set of Pabcde stereoisomers and pseudorotation steps is conveniently depicted as a graph in which 20 vertices represent stereoisomers and 30 edges represent pseudorotation steps.[33]

Reaction sequences in which a compound is converted to its enantiomer without racemization have long been known. The first example of such a reaction sequence, the conversion of (+)- into (−)-isopropylmalonamic acid, was reported in 1914.[34] The individual steps are shown in Fig. 12, along with configurational descriptors *S* and *R* for each molecule; although the actual configurations are still unknown, those shown in Fig. 12 represent an "enlightened guess."[35] Judging by the configurational descriptors, it might appear as though the three compounds at the top of Fig. 12 belong to one homochirality class (*S* = **L**), whereas the three compounds at the bottom belong to the other (*R* = **R**). This impression would, however, be mistaken. The six compounds shown in

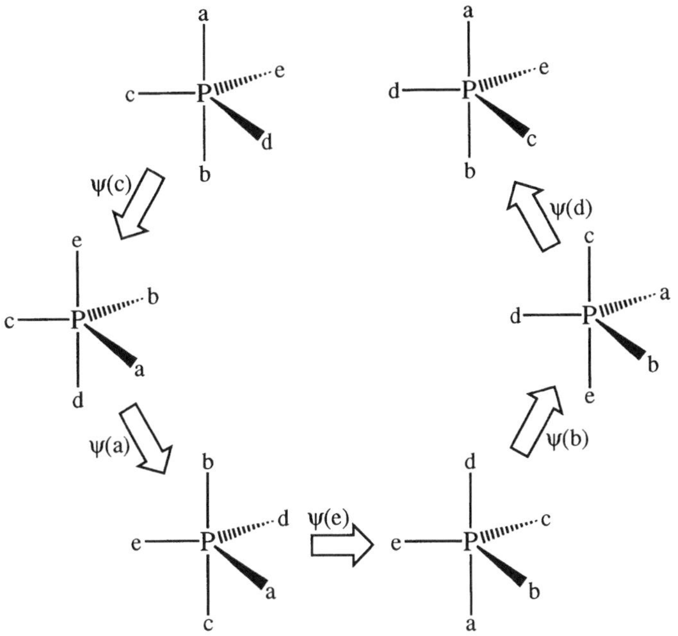

FIGURE 11 Steps in the conversion of an asymmetric phosphorane (top left) into its enantiomer (top right) by a Berry pseudorotation mechanism. In each step, labeled $\Psi(x)$, one of the three equatorial ligands (the pivot ligand x) remains equatorial and is at the apex of the square pyramid in the transition structure.

Fig. 12 represent particular stationary points on the multidimensional potential energy surface and are chirally connected through a continuous reaction pathway that involves no achiral intermediates; this is evidently the case because the product of the reaction sequence is not racemic. Hence, they cannot be partitioned into **R** and **L** homochirality classes.

Thus, even though the S and R forms of each individual molecule (for example, the enantiomeric phosphoranes in Fig. 11 and the enantiomeric isopropylmalonamic acids in Fig. 12) always belong to heterochiral classes in the restricted sense of Kelvin's definition, this classification cannot be consistently extended beyond individual enantiomeric pairs.

D. Pseudoscalar Properties and Chiral Zeroes

Chirality in the world of observables is characterized by pseudoscalar properties—properties that remain invariant under proper rotation but change sign under improper rotation. Enantiomers and, in general, enantiomorphous molecules, have identical scalar properties, such as melting points or dipole moments, and pseudoscalar properties that are identical in

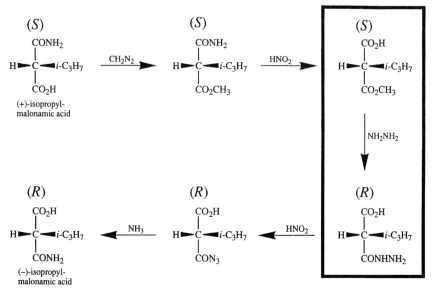

FIGURE 12   Conversion of (+)- into (−)-isopropylmalonamic acid.[34] The step at which the configuration changes from S to R is boxed.

magnitude but opposite in sign, such as optical rotation. Achiral systems have no pseudoscalar properties, or, to be more exact, the pseudoscalar properties of achiral systems are zero under all conditions of measurement. This is called an achiral zero. It is possible, however, that under certain conditions of measurement a pseudoscalar property of a chiral system is also exactly zero; a well-known example is the optical rotation at the crossover point in optical rotatory dispersion curves. This chiral zero is, however, the result of an accidental equivalence: in contradistinction to achiral zeroes, chiral zeroes disappear and nonzero pseudoscalar properties reappear when conditions of measurement (for example, the wavelength at which the rotation is observed) are changed.

Because enantiomers have oppositely signed pseudoscalar properties, chiral zeroes are unavoidable at some stage in the conversion of a molecule into its enantiomer along a chiral pathway. This is true of chirally connected enantiomeric conformations in chemically achiral molecules, such as $(1R)$-menthyl $(1S)$-menthyl $2, 2', 6, 6'$-tetranitro-$4, 4'$-diphenate, and of chirally connected enantiomers, such as (+)- and (−)-isopropylmalonamic acids. More generally, as previously noted, any chiral molecule composed of five or more atoms is in principle always capable of conversion into its enantiomer by chiral as well as by achiral pathways, provided that this is energetically feasible. Hence, unless it can be demonstrated that an achiral pathway for enantiomerization exists that is energetically much preferred, one may expect that, for all but the simplest molecules, chiral zeroes will inevitably be observed in the pseudoscalar properties that attend enantiomerization by chiral pathways. The existence of such chiral zeroes in turn implies that no sign-changing continuous function, and, in particular, no continuous pseudoscalar function, can be used as a chirality measure in 3-space because such a function necessarily has one or more chiral zeroes, which violates the essential condition that the value of any chirality measure can be zero if and only if the object is achiral.[13] This, then, is another aspect of the homochirality problem.

E. Homochirality Classes of Topological Constructions

Certain types of knots and links exist as topologically chiral enantiomorphs. Such enantiomorphs cannot be interconverted by continuous deformation ("ambient isotopy").[36] Homochirality classes can therefore be defined for this type of mathematical object.[37]

A variety of topologically chiral knots and links is found among circular DNA molecules,[38] and a few examples of topologically chiral knotted and linked substructures have recently been discovered among proteins.[39] Sauvage and co-workers have achieved the synthesis of topologically chiral molecules abstractly represented by the trefoil knot,[40] by the four-crossing two-component link,[41] and by the oriented two-crossing link[42] (Fig. 13). Given a criterion suitable for the characterization of a sense of chirality—such as writhe profiles in the case of knots[37]—and, if dictated by circumstances, an ancillary set of rules, it might be possible, in principle, to group these molecules into homochirality classes. The main difficulty in doing so, however, derives from the fuzziness that is introduced when a molecular constitution is translated into a molecular graph. The importance of this point warrants a brief recapitulation of some earlier remarks on this subject.[39, 43]

In a molecular graph that represents the constitutional formula of a molecule, there is no problem in identifying the vertex set because each vertex bears a one-to-one correspondence to an appropriately labeled atom in the molecule. The relationship of edges in the graph to bonds in the molecule is, however, far less well defined. The crucial question is "which bonds in the molecule are regarded as 'topologically significant'[44]?" Because the concept of a chemical bond cannot be sharply defined,[45] some arbitrariness is unavoidably built into the definition of a molecular graph, specifically with regard to membership in the edge set. Hence, whether or not a given geometrically chiral molecular model is considered to be topologically chiral depends crucially on which subset of bonds in the molecule is considered to be topologically significant. In short, the uncer-

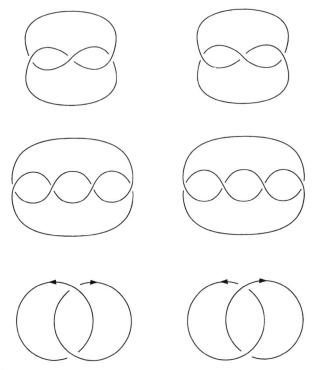

FIGURE 13 Enantiomorphs of topologically chiral constructions. Top: Trefoil knot. Center: Four-crossing two-component link. Bottom: Oriented two-crossing link.

tainty entailed in defining membership in the edge set renders uncertain the very existence of topological chirality in a molecule.

ACKNOWLEDGMENTS

I thank Chengzhi Liang and Noham Weinberg for helpful discussions. This work was supported by the National Science Foundation.

REFERENCES

1. W. T. Kelvin, "The Second Robert Boyle Lecture," *J. Oxford Univ. Junior Scientific Club* **18**, 25 (1894).
2. K. Mislow and P. Bickart, *Israel J. Chem.* **15**, 1 (1976/77).
3. W. H. Mills, *Chem. Ind.* (*London*), 750 (1932).
4. E. Baer and H. O. L. Fischer, *J. Biol. Chem.* **128**, 463, 475 (1939); H. O. L. Fischer and E. Baer, *Chem. Rev.* **29**, 287 (1941).

5. K. Mislow, R. Graeve, A. J. Gordon, and G. H. Wahl, Jr., *J. Am. Chem. Soc.* **86**, 1733 (1964).

6. H. Wynberg, G. L. Hekkert, J. P. M. Houbiers, and H. W. Bosch, *J. Am. Chem. Soc.* **87**, 2635 (1965).

7. M. Black, *Dialogue* **2**, 1 (1963).

8. L. A. Zadeh, *Inf. Control* **8**, 338 (1965).

9. L. A. Zadeh, *J. Cybern.* **2**, 4 (1972); G. Lakoff, *J. Philos. Logic* **2**, 458 (1973).

10. J. D. Morrison and H. S. Mosher, *Asymmetric Organic Reactions*, p. 5. Prentice-Hall, Englewood Cliffs, NJ, 1971.

11. M. Scriven, *J. Philos.* **56**, 857 (1959).

12. A. B. Buda, T. Auf der Heyde, and K. Mislow, *Angew. Chem. Int. Ed. Engl.* **31**, 989 (1992).

13. N. Weinberg and K. Mislow, *J. Math. Chem.* **17**, 35 (1995).

14. A. B. Buda, T. P. E. Auf der Heyde, and K. Mislow, *J. Math. Chem.* **6**, 243 (1991).

15. T. P. E. Auf der Heyde, A. B. Buda, and K. Mislow, *J. Math. Chem.* **6**, 255 (1991).

16. A. B. Buda and K. Mislow, *J. Mol. Struct.* **232**, 1 (1991); *Elem. Math.* **46**, 65 (1991).

17. (a) A. B. Buda and K. Mislow, *J. Am. Chem. Soc.* **114**, 6006 (1992); (b) N. Weinberg and K. Mislow, *J. Math. Chem.* **14**, 427 (1993).

18. A. Mead, *Topics Current Chem.* **49**, 1 (1974).

19. K. Mislow and P. Poggi-Corradini, *J. Math. Chem.* **13**, 209 (1993).

20. (a) E. Ruch and I. Ugi, *Theor. Chim. Acta* **4**, 287 (1966); (b) E. Ruch, A. Schönhofer, and I. Ugi, *Theor. Chim. Acta* **7**, 420 (1967); (c) E. Ruch and A. Schönhofer, *Theor. Chim. Acta* **10**, 91 (1968); (d) E. Ruch, *Theor. Chim. Acta* **11**, 183 (1968); (e) E. Ruch and I. Ugi, *Topics Stereochem.* **4**, 99 (1969); (f) E. Ruch and A. Schönhofer, *Theor. Chim. Acta* **19**, 225 (1970); (g) E. Ruch, *Acc. Chem. Res.* **5**, 49 (1972); (h) A. Mead, E. Ruch, and A. Schönhofer, *Theor. Chim. Acta* **29**, 269 (1973); (i) E. Ruch, *Angew. Chem. Int. Ed. Engl.* **16**, 65 (1977).

21. (a) L. Salem, J. Durup, G. Bergeron, D. Cazes, X. Chapuisat, and H. Kagan, *J. Am. Chem. Soc.* **92**, 4472 (1970); (b) L. Salem, *Acc. Chem. Res.* **4**, 322 (1971).

22. S. Wolfe, H. B. Schlegel, I. G. Csizmadia, and F. Bernardi, *J. Am. Chem. Soc.* **97**, 2020 (1975).

23. (a) K. Mislow, *Science* **120**, 232 (1954); (b) K. Mislow and R. Bolstad, *J. Am. Chem. Soc.* **77**, 6712 (1955); (c) K. Mislow, *Trans. New York Acad. Sci.* **19**, 298 (1957); (d) K. Mislow, *Introduction to Stereochemistry*, pp. 91–93. Benjamin, NY, 1965.

24. (a) J. Dugundji, J. Showell, R. Kopp, D. Marquarding, and I. Ugi, *Israel J. Chem.* **20**, 20 (1980); (b) I. Ugi, J. Dugundji, R. Kopp, and D. Marquarding, *Perspectives in Theoretical Stereochemistry*. Springer-Verlag, Berlin, 1984.

25. D. M. Walba, in *New Developments in Molecular Chirality* (P. G. Mezey, ed.), pp. 119–129. Kluwer, Dordrecht, 1991.

26. See for example: (a) G. W. Wheland, *Advanced Organic Chemistry*, 3rd ed., pp. 278–280. Wiley, New York, 1960; (b) F. A. Cotton, *Chemical Applications of Group Theory*, 3rd ed., pp. 38–39. Wiley, New York, 1990; (c) V. I. Sokolov, *Introduction to Theoretical Stereochemistry*, p. 278. Gordon and Breach, New York, 1991; (d) E. L. Eliel and S. H. Wilen, *Stereochemistry of Organic Compounds*, pp. 62, 92. Wiley, New York, 1994.

27. (a) H. Frei and H. H. Günthard, *Chem. Phys.* **15**, 155 (1976); (b) H. Frei, A. Bauder, and H. H. Günthard, *Topics Current Chem.* **81**, 1 (1979).

28. J. G. Nourse, *J. Am. Chem. Soc.* **102**, 4883 (1980).

29. (a) D. Gust and K. Mislow, *J. Am. Chem. Soc.* **95**, 1535 (1973); (b) K. Mislow, *Acc. Chem. Res.* **9**, 26 (1976); (c) R. Glaser, J. F. Blount, and K. Mislow, *J. Am. Chem. Soc.* **102**, 2777 (1980).

30. (a) W. D. Hounshell, C. A. Johnson, A. Guenzi, F. Cozzi, and K. Mislow, *Proc. Nat. Acad. Sci. U.S.A.* **77**, 6961 (1980); (b) A. Guenzi, C. A. Johnson, F. Cozzi, and K. Mislow, *J. Am. Chem. Soc.* **105**, 1438 (1983); (c) H. Iwamura and K. Mislow, *Acc. Chem. Res.* **21**, 175 (1988).

31. T. Hoye, personal communication.

32. E. L. Muetterties, *Inorg. Chem.* **6**, 635 (1967).

33. (a) P. C. Lauterbur and F. Ramirez, *J. Am. Chem. Soc.* **90**, 6722 (1968); (b) K. Mislow, *Acc. Chem. Res.* **3**, 321 (1970).

34. E. Fischer and F. Brauns, *Ber. Dtsch. Chem. Ges.* **47**, 3181 (1914).

35. E. L. Eliel, *Stereochemistry of Carbon Compounds*, p. 18. McGraw-Hill, New York, 1962; E. L. Eliel and S. H. Wilen, *Stereochemistry of Carbon Compounds*, p. 61. Wiley, New York, 1994.

36. R. H. Crowell and R. H. Fox, *Introduction to Knot Theory*. Springer-Verlag, New York, 1963; C. Livingston, *Knot Theory*. Mathematical Association of America, Washington, DC, 1993; C. C. Adams, *The Knot Book—An Elementary Introduction to the Mathematical Theory of Knots*. Freeman, New York, 1994.

37. (a) C. Liang and K. Mislow, *J. Math. Chem.* **15**, 35 (1994). (b) C. Liang, C. Cerf, and K. Mislow, *J. Math. Chem.* **19**, 241 (1996).

38. D. W. Sumners, *in Geometry and Topology* (C. McCrory and T. Schifrin eds.), pp. 297–318. Dekker, New York, 1987; C. O. Dietrich-Buchecker and J.-P. Sauvage, in *Bioorganic Chemistry Frontiers* (H. Dugas, ed.), Vol. 2, pp. 95–248. Springer-Verlag, Berlin, 1991; A. D. Bates and A. Maxwell, *DNA Topology*. Oxford Univ. Press, London, 1993.

39. C. Liang and K. Mislow, *J. Am. Chem. Soc.* **117**, 4201 (1995).

40. (a) C. O. Dietrich-Buchecker and J.-P. Sauvage, *Angew. Chem. Int. Ed. Engl.* **28**, 189 (1989); (b) C. O. Dietrich-Buchecker, J. Guilhem, C. Pascard, and J.-P. Sauvage, *Angew. Chem. Int. Ed. Engl.* **29**, 1154 (1990); (c) J.-P. Sauvage, *Acc. Chem. Res.* **23**, 319 (1990); (d) Ch. Dietrich-Buchecker and J. P. Sauvage, *New J. Chem.* **16**, 277 (1992); (e) C. O. Dietrich-Buchecker, J.-P. Sauvage, J.-P. Kintzinger, P. Maltèse, C. Pascard, and J. Guilhem, *New J. Chem.* **16**, 931 (1992); (f) C. Dietrich-Buchecker and J.-P. Sauvage, *Bull. Soc. Chim. Fr.* **129**, 113 (1992).

41. J.-F. Nierengarten, C. O. Dietrich-Buchecker, and J.-P. Sauvage, *J. Am. Chem. Soc.* **116**, 375 (1994).

42. J.-C. Chambron, D. K. Mitchell, and J.-P. Sauvage, *J. Am. Chem. Soc.* **114**, 4625 (1992).

43. K. Mislow, *Bull. Soc. Chim. Belg.* **86**, 595 (1977); C. Liang and K. Mislow, *J. Math. Chem.* **15**, 245 (1994); K. Mislow, *Croat. Chem. Acta*, **69**, 485 (1996).

44. D. M. Walba, *Tetrahedron* **41**, 3161 (1985).

45. V. Prelog and G. Helmchen, *Angew. Chem.* **94**, 614 (1982).

4

Fuzzy Classical Structures in Genuine Quantum Systems

ANTON AMANN

ETH Hönggerberg
CH-8093 Zürich, Switzerland

I. INTRODUCTION

The notions of chemistry and quantum theory are in a state of conflict. Indeed, it must be said that quantum mechanics and traditional chemistry contradict each other at a very fundamental level. Whereas traditional chemistry claims that each nucleus in a molecule is located at some particular (possibly fluctuating) position in space, quantum mechanics asserts that nuclei within a molecule need not have a dispersion-free position at all.

Consider the example of a molecule such as that shown in Figs. 1 and 2 having adjacent carbonyl and hydroxy groups (cf. ref. 35), with a hydrogen bond between the two oxygen nuclei. The question we might ask concerns the dynamics of the relevant proton. In traditional chemistry, one would expect the proton to move between two different positions (near the two oxygen nuclei) or to take an intermediate position (between the oxygens). In both cases, the chemist's notion is that the proton has an unambiguous position at every moment in time (see Fig. 1). Quantum mechanics offers another possible behavior, namely, that the proton's position is ambiguous—located near both oxygen nuclei (see Fig. 2). Quantum-mechanical states having this property can easily be constructed as *superpositions* of two (pure) states with respect to which the proton has a different position. A superposition of the two states shown in the upper part of Fig. 1, for example, gives the (pure) state illustrated in Fig. 2. This latter state is *strictly stationary*, that is to say, it is an eigenstate of the underlying molecular Hamiltonian. Consequently, in this state the proton between the oxygen nuclei does not move and its property of being located

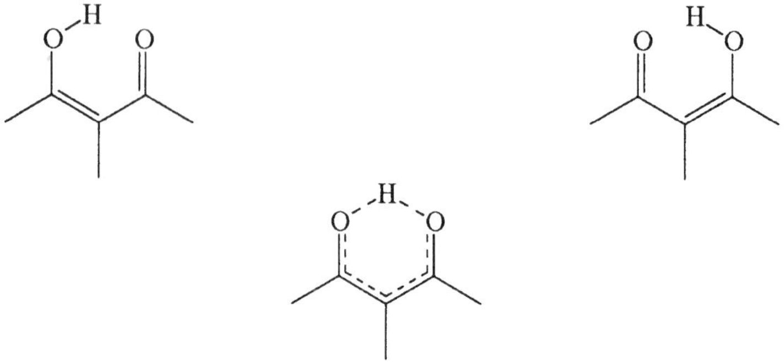

FIGURE 1 Hydrogen bond with the proton sitting at some fixed unambiguous position for every moment of time. The proton could either move back and forth (upper part of the figure) or sit at a stationary position in between the two oxygen nuclei (lower part of the figure).

near both oxygen nuclei should not be interpreted as a frequent change of position in time!

Such quantum-mechanical molecular states, with respect to which a given nucleus has an ambiguous position, will be referred to here as "strange states." This is done to stress the fact that such states are neither predicted by nor admissible in traditional chemistry. In the latter only different mesomeric forms of a given molecule are admitted, and it is not assumed that the molecule is in different mesomeric forms at the same time. In traditional chemistry, the nuclei of a molecule always have an unambiguous position in space and hence the molecule has a nuclear structure.

Most molecular species can be expected to possess a nuclear structure, although for some, e.g., hydrogen bonded species or ammonia, the issue is not so obvious. The ammonia-maser transition, for example, is thought to be a transition between "strange" molecular (pure) states that do not admit a nuclear structure.

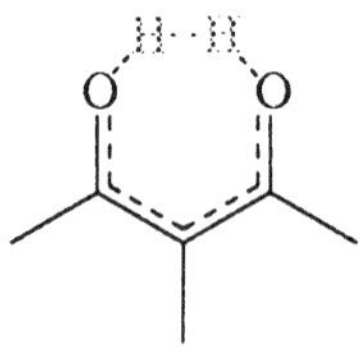

FIGURE 2 Sketch of a stationary quantum state for a hydrogen bond where the proton does not have an unambiguous position. This state is a superposition of the two states sketched in the upper part of Fig. 1.

Since one cannot sharply distinguish between molecular species that exhibit strange (pure) states and those that do not, it is plausible that all molecules can in principle be prepared in such strange (pure) states, but that the latter do not play an important role in most molecular species (with certain exceptions such as hydrogen-bonded species or ammonia, for example). This suggests that classical chemical concepts, e.g., chirality, knot type, and nuclear structure of molecules, are not strictly classical in the sense that superpositions between different structures are strictly forbidden. Such superpositions may still arise, but are expected to be unstable under external perturbations and therefore are usually not observed. The instability of strange states is thought to increase with increasing nuclear mass or with decreasing level splitting between low-lying eigenstates of the molecular Hamiltonian, as, for instance, in the sequence of species {monodeuteroaniline → ammonia → naphthazarin → ⋯ → sugar/amino acid}. In other words, classical chemical structures are fuzzy because strange superpositions—not compatible with classical notions—are still admissible. However, the fuzziness of such classical structures decreases with increasing nuclear mass because strange states may be said to die out.

The question then is: Can quantum theory predict whether a molecule possesses an approximate nuclear structure or not? In present-day quantum mechanics, this question is usually not posed. As a rule, it is assumed that molecules have an approximate nuclear structure, with exceptional cases (such as ammonia) being given a special treatment. This assumption of a nuclear molecular structure has its origin in traditional chemistry and is embedded in the Born–Oppenheimer scheme used in quantum chemistry.[18,63] In the Born–Oppenheimer scheme one initially computes the electronic energy for some given nuclear configuration and then includes nuclear motion as a second step. Thus, the Born–Oppenheimer way of looking at molecules starts with states having a nuclear structure. It does, of course, allow us to write down superpositions of such states (resulting in strange states, such as those necessary for ammonia[52]). However, no criterion tells us which (stationary or nonstationary) states actually arise for a given molecular species (cf. Section V), say, in a thermal situation. A single molecule can be prepared in *any* (stationary or nonstationary) quantum-mechanical state, but the molecules in an ensemble in contact with a thermal reservoir will assume differing states with higher or lower probability.

Much more attention has been given to the computation of molecular spectra (for ensembles of molecules); for this the actual structure of molecules is not really necessary. The difference between the spectroscopy of molecular ensembles and that of single molecules has been discussed in ref. 11. Here, only a simple example will be cited to illustrate that molecular structure constants (such as bond lengths or bond angles) are usually derived under the assumption that the molecular species under investigation definitely has a nuclear structure. Consider, for instance, trifluoronitrogen (NF_3). Its microwave spectrum shows a line at the frequency $\nu = 21{,}362{,}040{,}000$ s^{-1} (and at $\bar{\nu} = 21{,}258{,}880{,}000$ s^{-1} for the isotopically substituted version $^{15}NF_3$). Assuming now that NF_3 is a symmetric rigid rotor, it becomes possible to compute the bond length R_0 between nitrogen and fluorine as $R_0 = 1.37$ Å. This numerical value is conditional: under the condition that NF_3 has a nuclear structure, the bond length may be given as 1.37 Å. The spectrum by itself does not imply that NF_3 has a nuclear structure at all.

In the spectroscopy of molecular ensembles, there is much freedom in assigning virtual attributes to molecules, such as the (possibly) virtual bond length in NF_3. There is a deeper reason for this. To describe ensemble spectroscopy it is sufficient to have an appropriate so-called master equation[32,43,62] for the average expectation values of all the (pure) states describing the individual molecules in the ensemble. Many different assumptions for the individual molecules in the ensemble are compatible with the average of expectation values and with the average master equation. Hence, no conclusive evidence about the molecular structure can be derived from the spectroscopy of molecular ensembles at any nonzero temperature. One may either assume that the individual molecules in the ensemble admit a nuclear structure or assume the contrary: spectroscopic results may be compatible with both of these completely different assumptions (see Section V). In single-molecule spectroscopy, the situation is—at least in principle—different. A fully quantum-mechanical understanding of single-molecule spectroscopy (SMS) is therefore highly desirable.[11] An interesting point here is the stochastic behavior of migrating lines in the spectrum (see Fig. 9 and refs. 25, 37, 38, 47–51, 56, 57, 82, and 91).

Averaging over all different possible stochastic behaviors in SMS yields the master equation used in ensemble spectroscopy, but the averaged master equation does not determine the dynamics of the (pure) states of individual molecules. Certain attempts have been made to derive a proper theory of individual behavior of single quantum systems,[33,34,60,61] but a rigorous interpretation is still lacking.

Hence, the chemical structure, dynamics, and spectroscopy of *single* molecules cannot be rigorously discussed in the present formalism of quantum mechanics and the problem is to construct a quantum theory for individual molecules. One possible starting point is an averaged (e.g., thermal) description and an averaged dynamics over an ensemble of molecules. The average ranges over all the pure states of the molecules in the ensemble. In technical terms, an individual quantum theory is related to a decomposition of nonpure states into pure ones (for a precise definition, see subsequent text). This decomposition is not unique. One and the same averaged (nonpure, e.g., thermal) state can be decomposed into either pure states admitting a nuclear structure or strange pure states. Different decompositions give also rise to different stochastic dynamics on the level of pure states. It will be argued that this dynamics is stochastic due to the influence of the molecular environment.

There are good reasons to believe that a strictly isolated molecule can be prepared experimentally in any (pure) state.[21,22,64–66,69] However, not all molecular states are equally stable under (small) external stochastic perturbations. The molecular environment is thus of decisive importance in all these discussions on molecular structure. Examples of environments that cannot be completely screened out are, e.g., quantum radiation and gravitational fields. Here an attempt is made to incorporate external environmental stochastic perturbations and to look for decompositions that are stable under these perturbations. This opens up a possible route to enable us to:

(1) Understand the concept of chemical structure;
(2) Derive and study deviations from chemical structures in the sense that strange quantum-mechanical superpositions of different chemical structures appear for certain molecules like ammonia;
(3) Study the fuzziness of chemical concepts and the decrease of fuzziness (to zero) when going to some appropriate limit, e.g., the limit of infinite nuclear molecular masses;
(4) Gain a deeper understanding of the Born–Oppenheimer approximation; and
(5) Take the first steps towards a fully quantum-mechanical explanation of the stochastic aspects in single-molecule spectroscopy.

Point (3) in this list of desiderata needs further explanation. Consider a dilute gas of ammonia-type molecules NX_3 in a thermal situation. In the limit of infinite mass for each of the ligands X, one would expect completely classical behavior for the nuclei in all the molecules of the gas. The position of the nuclei, in particular, would be unambiguously determined. For large but finite masses of the ligands X, the behavior of individual molecules can still be expected to be more or less classical, but quantum corrections will come into play. The precise nature of these quantum corrections will depend decisively on the way in which quantum mechanics is introduced. Symmetry-adapted eigenfunctions of the molecular Hamiltonian (see Section II), for example, will never give rise to unambiguously determined nuclear positions for any chosen finite mass of the ligands, however large they may be. Hence bringing quantum mechanics in via eigenfunctions of the Hamiltonian leads to a discrepancy in the behavior in the limit of infinite masses. The physical reason for this discrepancy is the instability of the eigenfunctions under small external perturbations, which (heuristically) increases with increasing mass of the ligand. Letting quantum mechanics enter in accordance with stability requirements, as proposed here, leads to almost classical behavior for large but finite masses and hence shows the correct behavior in the limit. The respective quantum fluctuations for large but finite masses are thus trustworthy indicators and can in principle be computed. Finally, one could try to establish the percentage of (pure) states that allow an approximate nuclear structure against the percentage of (pure) strange states for species like ammonia or NF_3. The greater the percentage of strange states, the greater will be the fuzziness of the chemical concept "nuclear structure" for a particular molecular species. Heuristically, one expects that the nuclear structure of ammonia is fuzzier than that of NF_3 (because of the greater mass of F compared to H).

At the present time, a detailed numerical investigation of these molecular structure problems is not yet available. We shall therefore carry out a check on the simpler quantum-mechanical Curie–Weiss model of a magnet, which might be considered as a naive model for the generation of fuzzy classical structures in quantum systems. The question now is how "fast" the specific magnetization becomes a classical observable and how "fast" the superpositions of states with opposite permanent magnetization die out with an increasing number N of spins. This increasingly classical (or decreasingly fuzzy) behavior is described here in terms of a *large-deviation entropy*, which is at the heart of present studies on the fuzziness of classical concepts (see Section XI for precise definitions). It is this large-deviation entropy that numerically characterizes the fuzziness of classical structures in quantum systems. For finite N, the superposition principle is still universally valid and yet the fuzzy classical observable "magnetization" appears and becomes strictly classical in the limit $N \to \infty$.

Similar behavior might be expected for molecular structure problems, the limit $N \to \infty$ being replaced now by the limit of all molecular masses going to infinity. The respective large-deviation entropy would characterize the quantum fluctuations caused by nonclassical behavior of the molecular nuclei (strictly classical behavior occurs only in the limit of infinite nuclear masses). It may be argued that traditional statistical (algebraic) quantum mechanics imposes excessive conditions on symmetry breaking and classical structures (which arise only in the limit of infinitely many degrees of freedom). Only an individual formalism of quantum mechanics can cope with the problem of chemical structure.

In summary, there are several important points that need to be stressed:

(1) The molecules of chemists are usually not in strict eigenstates, since such states do not show chemical structure (handedness, isomerism, knot type, nuclear structure). In quantum mechanics,

on the other hand, the ground state of a molecule is always strictly stationary. Hence, from a quantum mechanical point of view, it is not clear why states showing chemical structure should arise.[2,4,5,7,8,23,24,26,27,55,58,59,79,84–90]

(2) Ammonia typifies a molecule without a chemical structure. Its ground state, in particular, does not admit a nuclear molecular framework. Moreover, for *all* molecules with at least two possible isomeric structures, the same strange conclusions could be drawn.

(3) An isolated molecule can be prepared in any stationary or nonstationary pure state. There is thus no reason to restrict oneself to eigenstates of the Hamiltonian, i.e., solutions of the time-independent Schrödinger equation. Hence isolated molecules do not show chemical structure. A derivation of chemical structure from the Schrödinger equation for an isolated molecule can work only by tricks or approximations hidden in the complicated mathematics.

(4) Without a consideration of the molecular environment, it is impossible to derive and understand chemical structure. Possible environments to be considered are the radiation or gravitational fields, collisions with other molecules, etc.

(5) The Born–Oppenheimer idea of chemical structure goes back to the traditional chemistry of the prequantum era and is likely to be essentially correct. The problem is to understand chemical structure in quantum-mechanical terms and to work out and discuss borderline cases that do not conform to traditional chemical thinking.

(6) Chemical concepts usually are considered to be strictly classical concepts. Here arguments will be given in favor of fuzzy classical chemical concepts, where the fuzziness decreases with increasing nuclear masses, etc.

(7) Some interesting behavior in single-molecule spectroscopy involves the stochastic migration of lines. Usual statistical quantum theory describes only mean values or dispersions of observables, but not the actual fluctuations in the dynamics of single quantum systems. In an individual formalism of quantum mechanics, such fluctuations are of great importance.

It should be emphasized that it is not our aim here to investigate some particular pure state (wave functions), such as the ground state, of a molecular species. The issue at stake is rather *which* pure states actually arise in a thermal situation? Are these pure states strange or do they admit a nuclear structure? What are the respective probabilities of finding strange states in a vessel containing 10^{20} molecules at a given temperature and pressure?

II. STRANGE STATES OF MOLECULES

The superposition principle of quantum mechanics immediately gives rise to what have been termed strange states. These can be illustrated by reference to the ammonia molecule.

Let us consider Fig. 3 in which a double-minimum function represents the Born–Oppenheimer potential for the electronic ground-state of ammonia. This is an energy versus internal inversion coordination diagram. Every value of the inversion coordinate corresponds to a particular nuclear framework for ammonia. The two minima, for example, correspond to pyramidal structures, whereas the maximum corresponds to a planar

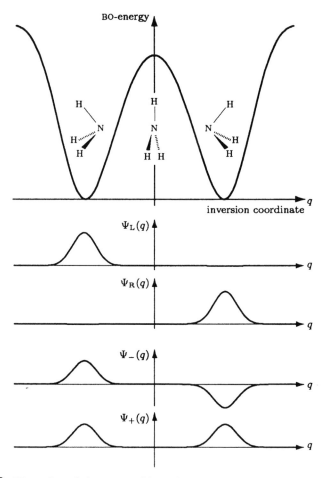

FIGURE 3   Illustration of the superposition principle using ammonia and ammonia-type molecules.

structure. At first sight, ammonia molecules roughly retain their traditional chemical structure even in a quantum description. The pyramidal forms, for example, are described by the rather peaked (nonstationary) wave functions Ψ_L and Ψ_R, and therefore permit at least an approximate nuclear framework. However, as soon as we consider superpositions of Ψ_L and Ψ_R, things change dramatically. The superposition

$$\Psi_+ \overset{\text{def}}{=} \frac{1}{\sqrt{2}}(\Psi_L + \Psi_R) \tag{1}$$

describes the proper ground state of ammonia and the superposition

$$\Psi_- \overset{\text{def}}{=} \frac{1}{\sqrt{2}}(\Psi_L - \Psi_R) \tag{2}$$

describes the first excited state of ammonia. (Here and in the following $=_{\text{def}}$ denotes a definition.) Both Ψ_+ and Ψ_- are eigenstates of the underlying molecular Hamiltonian and therefore do not change under the time evolution given by the time-dependent Schrödinger equation, apart from the phase factors $\exp\{-itE_\pm/\hbar\}$, where E_+ and E_- are the respective eigenvalues. This implies that all the expectation values $\langle \Psi_\pm | \hat{A}\Psi_\pm \rangle$ of the observables $\hat{A}$ are invariant under Schrödinger time evolution, and not just the expectation values of (say) the nuclear position operators. Since expectation values do not change, the states Ψ_+ and Ψ_- are *stationary*.

Figure 4 presents a depiction of the ground state Ψ_+ of ammonia. Although still shown here, the nitrogen−hydrogen bonds lose their meaning in this context. With respect to Ψ_+ and Ψ_-, a single ammonia

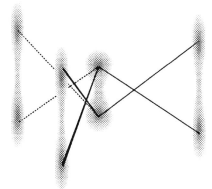

FIGURE 4 Sketched distribution of the nuclei in the ground state of an ammonia-type molecule.

molecule no longer has a nuclear framework. The nitrogen and hydrogen nuclei are not located at fixed positions; only probability distributions for their positions can be given, just as is usual with electrons. Hence, ammonia does not possess a nuclear framework with respect to the states Ψ_+ and Ψ_-. It is this fact that makes these states appear strange from the point of view of the chemist.

The transition between the two stationary states Ψ_+ and Ψ_- is the ammonia-maser transition with transition frequency $\nu = (E_- - E_+)/h = 23{,}870{,}110{,}000 \ \text{s}^{-1}$. The very existence of the ammonia-maser transition suggests that the states Ψ_+ and Ψ_- of ammonia do indeed exist in reality and not just in the quantum-mechanical formalism.

Incidentally, the states Ψ_L and Ψ_R could be written as the superpositions

$$\Psi_L \overset{\text{def}}{=} \frac{1}{\sqrt{2}}(\Psi_+ + \Psi_-), \tag{3}$$

$$\Psi_R \overset{\text{def}}{=} \frac{1}{\sqrt{2}}(\Psi_+ - \Psi_-) \tag{4}$$

of the ground and excited states of ammonia. These are nonstationary states with Schrödinger time evolution

$$\frac{1}{\sqrt{2}}[\Psi_+ \pm \Psi_-] \rightarrow \frac{1}{\sqrt{2}}\left[\Psi_+ \pm \exp\left\{-it\,\frac{(E_- - E_+)}{\hbar}\right\}\Psi_-\right]. \tag{5}$$

Recall that the time-evolved state on the right-hand side of Eq. (5) does not just differ by a phase factor from the initial state $(1/\sqrt{2})[\Psi_+ \pm \Psi_-]$. The Schrödinger time evolution (5) describes a tunneling process that transforms the state Ψ_L into Ψ_R and vice versa.

As far as the quantum-mechanical formalism is concerned, ammonia is certainly no exception. Similar situations can be constructed for all molecules. Consider a molecule with an internal Hamiltonian H_0 (the kinetic energy of the center of mass being subtracted) and let I denote the unitary operator that implements the space inversion

$$I\Psi(q_1,\ldots,q_L;Q_1,Q_2,\ldots,Q_M)$$

$$\overset{\text{def}}{=} \Psi(-q_1,\ldots,-q_L;-Q_1,-Q_2,\ldots,-Q_M). \tag{6}$$

Here the qs are electronic coordinates and the Qs are nuclear coordinates. If the Hamiltonian H_0 contains no weak neutral current terms,[67, 68] it will be invariant under space-inversion, i.e., we have

$$IH_0I^{-1} = H_0. \tag{7}$$

The eigenstates Ψ_n of H_0 can then be chosen to be symmetry-adapted; thus

$$I\Psi_n = \pm\Psi_n, \qquad n = 1, 2, 3, \ldots, \tag{8}$$

and these will always satisfy the condition

$$\langle \Psi_n \mid Q_j \Psi_n \rangle = \langle I\Psi_n \mid Q_j I\Psi_n \rangle = \langle \Psi_n \mid I^{-1} Q_j I \Psi_n \rangle = \langle \Psi_n \mid -Q_j \Psi_n \rangle = 0, \tag{9}$$

where $j = 1, 2, \ldots, M$. This is a strange state of affairs, for then it follows that (i) a nuclear molecular framework does not exist; (ii) different isomeric molecular forms do not arise; and (iii) a sequence of monomers in a macromolecule does not make sense. Because the notions of nuclear framework, isomerism, and sequential structure in a macromolecule need (at least approximately) localized nuclei, the preceding conclusions are incompatible with the result obtained in Eq. (9).

The symmetry-adapted eigenstates in Eq. (8) correspond to the states Ψ_+ and Ψ_- of ammonia. They do not fit into the molecular structure scheme of traditional chemistry and are therefore considered as strange states from a chemical point of view. Neither chemical structures nor chemical bonds exist any more.

III. CHEMICAL CONCEPTS ARE FUZZY CLASSICAL CONCEPTS

Chemical concepts such as handedness and the isomerism of molecules or the knot type of circular DNA molecules[75,77] are often considered to be strictly classical concepts, i.e., to have strictly dispersion-free expectation values. This means that:

(1) The handedness of a molecule, for example, is thought to be either left or right, without any intermediate forms such as those involving superpositions of the left- and right-handed states.

(2) The isomer type of a molecule is thought to be unequivocally determined, without any intermediate forms such as those based on superpositions of different isomers.

(3) The knot type of a circular DNA molecule is thought to be fixed without any intermediate forms such as superpositions of differently knotted circular DNA molecules (having the same monomer sequence).

Starting from the ammonia molecule NH_3, one can replace two of its hydrogen ligands by a deuterium and a tritium atom to end up with a potentially chiral derivative, NHDT. For this derivative, assumption (1) is not fulfilled. The stationary ground and excited states, Ψ_+ and Ψ_-

(analogous to those of ammonia), are actually states that do not fit into a left/right classification scheme because they are symmetric under space-inversion. Nevertheless, these stationary states seem to exist in reality, as suggested by the associated maser transition between them (with a frequency which is smaller than that for NH_3).[52]

For other ammonia-type molecules, such as naphthazarin[40,52,71,80] or properly chiral molecules (amino acids, sugars, sulfoxides,[39,45] etc.), the situation is not so clear; see Fig. 1. These molecular species admit an inversion coordinate and hence have the same structure of states as ammonia, i.e., stationary states Ψ_+ and Ψ_- and nonstationary states Ψ_L and Ψ_R (in a two-level approximation); the main difference between them is the differing level splitting ($E_- - E_+$). Handwaving arguments based on quantum-mechanical perturbation theory[70] lead to the conclusion that the stationary states Ψ_+ and Ψ_- become unstable when the level splitting is sufficiently small. This would mean that the states Ψ_+ and Ψ_-, although they are stationary under Schrödinger time evolution, would decay into more stable states under the influence of small external perturbations. Since the level splitting of properly chiral molecules such as amino acids or sugars is estimated to be very small indeed,[55] it comes as no surprise that the states Ψ_+ and Ψ_- cannot easily be prepared experimentally.[21, 22, 64, 65] In other words, although the strange states Ψ_+ and Ψ_- might exist for all isolated or reasonably well screened ammonia-type molecules, they usually do not arise in the case of small level splitting due to lability under external perturbations.

Seen from this point of view, no phase transition would be expected to take place in a sequence of molecular species with decreasing level splitting such as in the sequence {monodeuteroaniline → ammonia → ··· → asparagic acid} of Table I. The heuristic idea is that the lability of the strange states Ψ_+ and Ψ_- increases *continuously* with decreasing level splitting. The most unstable situation is expected to occur for properly chiral molecules with very low level splitting (below 10^{-50} J mol^{-1}). Therefore, even the concept of handedness need not be strictly classical although it could be *fuzzy* classical. We might expect the superposition principle to hold universally for all molecules in Table I such that strange states may arise or be prepared experimentally. However, such strange states would likely arise only with small but nonzero probability in thermal equilibrium (for species with low level splitting). This would mean that the fuzziness of the classical concept "chirality" decreases continuously with decreasing level splitting. These considerations will be substantiated in Section XI.

Another interesting example of a classical chemical concept is the *nuclear framework of a molecule*. The usual Born–Oppenheimer approximation for molecules shows that this is not a strictly classical structure. With respect to the states Ψ_L and Ψ_R (see Fig. 3), for example, the

TABLE I Ammonia-Type Molecules

	Accessible states	Level splitting (J mol^{-1})
Monodeuteroaniline[66]	All pure states	600
Ammonia	All pure states	9.5
Naphthazarin[80]		
	Ψ_L and Ψ_R and ?	≈ 0.02
Aspartic acid	Ψ_L and Ψ_R "only"	$\approx 10^{-60}$

positions of the nuclei in ammonia show a small but nonzero dispersion. (With respect to the states Ψ_+ and Ψ_-, the dispersions of the nuclear positions are substantially increased anyway.) Hence again there is no reason to expect the molecular nuclear framework to be a strictly classical concept, but rather a fuzzy classical one. From this point of view all molecular species exhibit strange states with no nuclear framework such as superpositions of two (or more) different isomers. Such strange states can be expected to survive only in isolated molecules and to decay under the influence of external perturbations.

The parameter "level splitting" in the case of the two-level ammonia-type molecules discussed previously can now be replaced by the molecules' nuclear masses. With increasing nuclear masses (M_j, $j = 1, 2, \ldots, K$) one would expect a smaller probability of finding strange states such as superpositions of different isomers (in a thermal situation). In the limit $M_j \to \infty$ the nuclei should behave entirely classically. Again, there is no reason to believe that some phase transition takes place at particular values for the nuclear masses.

IV. STATISTICAL VERSUS INDIVIDUAL FORMALISMS OF QUANTUM MECHANICS

In quantum theory, the expression "nonpure states" has at least two different meanings. These can arise as *mixtures of pure states* or as a *restriction of a pure state onto a subsystem*. In the latter case, the respective nonpure state is called an improper mixture[12,30] and a decomposition (for a precise definition, see subsequent text) into pure states does not necessarily make sense.

Some remarks and definitions concerning states in quantum mechanics are called for here. Usually, a *pure* state in quantum mechanics is introduced with the help of a wave function Ψ. In modern terms, such a wave function Ψ is called a state vector, because it can be viewed as an element in a mathematical vector space. In the case of a two-level system, Ψ is given as a vector in a two-dimensional complex vector space

$$\begin{pmatrix} c_1 \\ c_2 \end{pmatrix}, \qquad |c_1|^2 + |c_2|^2 = 1, \tag{10}$$

with c_1 and c_2 being complex scalars. For an N-level system, one needs an N-dimensional complex vector space. For the description of a real molecule with infinitely many levels one needs a vector space of infinite dimension. The infinite dimension of a vector space should not be confused with the infinite number of degrees of freedom in a system. One molecule, with finitely many nuclei and electrons, has finitely many degrees of freedom and yet is described by infinitely many levels, and hence an infinite-dimensional vector space. Here, for simplicity, we shall usually restrict attention to two-level systems.

In a two-level system, an *observable* is represented by a 2×2 matrix

$$\hat{A} = \begin{pmatrix} a_1 & a_2 \\ a_3 & a_4 \end{pmatrix}, \tag{11}$$

with complex scalars $a_1, \ldots, a_4$ as entries. The respective expectation value of the observable $\hat{A}$ with respect to the state vector Ψ in Eq. (10) will be given as the scalar product

$$\langle \Psi | \hat{A}\Psi \rangle = (\bar{c}_1, \bar{c}_2) \begin{pmatrix} a_1 & a_2 \\ a_3 & a_4 \end{pmatrix} \begin{pmatrix} c_1 \\ c_2 \end{pmatrix}. \tag{12}$$

An important point (already mentioned) is that if a state is restricted to a subsystem, one cannot represent it by a state vector. If, for example, one uses the state vector

$$\xi \stackrel{\text{def}}{=} \begin{pmatrix} \xi_1 \\ \xi_2 \\ \xi_3 \\ \xi_4 \end{pmatrix} \tag{13}$$

for a four-level system (containing four complex entries $\xi_1, \xi_2, \xi_3, \xi_4$) and considers only expectation values with respect to some two-level subsystem of the four-level system in question, then these expectation values cannot be represented by a vector (10) in a two-dimensional vector space again.

The respective expectation values can however be represented by a so-called density matrix

$$\begin{pmatrix} d_1 & d_2 \\ d_3 & d_4 \end{pmatrix} \tag{14}$$

in the sense that the expectation value

$$\mathrm{Tr}\left(\begin{pmatrix} d_1 & d_2 \\ d_3 & d_4 \end{pmatrix}\begin{pmatrix} a & b \\ c & d \end{pmatrix}\right) = \mathrm{Tr}(D\hat{A}) \stackrel{!}{=} \langle \xi | \hat{A}\xi \rangle \tag{15}$$

coincides with the expectation value previously computed using a vector of type (13). The trace of a matrix is defined as the sum of its diagonal elements and is denoted by the symbol Tr.

A *density matrix* or *density operator* is defined as a self-adjoint matrix (operator) having positive eigenvalues and trace 1. This implies, of course, certain restrictions on the entries $d_1, \ldots, d_4$ in Eq. (14). Here the term "operator" has essentially the same meaning as "matrix," but is appropriate also in the case of infinite-dimensional vector spaces.

Given an arbitrary density matrix D, the expectation values of an observable $\hat{A}$ can be represented as a sum (or an integral) of expectation values with respect to the pure states Ψ_j ($j = 1, 2, \ldots, M$); thus,

$$\mathrm{Tr}(D\hat{A}) = \sum_{j=1}^{M} \langle \Psi_j | \hat{A}\Psi_j \rangle. \tag{16}$$

The state corresponding to D is then called a *nonpure* state and the representation (16) of the expectation values $\mathrm{Tr}(\hat{A}D)$ is described as a decomposition of the nonpure state D into the pure states Ψ_j ($j = 1, 2, \ldots, M$). A nonpure state D for a two-level system can be decomposed into two or three or arbitrarily many pure states (see subsequent text). Hence, even for a two-level system, the number M in Eq. (16) can be arbitrarily large and possibly infinite. Note that we identify nonpure states with the density matrix D representing them.

We have here a mere mathematical formalism. Let us consider a two-level description of an ammonia-type molecule (as before). We now introduce 2×2 density operators D_+, D_-, D_L, D_R corresponding to the pure state vectors $\Psi_+, \Psi_-, \Psi_L, \Psi_R$ in the sense that

$$\mathrm{Tr}(D_+\hat{A}) = \langle \Psi_+ | \hat{A}\Psi_+ \rangle \tag{17}$$

holds true for all 2×2 matrices $\hat{A}$, and similarly for D_-, D_L, D_R. The density operators D_+, D_-, D_L, D_R describe pure states, of course, but we can also look at proper mixed states, such as

$$D_{rac} \overset{def}{=} \tfrac{1}{2} D_L + \tfrac{1}{2} D_R, \tag{18}$$

describing a mixture of 50% left- and 50% right-handed molecules. A thermal state at an inverse temperature β may be regarded as a mixture

$$D_\beta \overset{def}{=} \frac{e^{-\beta E_+} D_+ + e^{-\beta E_-} D_-}{Tr\left(e^{-\beta E_+} D_+ + e^{-\beta E_-} D_-\right)} \tag{19}$$

of the eigenstates Ψ_+ and Ψ_- with appropriate Boltzmann weightings. In quantum mechanics, thermal states are introduced as nonpure states that satisfy certain stability requirements under perturbations, namely, the so-called Kubo–Martin–Schwinger (KMS) boundary condition.[19,72] For N-level systems, such KMS states can be constructed as mixtures over all eigenstates with appropriate Boltzmann weightings.

Here the nonpure states D_{rac} and D_β have been constructed as mixtures of pure states, but could, of course, also be viewed as restrictions of some pure state on an N-level system which contains our two-level system as a subsystem. Hence Eq. (16), even if mathematically correct, does not necessarily imply that the two-level subsystem is indeed in some of the states Ψ_j. In all of the following, one must be extremely careful. It is a very delicate matter to decide which of the possible interpretations of a nonpure state—restriction or mixture—is correct in a given physical situation. The decision to be made is not mathematical, but rather physical.

In summary, we may conclude that a nonpure state (and its corresponding density operator) can have two different meanings, and it is often not easy to figure out which one is appropriate. It is, for example, not at all obvious that a thermal state D_β can be considered as a mixture of pure states as insinuated in Eq. (19) (apart from mathematical convenience).

These difficulties in the interpretation of nonpure states give rise to different versions and interpretations of quantum mechanics, namely, a statistical and an individual one. Quantum theory in its usual *statistical* version embodies the following claims:

(1) A quantum system is always intrinsically coupled to its quantum environment and therefore cannot be expected to be in a pure state. Subsystems of a quantum system are not objects in their own right.

(2) Ultimately, only the whole universe can be described by a pure state, whereas all subsystems require a description in terms of density operators.

(3) Consequently, a decomposition of a nonpure state into pure states does not make physical sense.

(4) Decomposition of a nonpure state (into pure ones) can be used as a mathematical tool, but different decompositions are entirely equivalent.

Quantum theory in its *individual* version, on the other hand, embodies the following claims:

(1) Quantum systems are often intrinsically coupled to their quantum environment and cannot generally be expected to be in a pure state. Nevertheless, in many cases the coupling is "weak" and a description in terms of pure states is appropriate.
(2) In the latter case, a decomposition of a nonpure state into pure states can make sense.
(3) Different decompositions of a nonpure state are not necessarily physically equivalent.[9, 15, 46]

The latter items concerning different decompositions of nonpure states into pure ones will be addressed in the next section. For the moment, we shall concentrate on the relevance of pure states and the meaning of nonpure states as such.

As can be seen from the vague formulation of individual quantum theory, things are not really clear. Molecules in a vessel at a pressure of 1 mbar are usually considered to be individual objects in their own right, to be described by pure states. An electron in an ammonia molecule, on the other hand, is definitely not in a pure state (as can be shown computationally, for instance, by starting from the overall ground state Ψ_+ of the ammonia molecule). A proper discussion of these problems is difficult and involves us in so-called *dressing procedures*.[10, 14] Here we shall simply accept the assumptions of individual quantum theory (which can be true only if the quantum system considered is large enough, with large parts of the environment incorporated, if necessary). Alternatively, we may assume that an appropriate dressing has already been made.

Treating a two-level system in an individual way is, of course, a delicate matter and can be quite misleading, as will be seen later in our discussion of the Curie–Weiss model. This point cannot be stressed often enough. It is only for reasons of simplicity that two-level systems are used here for purposes of illustration.

V. THE DECOMPOSITION OF A NONPURE STATE INTO PURE STATES IS NOT UNIQUE

The problem we turn to now is the *nonuniqueness* of decompositions of nonpure thermal states into pure ones. This nonuniqueness is nicely illustrated in the case of a two-level system, because there the state space is isomorphic to the sphere in three-dimensional Euclidean space, the so-called *Bloch sphere*.

Our discussion is prefaced by the remark that for every nonpure state of a two-level system with associated density operator D, there corresponds a uniquely determined vector **b** in three-dimensional Euclidean space defined by the equation

$$\begin{pmatrix} b_1 \\ b_2 \\ b_3 \end{pmatrix} \overset{\text{def}}{=} \begin{pmatrix} \text{Tr}(D\sigma_1) \\ \text{Tr}(D\sigma_2) \\ \text{Tr}(D\sigma_3) \end{pmatrix}, \tag{20}$$

with σ_j, $j = 1, 2, 3$, being the Pauli matrices. With this correspondence, pure states are mapped to the surface S_2 of the Bloch sphere, whereas nonpure states are mapped onto its inner points. *The mixing of pure states corresponds to the mixing of the vectors* **b**. If the density operators D_1 and D_2 are mapped onto the vectors $\mathbf{b}_1$ and $\mathbf{b}_2$, respectively, then the mixture

$$D \overset{\text{def}}{=} \lambda_1 D_1 + \lambda_2 D_2 \tag{21}$$

will be mapped onto the vector

$$\mathbf{b} \overset{\text{def}}{=} \lambda_1 \mathbf{b}_1 + \lambda_2 \mathbf{b}_2. \tag{22}$$

Representing the states Ψ_+ and Ψ_- as

$$\Psi_- = \begin{pmatrix} 1 \\ 0 \end{pmatrix}, \qquad \Psi_+ = \begin{pmatrix} 0 \\ 1 \end{pmatrix}, \tag{23}$$

one gets the north and south pole, respectively, as corresponding points on the Bloch sphere. A pure state

$$\Psi \overset{\text{def}}{=} c_+ \Psi_+ + c_- \Psi_- = \begin{pmatrix} c_- \\ c_+ \end{pmatrix} \tag{24}$$

corresponds to the point

$$\begin{pmatrix} 2\,\text{Re}(c_-^* c_+) \\ 2\,\text{Im}(c_-^* c_+) \\ c_-^* c_- - c_+^* c_+ \end{pmatrix}. \tag{25}$$

Recall again, that it is quite tricky to use two-level systems (instead of, say, an ammonia molecule) in individual quantum theory. Nevertheless, two-level systems can be quite instructive, precisely because simple visualization is possible by means of the Bloch sphere.

Let us devise a *Gedankenexperiment* for a gas of ammonia-type molecules in a vessel at the pressure $p = 1$ mbar and with a given inverse temperature β by making the following assumptions:

(1) All the molecules can be treated as two-level systems.
(2) All the molecules are in a pure state.
(3) The vessel has a small aperture, through which molecules may escape (at a rate of, say, 1 molecule per second). The pure state of each of these escaping molecules is measured by a "protective measurement," e.g., in the sense of Aharonov and Anandan.[1]
(4) The mixture of the molecules' pure states is given by the thermal density operator

$$D_\beta \approx \frac{1}{2} \begin{pmatrix} 1 & 0 \\ 0 & 1 \end{pmatrix}. \tag{26}$$

This will be the case if either the level splitting is low enough (as with chiral molecules; see Table I) or the temperature is high enough. The last assumption is made only for simplicity.

The question then becomes: What is the distribution of pure states found in such an experiment, when the ammonia molecules in the vessel are in thermal equilibrium?

The following distributions are possible to be found:

(1) A mixture of eigenstates Ψ_+ and Ψ_-, each with 50% probability. This would correspond to a decomposition of the thermal density operator D_β into the pure states as $D_\beta = \frac{1}{2} D_+ + \frac{1}{2} D_-$.
(2) A mixture of chiral states Ψ_L and Ψ_R, each with 50% probability. This would correspond to a decomposition of the thermal density operator D_β as $D_\beta = \frac{1}{2} D_L + \frac{1}{2} D_R$.
(3) A mixture of "alternative" chiral states

$$\tilde{\Psi}_L \stackrel{\text{def}}{=} \frac{1}{\sqrt{2}} (\Psi_+ + i\Psi_-), \tag{27}$$

$$\tilde{\Psi}_R \stackrel{\text{def}}{=} \frac{1}{\sqrt{2}} (\Psi_+ - i\Psi_-), \tag{28}$$

which again are transformed into one another under space-inversion. This would correspond to a decomposition of the thermal density operator D_β as $D_\beta = \frac{1}{2} \tilde{D}_L + \frac{1}{2} \tilde{D}_R$.
(4) An ensemble of 25% left-handed, 25% right-handed, 25% ground, and 25% excited states. This would correspond to a decomposition of D_β as $D_\beta = \frac{1}{4} D_+ + \frac{1}{4} D_- + \frac{1}{4} D_L + \frac{1}{4} D_R$.

In Fig. 5 some of these different decompositions are visualized with the aid of the Bloch sphere. The thermal density operator $D_\beta = \frac{1}{2} \mathbf{1}$ from

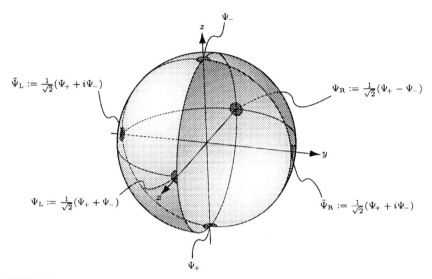

The following labels appear around the Bloch sphere:

$$\tilde{\Psi}_L := \tfrac{1}{\sqrt{2}}(\Psi_+ + i\Psi_-)$$

$$\Psi_R := \tfrac{1}{\sqrt{2}}(\Psi_+ - \Psi_-)$$

$$\Psi_L := \tfrac{1}{\sqrt{2}}(\Psi_+ + \Psi_-)$$

$$\tilde{\Psi}_R := \tfrac{1}{\sqrt{2}}(\Psi_+ + i\Psi_-)$$

FIGURE 5 A thermal density operator can be decomposed into pure states in infinitely many different ways. Mixing the vectors corresponding to eigenstates or chiral states or alternative chiral states with 50% probability always leads to the zero vector, i.e., the center of the Bloch sphere, corresponding to the density operator $D_\beta = \tfrac{1}{2}\mathbf{1}$.

Eq. (26) corresponds to the zero vector in three-dimensional space, i.e., to the center of the Bloch sphere. Statements about the decompositions of D_β made earlier can be verified by computing the vectors $\mathbf{b}_+$, $\mathbf{b}_-$, $\mathbf{b}_L$, $\mathbf{b}_R$, $\tilde{\mathbf{b}}_L$, and $\tilde{\mathbf{b}}_R$ which correspond to the respective pure state vectors Ψ_+, Ψ_-, Ψ_L, Ψ_R, $\tilde{\Psi}_L$, and $\tilde{\Psi}_R$. This can be done by using Eq. (25). The chiral states, for example, correspond to points on the Bloch sphere having zero y and x coordinates. Then a mixture of 25% $\mathbf{b}_L$, 25% $\mathbf{b}_R$, 25% $\mathbf{b}_+$, and 25% $\mathbf{b}_-$ obviously gives the vector $\mathbf{0}$, corresponding to our thermal density operator $D_\beta = \tfrac{1}{2}\mathbf{1}$. This verifies one of the preceding claims [cf. Eq. (22)].

In summary, one has infinitely many different ways of decomposing a thermal density operator into pure states. Different decompositions can refer to entirely different physical or chemical points of view; viz.:

(1) The decomposition of $D_\beta = \tfrac{1}{2}\mathbf{1}$ into chiral states Ψ_L and Ψ_R gives rise to an approximate nuclear molecular structure (simply because we assume that the molecule is in either of these states with equal probability and both Ψ_L and Ψ_R admit an approximate nuclear structure; see Fig. 3).

(2) The decomposition of $D_\beta = \tfrac{1}{2}\mathbf{1}$ into the eigenstates Ψ_+ and Ψ_- does not admit an approximate nuclear molecular structure (simply because we assume that the molecule is in either of these states with equal probability and both Ψ_+ and Ψ_- will not permit an approximate nuclear structure; see Fig. 3).

(3) The decomposition of $D_\beta = \frac{1}{2}\mathbf{1}$ into alternative chiral states $\tilde{\Psi}_L$ and $\tilde{\Psi}_R$ does not admit an approximate nuclear molecular structure (simply because we assume that the molecule is in either of these states with equal probability and both $\tilde{\Psi}_L$ and $\tilde{\Psi}_R$ do not admit an approximate nuclear structure).

We stress again that different decompositions of a molecular thermal state D_β can refer to entirely different physical situations! If all possible decompositions are considered as being equivalent, one cannot infer from the thermal state alone whether the molecule under discussion has a nuclear structure or not. Also, one cannot infer from the thermal state alone if the molecule admits strange states and so on.

VI. DECOMPOSITIONS OF A THERMAL STATE INTO CONTINUOUSLY MANY PURE STATES

It is, of course, not compulsory to decompose a thermal density operator into only two or finitely many pure states. One could equally well try to decompose a thermal density operator into a continuum of pure states, or even into all the pure states of the system in question. This possibility is illustrated here by use of a two-level system and the relevant Bloch sphere.

The physical background is as follows: Starting from a thermal state D_β, one wants to estimate the probabilities for certain pure states to arise. This is done by specification of some decomposition of D_β into pure states. If one could be sure that only finitely many pure states are admitted for the molecular species in question, restriction to these particular states is legitimate. Here is an example. If, in a two-level system, one can assert that only eigenstates of the Hamiltonian arise, then only the spectral decomposition used in Eq. (19) is possible. We have argued that molecules (such as ammonia) can be in any possible pure state (such as the handed states Ψ_L and Ψ_R, which are not eigenstates). Consequently, decompositions into all possible pure states should be considered. Even for a two-level system, continuously many possible pure states can arise (a result exemplified by the Bloch sphere).

This means that we cannot avoid introducing probability distributions of pure states; these are only slightly more complicated than those discussed in the last section. For most purposes, it will be sufficient to consider Fig. 6 and then skip the rest of this section. The essential point there is that decompositions into two states are not obligatory and that decompositions into continuously many pure states are natural. In the lower right part of Fig. 6, for example, the probability of finding the two-level molecule in some pure state is indicated by the shading (heavily shaded in the southern part of the sphere, including the ground state;

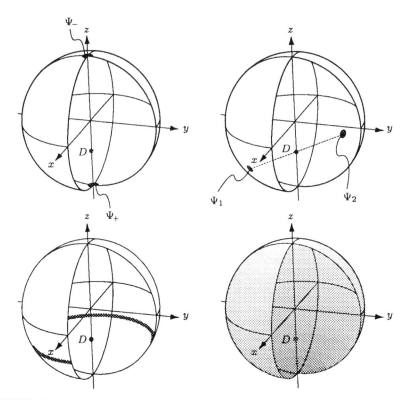

FIGURE 6 A nonpure state (with density operator D) can be decomposed in many different ways into pure states. Here the situation is sketched for 2×2 matrices. The upper left part of the figure shows a decomposition of a density operator D into its eigenstates. The upper right part of the figure shows a decomposition of D into two nonorthogonal pure states. The lower left part of the figure shows a decomposition into pure states all lying on the same parallel of latitude. The lower right part of the figure illustrates the decomposition of D into all possible pure states (of a two-level system). The probability density of pure states arising in this decomposition is indicated by the shading.

lightly shaded in the northern part of the sphere, including the first excited state). If the sphere were shaded uniformly, this would correspond to *equidistribution* of pure states: every pure state would then arise with equal probability. Note that "equidistribution" does not mean "equidistribution of eigenstates," but rather equidistribution of all possible pure states. Here, equidistribution of pure states will be used as a *reference distribution*. The physical distributions used will, of course, not be equidistributed.

The shading used to describe a continuous distribution of pure states can be described by a *shading function* $f = f(\vartheta, \varphi)$ on the sphere

$$f: S_2 \rightarrow [0, \infty], \tag{29}$$

whose precise definition follows. For equidistribution, the shading function f would take the constant value 1 for all points on the Bloch sphere.

Hence, instead of giving probabilities of finding the system in one of finitely many possible states, probabilities $\mu(\tilde{B})$ must be given for finding the system in question in a certain subset $\tilde{B}$ of the set of all pure states. In the case of a two-level system, the set of all pure states is isomorphic to the three-dimensional sphere S_2; hence, a subset $\tilde{B}$ of the set of all pure states can be specified by giving the corresponding subset B of S_2 (see Fig. 7).

As a typical example for such a probability distribution, consider the equidistribution of pure states, with probabilities

$$\mu_{\text{eqp}}(\tilde{B}) \overset{\text{def}}{=} \int_{B \subseteq S_2} \frac{\sin \vartheta \, d\vartheta \, d\varphi}{4\pi}, \tag{30}$$

where the spherical coordinates ϑ and φ are used. Here the subset $\tilde{B}$ of all the pure states of the two-level system in question corresponds to the subset B of the surface of the three-dimensional sphere. The probability distribution in Eq. (30) of pure states is called *equidistribution* of pure states since it is invariant under all symmetries of the two-level system, i.e.,

$$\mu(V\tilde{B}V^*) = \mu(\tilde{B}) \tag{31}$$

holds for all unitary 2×2 matrices V. Mixing all the pure states of the two-level system with probability μ_{eqp} corresponds [see Eq. (22)] to mixing all the vectors **b** according to

$$\int_{S_2} \mathbf{b}(\vartheta, \varphi) \frac{\sin \vartheta \, d\vartheta \, d\varphi}{4\pi}, \tag{32}$$

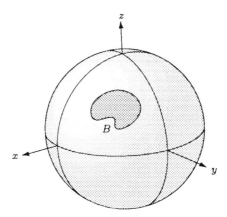

FIGURE 7 A subset $\tilde{B}$ of all the pure states of a two-level system can be specified by giving the corresponding subset B of the surface S_2 of the Bloch sphere.

which is clearly the zero vector **0** corresponding to the density operator $D_\beta = \frac{1}{2}\mathbf{1}$. Hence the mixture of all the pure states of the two-level system with probability μ_{eqp} results in the nonpure state with associated density operator $D_\beta = \frac{1}{2}\mathbf{1}$. Thus, this gives a new decomposition of $D_\beta = \frac{1}{2}\mathbf{1}$, specified by the equipartition probabilities in Eq. (30). Written in terms of density operators, this decomposition would read

$$\text{Tr}\left(\left(\frac{1}{2}\mathbf{1}\right)A\right) = \int_{S_2} \text{Tr}(D_{\vartheta,\varphi}A) \frac{\sin\vartheta\, d\vartheta\, d\varphi}{4\pi}, \tag{33}$$

where A is an arbitrary observable of the two-level system (i.e., an arbitrary 2×2 matrix) and where $D_{\vartheta,\varphi}$ is the density operator corresponding to the (pure) state in the Bloch-sphere description that has the spherical coordinates ϑ and φ. In more abstract notation, this decomposition could be denoted by

$$\text{Tr}\left(\left(\frac{1}{2}\mathbf{1}\right)A\right) = \int_{S_2} \text{Tr}(D_{\vartheta,\varphi}A)\mu_{\text{eqp}}(d\vartheta, d\varphi), \tag{34}$$

with μ_{eqp} being the equipartition measure on the pure states of the two-level system (with two-level systems only being used because of their simplicity).

In Fig. 6 again different decompositions of some thermal state D into pure states are shown: This time the density operator D is not chosen to be equal to $\frac{1}{2}\mathbf{1}$. Since every thermal density operator D can be written as a mixture of the eigenstates of the underlying Hamiltonian, the corresponding vector $\mathbf{b}_D$ must be some mixture of the north and south pole vectors corresponding to the eigenstates Ψ_+ and Ψ_-; hence $\mathbf{b}_D$ must be on the z-axis, as indicated in Fig. 6.

In the left upper part of Fig. 6 the spectral decomposition of D into orthogonal eigenstates is illustrated. Since it is not necessary to decompose a density operator into orthogonal states, another decomposition into two nonorthogonal pure states could also be chosen, as indicated in the upper right part of Fig. 6. As it is not obligatory to decompose into two pure states, decompositions into continuously many pure states are also possible; e.g., into all the pure states lying on the same parallel of latitude as D, indicated in the lower left part of the figure. Alternatively, decompositions into all the pure states of the two-level system under discussion can be made as illustrated in the lower right part of Fig. 6, where the shading indicates the probability of finding a pure state in the respective region of the set of all pure states.

A decomposition of a thermal state D_β into pure states is nothing other than the specification of probabilities $\mu(\tilde{B})$ that the system under discussion is in one of the pure states of the subset $\tilde{B}$ of all pure states possible. A typical example may be given as

$$\mu_f(\tilde{B}) \stackrel{\text{def}}{=} \int_{B \subseteq S_2} f(\vartheta, \varphi) \frac{\sin \vartheta \, d\vartheta \, d\varphi}{4\pi} \tag{35}$$

where B is the subset of the Bloch sphere corresponding to the chosen subset $\tilde{B}$ of pure states and where $f: S_2 \to [0, \infty[$ is now some fixed function from the Bloch sphere into the positive real numbers. The respective decomposition would then read as

$$\text{Tr}(D_\beta A) = \int_{S_2} \text{Tr}(D_{\vartheta, \varphi} A) f(\vartheta, \varphi) \frac{\sin \vartheta \, d\vartheta \, d\varphi}{4\pi} \tag{36}$$

or, in more abstract notation, as

$$\text{Tr}(D_\beta A) = \int_{S_2} \text{Tr}(D_{\vartheta, \varphi} A) \mu(d\vartheta, d\varphi). \tag{37}$$

The probability of finding the system in any pure state must be equal to 1, i.e.,

$$\int_{S_2} f(\vartheta, \varphi) \frac{\sin \vartheta, d\vartheta, d\varphi}{4\pi} = 1. \tag{38}$$

The function f is a probability density, indicated by the shading of the lower right part of Fig. 6. To bring about a decomposition of D_β into eigenstates, one must choose the "function" f as a linear combination of Dirac delta functions (i.e., as an appropriate distribution).

VII. CHEMICAL VERSUS QUANTUM-MECHANICAL POINT OF VIEW

A given thermal density operator can be decomposed into *stationary* or *nonstationary* pure states. Depending on this choice, two different points of view are adopted, which for simplicity will be referred to as the spectroscopist's and the chemist's point of view, respectively.

Let us again look at ammonia molecules: (i) From a spectroscopist's point of view, ammonia molecules are in the stationary eigenstates Ψ_+ and Ψ_- with stochastically occurring "quantum jumps" between them (corresponding to the maser-transition). Note that the expectation value of the dipole moment operator $\hat{\mu}$ is zero with respect to these stationary states. (ii) From a traditional chemist's point of view, ammonia molecules have a (dispersion-free) nuclear structure, oscillating back and forth between the two pyramidal forms, preserving a nuclear structure during the whole oscillation process and with a planar transition state. (iii) Matching the traditional chemist's point of view with that of quantum mechanics, one arrives at what we shall call the chemist's point of view, omitting the epithet "traditional." Ammonia molecules are in nonstationary (pyramidal) states $\Psi_L = (1/\sqrt{2})[\Psi_+ + \Psi_-]$ and $\Psi_R = (1/\sqrt{2})[\Psi_+ - \Psi_-]$ with Schrödinger time evolution

$$\frac{1}{\sqrt{2}}[\Psi_+ \pm \Psi_-] \rightarrow \frac{1}{\sqrt{2}}\left[\Psi_+ \pm \exp\left\{-it\,\frac{(E_- - E_+)}{\hbar}\right\}\Psi_-\right]. \quad (39)$$

During this tunneling process, the nuclear molecular framework is not conserved; in between the "alternative chiral" states $(1/\sqrt{2})[\Psi_+ \pm i\Psi_-]$ arise, which do not possess a nuclear structure. Incidentally, for small level splitting $(E_- - E_+)$, the tunneling process is very slow and so we need to ask which of the available chiral states (on the equator of the Bloch sphere) actually arise in a properly chiral molecule.

The example of hydrogen bond formation between two oxygen atoms (one to a carbonyl group and the other to a hydroxy group, as in naphthazarin, citrinin, and dicarboxylic acids; see Figs. 1 and 2) was considered in Section I. These have a completely analogous structure. The questions again are: Is the hydrogen located near one of the oxygens, does it oscillate back and forth, or should one consider a stationary state (analogous to the ground state of ammonia) without a nuclear structure, i.e., without a dispersion-free position of the hydrogen atom (and other nuclei)?

In such situations either of the different descriptions mentioned can be used:

(1) A "quantum" description in terms of some stationary state Ψ_0 or various stationary states with additional stochastic dynamics (imposed on the usual Schrödinger dynamics).

(2) A "chemical" description by nonstationary states whose superposition is Ψ_0.

This dichotomy reflects itself in the different possible choices for decompositions of the thermal density operator D_β into pure states, viz.: (i) A decomposition of the thermal density operator D_β into (symmetry-adapted) eigenstates of the Hamiltonian. If superpositions of these eigenstates are not considered, one obtains a *classical energy observable*. (ii) A decomposition of the thermal density operator D_β into pure handed states. If superpositions of these handed states are not considered, one gets a *classical chirality observable*. (iii) A decomposition μ of the thermal density operator D_β into pure states Ψ such that the average dispersion

$$\langle \Psi \mid \hat{Q}_j^2 \Psi \rangle - \langle \Psi \mid \hat{Q}_j \Psi \rangle^2 \quad (40)$$

for the expectation values of the nuclear position operators $\hat{Q}_j$ ($j = 1, 2, \ldots, K$) is minimal. Using the nomenclature of two-level systems [compare Eq. (37)], we may define this average dispersion to be

$$\int_{S_2} \left\{ \text{Tr}\left(D_{\vartheta,\varphi} \hat{Q}_j^2 \right) - \text{Tr}\left(D_{\vartheta,\varphi} \hat{Q}_j \right)^2 \right\} \mu(d\vartheta, d\varphi). \tag{41}$$

If superpositions of these minimal-dispersion states are not considered, one obtains an approximately *classical nuclear structure*, which is the best possible nuclear structure compatible with the thermal density operator D_β.

This way of imposing classical structures "by hand" by simply omitting the superpositions between certain chosen states is a very common procedure. Often, the same phenomenon (e.g., the spectrum of ammonia) can be explained in different ways (with, e.g., pyramidal nonstationary states instead of eigenstates), giving a chemical or a quantum-mechanical explanation in our sense. Nevertheless, such differing explanations are not mutually compatible (and may sometimes even be complementary), since either energy or handedness may be a classical observable, but not both together.

Here another approach will be advocated by assuming that a quantum object is indeed in some pure state (without claiming that this pure state can readily be determined). Pure states in quantum mechanics are thus not to be considered as artifacts that can be interchanged ad libitum. We shall seek a canonical decomposition of thermal density operators D_β into pure states. Here "canonical" decomposition means a decomposition that is uniquely determined by stability considerations (for details, see subsequent text). This canonical, i.e., preferred, decomposition should depend on the particular situation. For ammonia, the eigenstates might be expected to play an important role, whereas for a sugar molecule the handed states might be expected to be important. By important is meant here that they will have a high probability with respect to the canonical decomposition.

With a canonical decomposition, the fuzziness of any observable $\hat{A}$ can be unambiguously described. This could yield, for instance, the probability of finding the eigenstates of $\hat{A}$ with respect to the canonical decomposition of the thermal density operator D_β. If different decompositions are admitted, one cannot say anything about the fuzziness of some observable: the energy, for example, is dispersion-free with respect to the decomposition into eigenstates, but not with respect to other decompositions of D_β into pure states. Consequently, the distribution of the expectation values of the energy or the distribution of the dispersions of the molecular energy is very different depending on the particular decomposition of D_β into pure states.

VIII. EFFECTIVE THERMAL STATES

Consider two or more different isomers of some molecular species. Since different isomers have the same underlying Hamiltonian H_0, the respective thermal density operator

$$D_\beta \overset{\text{def}}{=} \frac{\exp(-\beta H_0)}{\text{Tr}(\exp(-\beta H_0))} \tag{42}$$

is identical for different isomers. Consequently, the expectation values

$$\text{Tr}\left(D_\beta \hat{Q}_j\right), \qquad (j = 1, 2, \ldots, K), \tag{43}$$

of the nuclear position operators $\hat{Q}_j$ are identical for different isomers. However, this result is in obvious contradiction with chemical experience, which indicates that in different isomers the nuclei are located at different (relative) positions.

The conclusion is that the prescriptions of statistical quantum mechanics, e.g., that governing the way a thermal state is defined in Eq. (42), cannot explain chemical phenomena without taking over concepts from traditional chemistry in an ad hoc manner. These prescriptions do not give rise to (i) molecular isomers, (ii) handed molecules, (iii) monomer sequences in a macromolecule, or (iv) differently knotted macromolecules. For all these chemically well-known concepts, different expectation values of the nuclear position operators are necessary.

The main reason for this bewildering observation is that the thermal density operators D_β describe a strictly stationary situation. To make this point clear, consider a "thermal state" $D_{\beta, L}$ describing an ensemble of left-handed molecules of some given species. Since handed molecules racemize—albeit slowly—one may expect that $D_{\beta, L}$ evolves into $\frac{1}{2}(D_{\beta, L} + D_{\beta, R})$ for large times. Consequently, the density operator $D_{\beta, L}$ is not strictly stationary and therefore cannot be a thermal state in the strict sense.

"Thermal state in the strict sense" means here that the state fulfills the Kubo–Martin–Schwinger (KMS) boundary condition, which is a stability requirement[19,72] under a specified class of external perturbations. In a quantum system having finitely many degrees of freedom there exists precisely one KMS state for every positive temperature, namely, the state defined in Eq. (42).

In summary, the usual prescriptions of statistical quantum mechanics cannot explain chemical phenomena such as isomerism or the handedness of molecules. The question is then how to introduce effective thermal states for different isomers or differently handed molecules, etc.

With the Born–Oppenheimer (BO) idea in mind, one might try to restrict oneself to a particular minimum of the BO potential. In Fig. 3, for example, one could consider only those state vectors associated with the left or the right minimum and construct corresponding effective thermal states by some ad hoc procedure [and decompose the overall thermal state in Eq. (42) accordingly, as $D_\beta = \frac{1}{2}(D_{\beta,L} + D_{\beta,R})$, for example]. Then the original question remains unanswered: Why should other state vectors, like the ground state vector Ψ_+ or the alternative handed states $\tilde{\Psi}_L$ and $\tilde{\Psi}_R$ of an ammonia-type molecule, be excluded? Any ad hoc procedure that introduces effective thermal states is thus based on a particular choice of state vectors or on a particular choice for decomposition of the overall thermal state D_β of Eq. (42).

We therefore end up again with the problem of finding a canonical decomposition of the "overall" thermal density operator D_β of some molecular species into pure states. Based on such a canonical decomposition of D_β (and dynamical arguments; see subsequent text), one can introduce effective thermal states in some ad hoc manner.

Another option that has been proposed in algebraic quantum mechanics stresses that[19, 20, 58, 72, 76] (i) a quantum system with finitely many degrees of freedom allows one thermal state D_β for every temperature, whereas (ii) a quantum system with infinitely many degrees of freedom may admit several different thermal states $D_{\beta,j}$ ($j = 1, 2, \ldots$) for a given temperature. Our ultimate intention, of course, is to find several different even strict, i.e., KMS thermal, states for different coexisting phases, such as water and steam at 100°C, or for differing molecular isomers or for differently handed molecules, etc.

Let us examine this option for another, perhaps more transparent example, namely, a Curie–Weiss magnet below the Curie temperature, i.e., one with the option of having two different phases with positive and negative permanent magnetization (cf. refs. 16, 17, and 81 for mean-field models). Then, by considering finitely many spins, we are always led to one thermal state D_β^N only (N being the number of spins). The Curie–Weiss model, however, consisting of infinitely many spins, admits two different (strict, i.e., KMS) thermal states $D_{\beta,+}$ and $D_{\beta,-}$ with positive and negative magnetization. In this case the "overall" thermal state D_β as defined in Eq. (42) is the mixture $D_\beta = \frac{1}{2}(D_{\beta,+} + D_{\beta,-})$. This result is somewhat frustrating: Why is it necessary to consider a limit of infinitely many spins to derive two thermal states $D_{\beta,+}$ and $D_{\beta,-}$ for a Curie–Weiss magnet?

Interestingly enough, in the infinite limit, one gets a superselection rule excluding superpositions between the two mentioned thermal states $D_{\beta,+}$ and $D_{\beta,-}$. For any finite number of spins, on the other hand, one always retains full validity of the superposition principle! Is it legitimate to ask how fast superpositions of states with positive and negative magnetization disappear with increasing number of spins?

The particular example of a Curie–Weiss magnet will be studied in Section XI in more detail. The interesting point in this example is that in the limit of infinitely many spins, the overall thermal state decomposes (as can be shown by theorem) into the (nonpure) states $D_{\beta,+}$ and $D_{\beta,-}$. This is because the specific magnetization operator becomes a strictly classical observable and takes only two different values of specific magnetization, denoted here as $+m_\beta$ and $-m_\beta$. The pure states Ψ arising in any further decomposition of $D_{\beta,+}$ and $D_{\beta,-}$ into pure states will have the same expectation values $+m_\beta$ and $-m_\beta$, respectively. On the other hand, in the case of finitely many spins, it is not clear at all how to decompose the unique thermal state D_β^N into the states $D_{\beta,+}^N$ and $D_{\beta,-}^N$ corresponding to the states $D_{\beta,+}$ and $D_{\beta,-}$ in the infinite limit. It is not at all necessary to decompose D_β^N into two such effective thermal states. Moreover, it is not necessary that the pure states Ψ in a decomposition of D_β^N into pure states lead to the expectation values $+m_\beta$ and $-m_\beta$ of the specific magnetization operator. Actually, a canonical decomposition of D_β^N might be expected to do the job, i.e., to lead to a definition of $D_{\beta,+}^N$ and $D_{\beta,-}^N$.

Whenever we introduce a canonical decomposition of thermal states into pure ones, it is therefore important to check compatibility with these limit considerations. Expectation values of the specific magnetization operator (with respect to the pure states Ψ in a canonical decomposition) other than $+m_\beta$ and $-m_\beta$ must disappear with an increasing number of spins in the Curie–Weiss model. The specific magnetization operator thus becomes "more and more classical," because the probability of finding expectation values other than $+m_\beta$ and $-m_\beta$ in the canonical decomposition will get ever smaller with increasing number of spins (see Section XI). In this and similar situations we shall speak of a *fuzzy classical observable* or *fuzzy classical structure*. Getting "more and more classical" means that the fuzziness disappears, ultimately yielding a strictly classical observable in the infinite limit having dispersion-free expectation values.

In summary, statistical quantum mechanics permits us to derive strictly classical observables (such as the classical specific magnetization operator) by appropriate limit considerations (such as a limit of infinitely many spins in case of the Curie–Weiss model). However, statistical quantum mechanics cannot cope with fuzzy classical observables (for finitely many degrees of freedom) since different decompositions of a thermal state D_β are considered to be equivalent. The introduction of a canonical decomposition of D_β into pure states will give rise to an individual formalism of quantum mechanics in which fuzzy classical observables can be treated in a natural way.

IX. STOCHASTIC DYNAMICS ON THE LEVEL OF PURE STATES

We have argued heuristically that certain molecular states are unstable under external perturbations. The symmetric ground state Ψ_+ of an ammonia-type molecule, for example, seems to be unstable under external perturbations if the level splitting $(E_- - E_+)$ is very small (see Table I). Alternatively, superpositions of states describing different molecular isomers would seem to be unstable. This is not to say that such strange states, as we have called them, cannot be prepared experimentally. On the contrary, it is reasonable to argue that a molecule can be prepared in any pure state available in the quantum-mechanical formalism as long as it is screened well enough from external influences.

External perturbations can be of various sorts. Examples include (i) quantum radiation and gravitational fields, (ii) collisions with neighbor particles, and (iii) coupling to some heat bath. Such external perturbations cannot easily be treated theoretically in a rigorous manner. The main problem here is that pure states of the joint quantum system {molecule & quantum environment} are usually not product states and therefore their restriction to the molecular part is no longer a pure state (see Section IV). Nevertheless, we may accept here the idea that appropriate dressing transformations[14] can resolve this situation (almost at least, for some effects due to nonproduct states will survive even the best possible dressing[14]). Hence, molecule and perturbing environment are both thought to be in pure states.

The pure state of the molecular environment is never precisely known. Since the dynamics of the molecular pure state depends on this unknown environment's (pure) state, one gets a stochastic dynamics for the molecular pure states. For a two-level system—with its pure states describable by a Bloch sphere—the situation is illustrated in Fig. 8. The dynamics of some given molecular initial state is governed not only by the Schrödinger equation, but also by external influence. Depending on the (pure) state of the environment, we reach different final molecular states. Usually only probabilistic predictions can be given (and no information about the precise trajectory of pure states, i.e., the trajectory on the Bloch sphere).

Stochastic dynamics for the pure states of a quantum system have been proposed or derived in a variety of different contexts.[33,34,36,53,60,61] Also, the transition rates of Fermi's golden rule,[31]

$$W(n \to m) = \frac{\pi}{2\hbar^2} |\langle \Psi_n | B\Psi_m \rangle|^2 \delta(\omega - |\omega_{nm}|), \qquad (44)$$

between the eigenstates of a molecular Hamiltonian lead to stochastic dynamics for pure states (with B being the coupling operator between molecule and radiation field, e.g., the dipole moment operator).

Unfortunately, a rigorous derivation of stochastic pure-state dynamics is still lacking. Nevertheless it is gratifying that such stochastic dynamics are important and in fact form the basis of a quantum theory of individual (quantum) objects. One hint in this direction comes from *single-molecule spectroscopy* (SMS), where "single molecule" always is to be understood as a single molecule embedded into a polymorphic matrix or a crystal.[13,25,37,38,47–51,56,57,82,83,91] The example used in Fig. 9 is a single

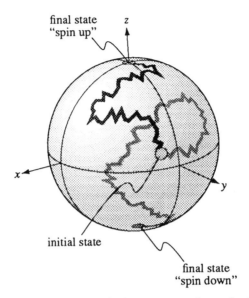

FIGURE 8   Sketch of a stochastic dynamics for pure states of a two-level molecule: Starting from an arbitrary initial state, one ends up with certain states such as, for example, eigenstates of some operator. Which final state is reached depends on the (pure) state of the molecular environment.

terylene defect molecule embedded in a hexadecane matrix; the actual individual quantum object therefore is the joint quantum system {terylene defect & hexadecane matrix}. Another well investigated example is a pentacene defect in a *p*-terphenyl crystal.[13] Investigation is done in these examples by fluoresence excitation spectroscopy, i.e., by exciting the system under discussion with a laser and measuring, at the same time, the overall emission of radiation (excluding the frequency of the laser itself). The interesting point with these SMS experiments is the stochastically migrating behavior of some of the spectral lines. This migrating behavior can be investigated either (i) by changing the laser frequency over some range (say 2000 MHz with one sweep per second) and checking at which frequency absorption (and reemission) takes place (as a function of time) or (ii) by leaving the laser frequency fixed and recording if absorption and reemission takes place at all (as a function of time).

   The second method has been used in Fig. 9, which shows very different stochastic behavior for different terylene molecules in a hexadecane matrix. In the corresponding *ensemble* spectroscopy, involving the investigation of a large number of terylene defects at the same time, all the possible spectral lines appear at the same time and hence a migrating behavior cannot be observed.

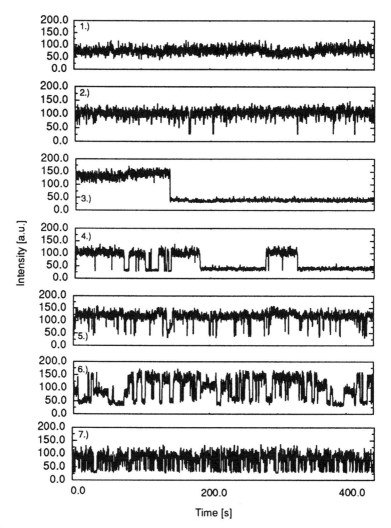

FIGURE 9 Migrating absorption frequencies attributed to single terylene defects in a hexadecane matrix. The examples shown here change between roughly two possible absorption frequencies. Depending on the particular situation, one observes a more or less pronounced stochastic behavior of spectral migration. This particular experiment was performed in the group of Professor Urs Wild at ETH-Zürich. Interested readers are invited to consult http://www.chem.ethz.ch/sms/ on the Internet.

In summary, one has a regular dynamics at the level of density operators (not differing much from the thermal state), whereas the dynamics at the level of pure states is stochastic. Averaging over all possible stochastic "paths" of pure states results in regular density-operator dynamics. To gain insight into single-molecule spectroscopy, for example, it is vital to investigate stochastic dynamics in the pure-state space of large quantum systems (and not only two-level systems, of course).

X. A CANONICAL DECOMPOSITION OF THERMAL STATES INTO PURE STATES

We ask now whether it is possible to give a canonical decomposition of thermal states that would be:

(1) Not necessarily concentrated on eigenstates of the underlying Hamiltonian and therefore able to give a "chemical" description in the sense of Section VII.

(2) (Roughly) concentrated on eigenstates of the Hamiltonian in the case of ammonia and (roughly) concentrated on handed states for properly chiral molecules (compare Table I).

(3) Compatible with limits (number of spins going to infinity or nuclear masses in a molecule going to infinity, etc.) in the sense of Section VIII.

(4) Dynamically *stable* under small external perturbations?

Let us consider first the requirement of stability and assume that a decomposition μ_0 of a thermal state D_β has been chosen arbitrarily [cf. Eq. (37)]. Assume now that the pure states arising from this decomposition are exposed to some stochastic dynamics (stemming from an external perturbation). The decomposition μ_0 will then develop in time $t \to \mu_t$ and normally converge to some limit μ_∞.

Continuous decompositions of D_β can be represented by shading functions (see Section VI). By starting with a particular shading function f_0, the shading will change in the course of time (with shading functions f_t) and converge to some equilibrium shading f_∞ for large times t. In the following text this behavior is described in more precise mathematical terms.

To every probability distribution μ_t of pure states (i.e., to every shading function f_t) there corresponds a density operator D_t [cf. Eq. (37)] defined by averaging

$$\text{Tr}(D_t A) = \int_{S_2} \text{Tr}(D_{\vartheta,\varphi} A)\mu_t(d\vartheta, d\varphi). \tag{45}$$

The density operator D_0 corresponding to μ_0 is the thermal state D_β. If no energy is transferred to the system from outside (on the average, and after large times t), the inverse temperature is again β after long times and we have $D_\infty = D_\beta$. However, the respective probability distribution μ_∞ of pure states might be different from the original μ_0 (if the latter was unstable under small external perturbations). In contrast to μ_0, the limit probability distribution μ_∞ does not change under the particular perturbation (i.e., the specific stochastic dynamics) considered, and hence is a suitable candidate for a stable distribution of pure states.

Let us use as an example a property chiral molecule, such as a sugar or an amino acid. Take μ_0 to be the decomposition of the thermal state D_β into its eigenstates Ψ_+ and Ψ_-. These eigenstates are unstable under external perturbations and could decay, for example, into the chiral states Ψ_L and Ψ_R, the final decomposition μ_∞ being concentrated on these chiral states.

The stochastic dynamics applied to the pure states of a given quantum system cannot easily be discussed. Hence, it would be desirable to have some criterion to estimate the "most stable" decomposition of a given thermal state D_β.

A very useful criterion in this respect is given by the *maximum entropy principle* in the sense of Jaynes.[41] The ingredients of the maximum entropy principle are (i) some reference probability distribution on the pure states and (ii) a way to estimate the quality of some given probability distribution μ on the pure states with respect to the reference distribution. As our reference probability distribution, we shall take the equidistribution μ_{eqp}, defined in Eq. (30), for a two-level system (this definition of equipartition can be generalized to arbitrary $d \times d$ matrices, being the canonical measure on the d-dimensional complex projective plane[42,54]). The relative entropy of some probability distribution μ_f [see Eq. (35)] with respect to μ_{eqp} is defined as

$$S(\mu_f \mid \mu_{eqp}) = -\int_{S_2} f(\vartheta, \phi) \ln[f(\vartheta, \phi)] \frac{\sin \vartheta \, d\vartheta \, d\phi}{4\pi}. \tag{46}$$

Note that the entropy as defined in Eq. (46) is not necessarily related to the von Neumann entropy $\mathrm{Tr}(D_\beta \ln D_\beta)$ of the thermal state. The entropy in Eq. (46) estimates only the information contained in the particular decomposition μ.

The maximum entropy principle can then be formulated as follows: Assume that a given quantum system is in a pure state but that this pure state is unknown. Assume furthermore that the only knowledge about the system is some nonpure state with density operator D_β. Then the probability of finding the pure state in some subset of all pure states is described by the probability distribution μ_{max} of pure states having maximum entropy with respect to equipartition. The probability distribution μ_{max} is chosen from all the probability distributions μ of pure states to yield the given density operator D_β via mixing in the sense of Eq. (37).

We stress again that different decompositions of a molecular thermal state D_β into pure states refer to entirely different physical or chemical descriptions, namely:

(1) A thermal state D_β can be decomposed into eigenstates of the underlying Hamiltonian. This so-called *spectral decomposition* may be unstable under external perturbations, in particular, for high energies or low level splitting. Furthermore, (strict) eigenstates such as the states Ψ_+ and Ψ_- for properly chiral molecules are eliminated that do not accord with chemical intuition.

(2) A molecular thermal state D_β can be decomposed into pure states in such a way that the (average) dispersions of the nuclear position operators Q_j $(j = 1, 2, \ldots, K)$ are minimal [see Eq. (41)]. In this case the question is why should one single out the nuclear *position* operators from the very beginning?

(3) A thermal state D_β can be decomposed according to the maximum entropy principle. The advantages are that (a) this decomposition conforms with dynamical stability under external perturbations,[11] (b) it may lead to quantum or chemical descriptions in the sense of Section VII, depending on the particular situation (i.e., the level splitting or the number of spins considered, etc.), (c) no particular operators such as the nuclear position operators are distinguished, and (d) the maximum entropy decomposition μ_{max} is uniquely determined.

Here, the maximum entropy decomposition μ_{max} of a given thermal state D_β will be defined to be its canonical decomposition. At a later stage of development, one could perhaps consider very specific mechanisms of external perturbation (i.e., particularly chosen stochastic dynamics on the pure states of the system in question). This might lead to a refined definition for the canonical decomposition of thermal states. There is certainly no reason to insist dogmatically on the maximum entropy decomposition. The maximum entropy decomposition should be considered only as a serious candidate for a canonical decomposition, the only one at hand for the moment.

XI. FUZZY CLASSICAL OBSERVABLES AND LARGE DEVIATION THEORY

The overall picture of an individual quantum object developed here may be summed up as follows:

(1) The individual quantum object is in a pure state, which may be unstable under external influences. The dynamics is not only given by the Schrödinger equation, but specified by additional stochastic terms (cf. ref. 11). The probability distribution of pure states in a

thermal situation is given by the maximum entropy decomposition μ_{max} of the respective thermal density operator D_β.

(2) All statistical results in terms of thermal density operators D_β are preserved. Consideration of the canonical decomposition μ_{max} allows us to discuss molecular structure or other fuzzy classical concepts, which cannot be done by considering thermal density operators alone.

It is, unfortunately, not simple to compute the maximum entropy decomposition in molecular situations. We shall therefore consider again the simpler example of the (quantum-mechanical) Curie–Weiss model with the Hamiltonian:

$$H_N = -\frac{J}{2N} \sum_{i,j=1}^{N} \sigma_{z,i}\sigma_{z,j}. \tag{47}$$

Here $\sigma_{z,j}$ are the Pauli matrices (spin $= \frac{1}{2}$) in the z direction for the particles $1, 2, \ldots, N$. The corresponding thermal state $D_{\beta,N}$ is defined as in Eq. (42). The Hamiltonian H_N is invariant under the symmetry transformation defined by

$$\sigma_{x,i} \rightarrow \sigma_{x,i}, \tag{48}$$

$$\sigma_{y,i} \rightarrow -\sigma_{y,i}, \tag{49}$$

$$\sigma_{z,i} \rightarrow -\sigma_{z,i} \tag{50}$$

for all spins i $(i = 1, 2, \ldots, N)$ in the model. The specific magnetization operator $\hat{m}_N$ is defined as

$$\hat{m}_N \overset{def}{=} \frac{1}{N} \sum_{j=1}^{N} \sigma_{z,j}. \tag{51}$$

We now wish to understand whether the specific magnetization becomes a "more and more classical observable" and how fast the fuzziness of the specific magnetization operator goes to zero with increasing number of spins in the magnet, to end up with a strictly classical observable in the limit of infinitely many spins. This issue is analogous to asking how fast the fuzziness of the nuclear molecular framework disappears with increasing mass of the nuclei, to lead to a strictly classical nuclear framework in the limit of infinite nuclear masses.

The following discussion is based on remarks in Section VIII. Recall again that the thermal state can be decomposed in many different ways into pure states.

One could, for example, decompose $D_{\beta,N}$ into symmetry-adapted eigenstates Ψ_n $(n = 1, 2, \ldots, 2^N)$ of the Hamiltonian H_N. Similarly, as in Eq. (9), it can be shown that the expectation values of $\hat{m}_N$ with respect to these eigenstates Ψ_n are zero, $\langle \Psi_n | \hat{m}_N \Psi_n \rangle = 0$. Hence the probability $P_{\beta, N, \mathrm{spectral}}[m_1, m_2]$ of finding an expectation value $\langle \Psi | \hat{m}_N \Psi \rangle$ in an interval $[m_1, m_2]$ (with respect to the spectral decomposition) is

$$P_{\beta, N, \mathrm{spectral}}[m_1, m_2] = 1, \quad \text{if } 0 \in [m_1, m_2] \tag{52}$$

$$= 0, \quad \text{otherwise.} \tag{53}$$

Consequently, the relevant probability density $p_{\beta, N, \mathrm{spectral}}$ for the distribution of the expectation values $\langle \Psi | \hat{m}_N \Psi \rangle$ defined by

$$P_{\beta, N, \mathrm{spectral}}[m_1, m_2] = \int_{m_1}^{m_2} p_{\beta, N, \mathrm{spectral}}(m) \tag{54}$$

is a Dirac δ-function at $m = 0$, i.e., $p_{\beta, N, \mathrm{spectral}}(m) = \delta_0(m)$. If the number N of spins in the model is increased, this probability density $p_{\beta, N, \mathrm{spectral}}$ does not change. In particular, the expectation values $\langle \Psi | \hat{m}_N \Psi \rangle$ with respect to the pure states Ψ in the spectral decomposition do not converge to the values $+m_\beta$ and $-m_\beta$ expected in the infinite limit (cf. Section VIII). Alternatively, one could decompose $D_{\beta, N}$ according to the maximum entropy principle. This results in a different picture (cf. ref. 11 and the large-deviation discussion that follows): The relevant probability density $p_{\beta, N, \max}$ now has two peaks (for high enough N, see the depiction in Fig. 10), i.e., the expectation values $\langle \Psi | \hat{m}_N \Psi \rangle$ concentrate around two different values. The interesting point is that the densities $p_{\beta, N, \max}$ get more sharply peaked with increasing number N of spins. Therefore, the expectation values $\langle \Psi | \hat{m}_N \Psi \rangle$ converge to the values $+m_\beta$ and $-m_\beta$ in the limit of infinitely many spins (cf. Section VIII and ref. 11).

With respect to the maximum entropy decomposition, the specific magnetization operator $\hat{m}_N$ becomes more and more classical, until—in the limit of infinitely many spins—one has a strictly classical observable (see Section VIII) with only two possible expectation values of $\hat{m}_N$, namely, $+m_\beta$ and $-m_\beta$. Nevertheless the superposition principle remains

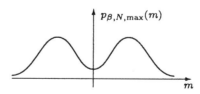

FIGURE 10 Sketch of the probability density $p_{\beta, N, \max}$ for the expectation values of the specific magnetization operator $\hat{m}_N$ with respect to the maximum entropy decomposition of the thermal state D_β of the Curie–Weiss model. This figure refers to some fixed inverse temperature. The expectation values of $\hat{m}_N$ concentrate around two possible values for positive and negative magnetization, respectively. For increasing N, the density $p_{\beta, N, \max}$ becomes more sharply peaked, converging to a sum of two Dirac δ-functions in the limit $N \to \infty$.

universally valid for every finite N. The only new aspect here is that certain pure states, such as symmetry-adapted eigenstates, become more and more unstable with increasing number N of spins in the magnet.

Let us return to our main question concerning how fast such symmetry-adapted eigenstates (and, more generally, pure states with expectation values of $\hat{m}_N$ equal to zero) die out with increasing N. The answer is "exponentially fast." The probabilities $P_{\beta, N, \max}[m_1, m_2]$ go either exponentially fast to zero or exponentially fast to 1 with increasing N, depending on the particular interval $[m_1, m_2]$ of expectation values considered. Furthermore the decay constant is given by an entropy in the sense of large-deviation theory [for general introductions, see refs. 28, 29, and 44; the concepts of a large-deviation entropy is different from an entropy in the sense of Jaynes, as used in Eq. (46)]:

$$\int_{[m_1, m_2]} P_{\beta, N, \max}(m) \sim \exp\left\{-N \inf_{m \in [m_1, m_2]} \left(s_{\beta, \text{mean}}(m)\right)\right\}. \quad (55)$$

To understand this formula, consider Fig. 11, which shows the large-deviation entropy $s_{\beta, \text{mean}}$. This figure is based on a computation (with some approximations) in ref. 11, the temperature now being chosen as one third of the Curie temperature. For a given interval $[m_1, m_2]$, one has to find the respective infimum of $s_{\beta, \text{mean}}$ on $[m_1, m_2]$ and this is the decay constant used in Eq. (55). Two different situations can occur, viz., (i) the interval $[m_1, m_2]$ does not contain one of the special values $+m_\beta$ or $-m_\beta$ (at which the function $s_{\beta, \text{mean}}$ takes its minimum zero). In this case one obtains a strictly positive decay constant for Eq. (55) and hence an

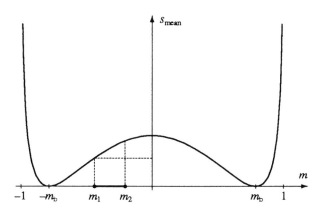

FIGURE 11 An entropy function s_{mean} in the sense of fluctuation (i.e., large-deviation) theory, describing how fast the mean magnetization of a spin system gets classical with an increasing number of spins. The figure is based on an approximate calculation for the Curie–Weiss model. The temperature is fixed and has been taken here as one third of the critical (Curie) temperature. Above the Curie temperature the respective entropy s_{mean} would only have one minimum, namely, at $m = 0$.

exponential decay of the probabilities $P_{\beta, N, \max}[m_1, m_2]$ to 0. (ii) Alternatively, the interval $[m_1, m_2]$ contains one of the special values $+m_\beta$ or $-m_\beta$ (at which the function $s_{\beta, \mathrm{mean}}$ takes its minimum zero). In this case one obtains a zero decay constant for Eq. (55) and hence an exponential decay of the probabilities $P_{\beta, N, \max}[m_1, m_2]$ to 1.

The large-deviation entropy $s_{\beta, \mathrm{mean}}$ describes how fast the fuzziness of the fuzzy classical observable $\hat{m}_N$ goes to zero with increasing number N of spins in a Curie–Weiss magnet.

(1) For a small and fixed values of N, the system behaves quantum mechanically, i.e., pure states Ψ with zero expectation value $\langle \Psi | \hat{m}_N \Psi \rangle = 0$ arise with relatively high probability in the thermal ensemble $\mu_{\beta, N, \max}$ of pure states.

(2) With increasing number N of spins, such pure states die out quickly, determined by the value $s_{\beta, \mathrm{mean}}(m = 0)$ of the large-deviation entropy at $m = 0$ in Eq. (55). In the infinite limit $N \to \infty$, only the expectation values $+m_\beta$ and $-m_\beta$ survive. This conforms with the result of algebraic quantum mechanics, that

$$\hat{m} \stackrel{\mathrm{def}}{=} \lim_{N \to \infty} \hat{m}_N \qquad (56)$$

is a strictly classical observable, i.e., superpositions of pure states Ψ_1 and Ψ_2 with differing expectation values

$$\langle \Psi_1 | \hat{m}_N \Psi_1 \rangle = +m_\beta, \qquad (57)$$

$$\langle \Psi_2 | \hat{m}_N \Psi_2 \rangle = -m_\beta \qquad (58)$$

are forbidden.

In summary, the specific magnetization operators $\hat{m}_N$ are fuzzy classical observables, and the fuzziness decreases continuously with increasing number N of spins in the Curie–Weiss magnet. This decrease of fuzziness is described by the large-deviation entropy $s_{\beta, \mathrm{mean}}$ in Eq. (55). In statistical mechanics, large deviations are "large" fluctuations around some mean value, around the internal energy, for example. In the present context, large deviations are fluctuations around "classical values" of certain observables. In the infinite limit (e.g., with the number of spins going to infinity), only classical values are admitted, whereas "before the limit" (i.e., for $N < \infty$) other expectation values (with respect to pure states) also arise.

XII. THE STRUCTURE OF SINGLE MOLECULES

Let us now again consider single ammonia-type molecules and first review the problem of the usual statistical formalism of quantum mechanics (which uses density operators and no particular decompositions of thermal density operators into pure states; cf. Section IV). We may conclude that

(1) Thermal states, as defined by the usual prescription of statistical quantum mechanics [see Eq. (42)] are in contradiction with traditional chemistry. Neither isomers, nor handed states, nor any other structural chemical concepts can be described or explained thereby.

(2) Fuzzy classical observables cannot be described; only strictly classical observables can be derived, but only in some appropriate limit (as, e.g., with the number of spins or molecular nuclear masses going to infinity).

(3) These limits are somewhat dubious because the superposition principle holds universally before the limit (e.g., for every finite number N of spins) and is broken afterward.

(4) The infinite limit should be the zeroth term in some expansion of the variable $1/N$. The interesting correction terms for large but finite N cannot be written down in the language of usual statistical quantum theory.

A typical question which makes no obvious sense in the usual statistical formalism of quantum mechanics is whether there exists some change in behavior in the sequence {monodeuteroaniline → ammonia → $\cdots$ → asparagic acid} of Table I such as a phase transition. As outlined in Section III, some possibility for such a phase transition to occur would be that there exists only one thermal state with density operator D_β for large level splitting $(E_- - E_+)$ (for a given inverse temperature β), whereas for small level splitting two different thermal states $D_{\beta,L}$ and $D_{\beta,R}$ exist for left- and right-handed molecular species, respectively. Change of behavior between the two different regimes would occur at some critical level splitting. Two arguments—one experimental and the other theoretical —show that a phase transition in this strict sense cannot exist, namely:

(1) Racemization implies that strict thermal states $D_{\beta,L}$ and $D_{\beta,R}$ (in the sense of KMS states) cannot exist (cf. Section VIII).

(2) Rigorous Araki perturbation theory[19] shows that a change in level splitting $(E_- - E_+)$ (mathematically speaking, this is a perturbation of finite norm) cannot lead to a change in the structure of the β-KMS states (for a given inverse temperature β). Hence, if ammonia admits only one KMS state, all the other species in Table I also admit only one KMS state, and conversely. This remark holds, in particular, for systems with infinitely many degrees of freedom such as the joint system {molecule & radiation field}. For finitely many degrees of freedom there exists only one thermal state (for every temperature), anyway. (This result does not apply to a Curie–Weiss magnet consisting of infinitely many spins,

because a change in the coupling constant J is not a perturbation of finite norm.)

If therefore the overall strict thermal state D_β, as defined in Eq. (42), is decomposed into the thermal states $D_{\beta,L}$ and $D_{\beta,R}$,

$$D_\beta = \tfrac{1}{2}(D_{\beta,L} + D_{\beta,R}), \tag{59}$$

then these states $D_{\beta,L}$ and $D_{\beta,R}$ can only be effective ones and not strictly thermal states, and similarly for different isomers of some molecular species. Moreover, since decompositions such as those in Eq. (59) are not acceptable in usual statistical quantum mechanics, such effective thermal states do not make sense there (and are, by the way, not uniquely defined, since different decompositions, if ever accepted, are considered as being equivalent; see Section IV).

Hence, starting from usual statistical quantum mechanics, we find it impossible to understand traditional chemical theories about single molecules.

Let us now turn to the individual formalism of quantum mechanics again, where thermal states D_β are decomposed in a canonical way according to the maximum entropy principle. We cannot compute yet maximum entropy decompositions and large-deviation entropies for molecular situations. Nevertheless, the (simpler) example of the Curie–Weiss magnet suggests that even in molecular situations:

(1) It is sufficient to consider large but finite nuclear molecular masses M_j ($j = 1, 2, \ldots, K$) and that it is not necessary to go to the limit $M_j \to \infty$.

(2) Fuzzy classical structures like handedness or the nuclear structure of molecules can arise.

(3) Distributions of expectation values are well defined in thermal situations.

An example of the latter can be given by some internal coordinate operator, such as the inversion coordinate operator $\hat{q}$ for an ammonia-type molecule (cf. Fig. 3). The decomposition of the thermal molecular state D_β into (strictly) symmetry-adapted eigenstates such as Ψ_+ and Ψ_- leads to expectation values of $\hat{q}$ that are equal to zero:

$$\langle \Psi_\pm | \hat{q} \Psi_\pm \rangle = 0; \tag{60}$$

cf. Eq. (9). Consequently, the probability density $p_{\beta,\text{spectral}} = p_{\beta,\text{spectral}}(q)$ of the expectation values q of $\hat{q}$ [defined analogously to those for the specific magnetization in Eq. (54)] is a Dirac δ-function $\delta_0(q)$. On the other hand, the decomposition of the thermal molecular state D_β according to the maximum entropy principle would be expected to lead to a probability density $p_{\beta,\text{max}} = p_{\beta,\text{max}}(q)$ with two or more peaks (see Fig. 12), analogously to what was shown for specific magnetization in the Curie–Weiss model.

Bearing in mind the large-deviation considerations for the Curie–Weiss model, one could try to characterize molecules by some large-deviation entropy that describes how fast a nuclear molecular structure appears with increasing molecular nuclear masses. Such a large-deviation entropy would describe the decrease in fuzziness of the molecular nuclear structure when the nuclear masses increase. In the limit of infinite nuclear masses one expects a strictly classical nuclear framework, this not being fuzzy anymore at all. Such a large-deviation entropy would also nicely describe the quantum fluctuations round the strictly classical nuclear structure.

Let us now assume that the probability density $p_{\beta,\max} = p_{\beta,\max}(q)$ does indeed show a peaked structure (see Fig. 12) similar to the probability density $p_{\beta,N,\max}$ of the Curie–Weiss magnet in Fig. 10. Then effective thermal states can be introduced in an ad hoc way. To this end, let

$$\text{Tr}(D_\beta T) = \int_{\substack{\text{all pure states} \\ \text{with state vector } \Psi}} \langle \Psi \mid T\Psi \rangle \mu_{\beta,\max}(d\Psi) \tag{61}$$

be the maximum entropy decomposition of a thermal molecular state D_β. [The only difference from Eq. (37) is that now the molecule is not written as a two-level approximation, and that therefore the probability distribution $\mu_{\beta,\max}$ is concentrated on all the pure states with state vectors Ψ and not on S_2. The notation $\mu_{\beta,\max}(d\Psi)$ is not entirely correct, since the probability distribution $\mu_{\beta,\max}$ is concentrated on pure states and not on the state vectors.] The density operators $D_{\beta,L}$ and $D_{\beta,R}$ may then be defined as

$$\text{Tr}(D_{\beta,L}T) = \frac{\int_{\{\Psi \mid \langle \Psi \mid \hat{q}\Psi \rangle \leq 0\}} \langle \Psi \mid T\Psi \rangle \mu_{\beta,\max}(d\Psi)}{\int_{\{\Psi \mid \langle \Psi \mid \hat{q}\Psi \rangle \leq 0\}} \mu_{\beta,\max}(d\Psi)}, \tag{62}$$

$$\text{Tr}(D_{\beta,R}T) = \frac{\int_{\{\Psi \mid \langle \Psi \mid \hat{q}\Psi \rangle \geq 0\}} \langle \Psi \mid T\Psi \rangle \mu_{\beta,\max}(d\Psi)}{\int_{\{\Psi \mid \langle \Psi \mid \hat{q}\Psi \rangle \geq 0\}} \mu_{\beta,\max}(d\Psi)}. \tag{63}$$

The denominators in these equations (being equal to $\frac{1}{2}$) are introduced to obtain normalized density operators, i.e., density operators D with $\text{Tr}(D) = 1$. Hence we simply dissect the set of pure molecular states into the set of

(1) Pure states $S_L =_{\text{def}} \{\Psi \mid \langle \Psi \mid \hat{q}\Psi \rangle \leq 0\}$ with negative expectation value of $\hat{q}$ and
(2) Pure states $S_R =_{\text{def}} \{\Psi \mid \langle \Psi \mid \hat{q}\Psi \rangle \geq 0\}$ with a positive expectation value of $\hat{q}$.

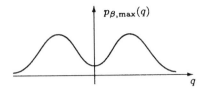

FIGURE 12 Sketch of the expected probability density $p_{\beta,\max} = p_{\beta,\max}(q)$ for the expectation values of the inversion coordinate operator $\hat{q}$ with respect to the maximum entropy decomposition of the molecular thermal state D_β of an ammonia-type molecule. This figure refers to some fixed inverse temperature. The expectation values of $\hat{q}$ concentrate around two possible values for positive and negative inversion coordinate, respectively. For increasing nuclear masses M_j, $j = 1, 2, \ldots, K$, the density $p_{\beta,\max}$ is expected to become more sharply peaked, converging to a sum of two Dirac δ-functions in the limit $M_j \to \infty$.

This procedure is ad hoc because effective thermal states can never really be defined in a completely unique way: The question, for example, whether some state Ψ fulfilling $\langle \Psi \mid \hat{q}\Psi \rangle = 0$ belongs to the set of left- or right-handed states, can never be conclusively answered or defined in an unambiguous way. A similar problem arises when one partitions the set of all pure states into the sets S_j $(j = 1, 2, \ldots)$ for the different isomers of some molecular species. The deeper reason for the ad hoc character of such partitioning procedures is that:

(1) Handed states and isomers are rigorously defined only in some classical limit, as, e.g., in the limit of infinite nuclear molecular masses;

(2) Before the limit, many quantum-mechanical pure states, called strange pure states in Section II, arise which simply do not conform at all to a particular classical classification; and

(3) These strange pure states are forced into some classical classification.

For strange states—such as the ground state Ψ_+ of ammonia—it is, of course, senseless to assign them to one of the sets S_L or S_R, because Ψ_+ is definitely neither left- nor right-handed. Hence, the main motivation for defining the effective thermal states $D_{\beta,L}$ and $D_{\beta,R}$ in Eqs. (62) and (63) is that the probability of finding strange states with respect to the maximum entropy decomposition is expected to get smaller and smaller with increasing nuclear molecular masses or with decreasing level splitting for ammonia-type molecules.

Consequently, the definitions of effective thermal states $D_{\beta,L}$ and $D_{\beta,R}$ in Eqs. (62) and (63) make sense if the respective probability density $p_{\beta,\max}$ shown in Fig. 12 is peaked sharply enough around nonstrange pure states that conform to the classical classification in the infinite limit (e.g., the classification into left- and right-handed states). Similar remarks apply to the problem of introducing different isomers of some molecular species.

Note again that a decomposition of the thermal state D_β into symmetry-adapted eigenstates leads to a situation where one has zero expectation values of the inversion coordinate operator $\hat{q}$, $\langle \Psi | \hat{q} \Psi \rangle = 0$, with probability 1. With this decomposition it is not possible to introduce effective thermal states $D_{\beta, L}$ and $D_{\beta, R}$ as in Eqs. (62) and (63).

XIII. CONCLUDING REMARKS

We have not presented here a fully worked out theory. All our material is in preliminary form, and a precise connection to mathematical fuzzy set theory is still lacking. Also, many of the old unsolved problems of quantum mechanics come into play, and there is no hope of eliminating these without a major effort.

The questions posed refer to many different fields:

(1) Can quantum mechanics give a full description of the stochastic behavior observed in single-molecule spectroscopy?
(2) Is it possible to explain the handedness, nuclear structure, and isomerism of molecular species?[11]
(3) Can fuzzy classical observables be properly defined?
(4) Can effective thermal states (for the handed molecules or chemical isomers) be defined?
(5) Can one understand fuzzy classical behavior in chemistry by specifying some strictly classical behavior at an appropriate limit (corresponding to crisp sets in fuzzy set theory) and giving quantum deviations therefrom for finitely many degrees of freedom or finite nuclear molecular masses (corresponding to proper fuzzy sets).

There are, of course, possibilities of applying mathematical fuzzy set theory directly to the problems discussed here. The subsets S_L or S_R used for the definition of the effective thermal states (62) and (63) should not be defined as crisp sets but as fuzzy sets. The respective compatibility functions of such fuzzy set definitions for S_L and S_R would assume very low values for the pure states Ψ, which satisfy the condition

$$\langle \Psi | \hat{q} \Psi \rangle < \varepsilon \quad (\varepsilon \text{ being small}), \qquad (64)$$

and zero values for pure states Ψ, which satisfy $\langle \Psi | \hat{q} \Psi \rangle = 0$. These matters have not been considered in detail because no more than the obviously trivial definitions could be given. Here the physical background of the fuzziness of chemical structures has been emphasized; the description of decreasing fuzziness (with increasing number of spins in a magnet or increasing nuclear molecular masses) by a large-deviation entropy (see Fig. 11) is of key importance.

ACKNOWLEDGMENTS

"The author gratefully acknowledges an APART-grant [Austrian Programme for Advanced Research and Technology] from the Austrian Academy of Sciences [Österreichische Akademie der Wissenschaften]." He is grateful to Professor Hans Primas for discussions on the individual interpretation of quantum mechanics, to Professor Martin Quack for discussions on the chirality problem and the decompositions of mixed states, to Dr. Don Travlos for discussions concerning naphthazarin, and to Professor Urs Wild for explaining his results on single-molecule spectroscopy and for allowing me to reproduce Fig. 9. Dr. Pitt Funck wonderfully prepared the figures and assisted with LaT$_E$X handling.

REFERENCES

1. Y. Aharonov, J. Anandan, and L. Vaidman, *Phys. Rev. A* **47**, 4616 (1993).
2. A. Amann, *J. Math. Chem.* **6**, 1 (1991).
3. A. Amann, *Ann. Phys.* **208**, 414 (1991).
4. A. Amann, in *Large-Scale Molecular Systems: Quantum and Stochastic Aspects— Beyond the Simple Molecular Picture* (W. Gans, A. Blumen, and A. Amann, eds.), p. 3, NATO ASI Series B 258. Plenum, London, 1991.
5. A. Amann, in *Large-Scale Molecular Systems: Quantum and Stochastic Aspects— Beyond the Simple Molecular Picture* (W. Gans, A. Blumen, and A. Amann, eds.), p. 23, NATO ASI Series B 258. Plenum, London, 1991.
6. A. Amann, *J. Chem. Phys.* **96**, 1317 (1992).
7. A. Amann, *South African J. Chem.* **45**, 29 (1992).
8. A. Amann, *Synthese* **97**, 125 (1993).
9. A. Amann, in *Symposium on the Foundations of Modern Physics 1993—Quantum Measurement, Irreversibility, and the Physics of Information* (P. Busch, P. Lahti, and P. Mittelstaedt, eds.), p. 3. World Scientific, Singapore, 1994.
10. A. Amann, *Proceedings of the Fourth Winter School on Measure Theory.* Tatra Mountains Mathematical Publication 10, 1997. To appear.
11. A. Amann, *J. Math. Chem.* **18**, 247 (1995).
12. A. Amann and H. Primas, in *Experimental Metaphysics—Quantum Mechanical Studies in Honor of Abner Shimony* (R. S. Cohen and J. Stachel, eds.). 1996. To appear.
13. W. P. Ambrose and W. E. Moerner, *Nature* **349**, 225 (1991).
14. H. Atmanspacher, G. Wiedenmann, and A. Amann, *Complexity* **1**, 15 (1995).
15. E. Beltrametti and S. Bugasjki, *J. Phys. A* **28**, 3329 (1995).
16. P. Bóna, *J. Math. Phys.* **29**, 2223 (1988).
17. P. Bóna, *J. Math. Phys.* **30**, 2994 (1989).
18. M. Born and J. R. Oppenheimer, *Ann. Phys.* **84**, 457 (1927).
19. O. Bratteli and D. W. Robinson, *Operator Algebras and Quantum Statistical Mechanics,* Vol. 2. Springer-Verlag, New York, 1981.
20. O. Bratteli and D. W. Robinson, *Operator Algebras and Quantum Statistical Mechanics,* Vol. 1, 2nd revised ed. Springer-Verlag, New York, 1987.
21. J. A. Cina and R. A. Harris, *J. Chem. Phys.* **100**, 2531 (1994).
22. J. A. Cina and R. A. Harris, *Science* **267**, 832 (1995).
23. P. Claverie and S. Diner, *Israel J. Chem.* **19**, 54 (1980).
24. P. Claverie and G. Jona-Lasinio, *Phys. Rev. A* **33**, 2245 (1986).
25. M. Croci, H.-J. Müschenborn, F. Güttler, A. Renn, and U. P. Wild, *Chem. Phys. Lett.* **212**, 71 (1993).
26. E. B. Davis, *Ann. Inst. H. Poincaré* **35**, 149 (1981).
27. E. B. Davis, *J. Phys. A* **28**, 4025 (1995).

28. J.-D. Deuschel and D. W. Stroock, *Large Deviations*. Academic Press, San Diego, 1989.
29. R. S. Ellis, *Entropy, Large Deviations, and Statistical Mechanics*. Springer-Verlag, New York, 1985.
30. B. d. Espagnat *in Preludes in Theoretical Physics* (A. De-Shalit, H. Feshbach and L. van Hove, eds.). North-Holland, Amsterdam, 1966.
31. E. Fermi, *Notes on Quantum Mechanics*. Univ. of Chicago Press, Chicago, 1961.
32. E. Fick and G. Sauermann, *Quantenstatistik Dynamischer Prozesse. Band IIa: Antwort- und Relaxationstheorie*. Deutsch, London, 1986.
33. G. C. Ghirardi, P. Pearle, and T. Weber, *Phys. Rev. A* **42**, 78 (1990).
34. G. C. Ghirardi, A. Rimini, and T. Weber, *Phys. Rev. D* **34**, 479 (1986).
35. P. Gilli, V. Ferretti, and G. Gilli, *in Proceedings of the Conference on Fundamentals of Molecular Modeling* (W. Gans, J. Boeyens, and A. Amann, eds). Plenum, New York, 1996.
36. N. Gisin, *Phys. Rev. Lett.* **52**, 1657 (1984).
37. F. Güttler, T. Irngartinger, T. Plakhotnik, A. Renn, and U. P. Wild, *Chem. Phys. Lett.* **217**, 393 (1994).
38. F. Güttler, J. Sepiol, T. Plakhotnik, A. Mitterdorfer, A. Renn, and U. P. Wild, *J. Luminescence* **56**, 29 (1993).
39. G. S. Hammond, H. Gotthardt, L. M. Coyne, M. Axelrod, D. R. Rayner, and K. Mislow, *J. Am. Chem. Soc.* **87**, 4959 (1965).
40. F. H. Herbstein, M. Kapon, G. M. Reisner, M. S. Lehman, R. B. Kress, R. B. Wilson, W.-I. Shiau, E. N. Duesler, I. C. Paul, and D. Y. Curtin, *Proc. Roy. Soc. London Ser. A* **399**, 295 (1985).
41. E. T. Jaynes, *Phys. Rev.* **106**, 620 (1957).
42. S. Kobayashi and K. Nomizu, *Foundations of Differential Geometry*, Vol. I. Wiley, New York, 1963.
43. R. Kubo, M. Toda, and N. Hashitsume, *Statistical Physics*, Vol. II. Springer-Verlag, Berlin, 1985.
44. O. E. Lanford, *in Statistical Mechanics and Mathematical Problems* (A. Lenard, ed.), p. 1. Springer-Verlag, Berlin, 1973.
45. K. Mislow, M. Axelrod, D. R. Rayner, H. Gotthardt, L. M. Coyne, and G. S. Hammond, *J. Am. Chem. Soc.* **87**, 4958 (1965).
46. B. Misra, *in Physical Reality and Mathematical Description* (C. P. Enz and J. Mehra, eds.), p. 455. Reidel, Dordrecht, 1974.
47. W. E. Moerner, *Science* **265**, 46 (1994).
48. W. E. Moerner and T. Basché, *Angew. Chem. Int. Ed. Engl.* **32**, 457 (1993).
49. W. E. Moerner, T. Plakhotnik, T. Irngartinger, M. Croci, V. Palm, and U. P. Wild, *J. Phys. Chem.* **98**, 7382 (1994).
50. W. E. Moerner, T. Plakhotnik, T. Irngartinger, U. P. Wild, D. W. Pohl, and B. Hecht, *Phys. Rev. Lett.* **73**, 2764 (1994).
51. M. Orrit, J. Bernard, and R. L. Personov, *J. Phys. Chem.* **97**, 10256 (1993).
52. D. Papousek and V. Spirko, *in Topics in Current Chemistry*, Vol. 68, p. 59. Springer-Verlag, Berlin, 1976.
53. P. Pearle, *Phys. Rev. D* **13**, 857 (1976).
54. D. Petz and C. Sudár, *Geometries of Quantum States*, preprint (1995).
55. P. Pfeifer, *Chiral Molecules—a Superselection Rule Induced by the Radiation Field*, Thesis 6551, ok Gotthard S + D AG, ETH-Zürich, Zürich, 1980.
56. M. Pirotta, F. Güttler, H.-R. Gygax, A. Renn, J. Sepiol, and U. P. Wild, *Chem. Phys. Lett.* **208**, 379 (1993).
57. T. Plakhotnik, W. E. Moerner, V. Palm, and U. P. Wild, *Opt. Commun.* **114**, 83 (1995).
58. H. Primas, *Chemistry, Quantum Mechanics, and Reductionism, Perspectives in Theoretical Chemistry*. Springer-Verlag, Berlin, 1983.
59. H. Primas, *Chemie in unserer Zeit* **19**, 109 (1985).

60. H. Primas, *in Sixty-two Years of Uncertainty: Historical, Philosophical, and Physical Inquiries into the Foundations of Quantum Mechanics* (A. I. Miller, ed.), p. 259. Plenum, New York, 1990.
61. H. Primas, *in Quantum Theory without Reduction* (M. Cini and J.-M/ Lévy-Leblond, eds.), p. 49. IOP Publishing Ltd., Bristol, 1990.
62. H. Primas, Physikalische Chemie A, Course at the Swiss Federal Institute of Technology Zürich, Zürich, 1994.
63. H. Primas and U. Müller-Herold, *Elementare Quantenchemie*, Teubner, Stuttgart, 1984.
64. M. Quack, *Chem. Phys. Lett.* **132**, 147 (1986).
65. M. Quack, *Angew. Chem. Int. Ed. Engl.* **28**, 571 (1989).
66. M. Quack, *Jahrbuch der Akademie der Wissenschaften zu Berlin 1990–1992.* 1993.
67. M. Quack, *J. Mol. Struct.* **292**, 171 (1993).
68. M. Quack, *Chem. Phys. Lett.* **231**, 421 (1994).
69. M. Quack, *Faraday Disc. Chem. Soc.* (1995).
70. M. Reed and B. Simon, *Methods of Modern Mathematical Physics. Volume IV: Analysis of Operators.* Academic Press, New York, 1978.
71. P. M. Rentzepis and V. E. Bondybey, *J. Chem. Phys.* **80**, 4727 (1984).
72. G. L. Sewell, *Quantum Theory of Collective Phenomena.* Clarendon, Oxford, 1986.
73. H. Spohn, *Commun. Math. Phys.*, **123**, 277 (1989).
74. H. Spohn and R. Dümcke, *J. Statist. Phys.* **41**, 389 (1985).
75. A. Stasiak and T. Koller, *in New Theoretical Concepts in Physical Chemistry* (A. Amann, L. Cederbaum, and W. Gans, eds.), p. 207, NATO ASI Series C 235. Kluwer, Dordrecht, 1988.
76. F. Strocchi, *Elements of Quantum Mechanics of Infinite Systems.* World Scientific, Singapore, 1985.
77. D. W. Sumners, *in New Theoretical Concepts in Physical Chemistry* (A. Amann, L. Cederbaum, and W. Gans, eds.), p. 221, NATO ASI Series C 235. Kluwer, Dordrecht, 1988.
78. B. T. Sutcliffe, *in Theoretical Models of Chemical Bonding, Part 1: Atomic Hypothesis and the Concept of Molecular Structure* (Z. B. Maksić, ed.), p. 1. Springer-Verlag, Berlin, 1990.
79. B. T. Sutcliffe, *J. Mol. Struct.* **259**, 29 (1992).
80. J. R. de la Vega, J. H. Busch, J. H. Schauble, K. L. Kunse, and B. E. Haaggert, *J. Am. Chem. Soc.* **104**, 3295 (1982).
81. R. F. Werner, *in Quantum Probability and Related Topics* (L. Accardi, ed.), Vol. VII, p. 34. World Scientific, Singapore, 1992.
82. U. P. Wild, M. Croci, F. Güttler, M. Pirotta, and A. Renn, *J. Luminescence* **60–61**, 1003 (1994).
83. U. P. Wild, F. Güttler, M. Pirotta, and A. Renn, *Chem. Phys. Lett.* **193**, 451 (1992).
84. R. G. Woolley, *Adv. Phys.* **25**, 27 (1976).
85. R. G. Woolley, *J. Am. Chem. Soc.* **100**, 1073 (1978).
86. R. G. Woolley, *in Structures versus Special Properties*, p. 1. Springer-Verlag, Berlin, 1982.
87. R. G. Woolley, *Chem. Phys. Lett.* **125**, 200 (1986).
88. R. G. Woolley, *New Scientist* **22**, 53 (1988).
89. R. G. Woolley, *in Molecules in Physics, Chemistry and Biology* (J. Maruani, ed.), Vol. 1. Kluwer, Dordrecht, 1988.
90. R. G. Woolley, *J. Mol. Struct.* **230**, 17 (1991).
91. J. Wrachtrup, *Magnetische Resonanz an Einzelnen Molekülen und Kohärente ODMR-Spektroskopie an Molekularen Aggregaten in Festkörpern*, Thesis, FU Berlin, Berlin, 1994.

5

Fuzzy Measures of Molecular Shape and Size

PAUL G. MEZEY

Mathematical Chemistry Research Unit
Department of Chemistry
  and Department of Mathematics and Statistics
University of Saskatchewan
Saskatoon, Canada S7N 5C9

I. INTRODUCTION

All aspects of molecular shape and size are fully reflected by the molecular electron density distribution.[1] A molecule is an arrangement of atomic nuclei surrounded by a fuzzy electron density cloud. Within the Born–Oppenheimer approximation, the location of the maxima of the density function, the actual local maximum values, and the shape of the electronic density distribution near these maxima are fully sufficient to deduce the type and relative arrangement of the nuclei within the molecule. Consequently, the electronic density itself contains all information about the molecule. As follows from the fundamental relationships of quantum mechanics, the electronic density and, in a less spectacular way, the nuclear distribution are both subject to the Heisenberg uncertainty relationship. The profound influence of quantum-mechanical uncertainty at the molecular level raises important questions concerning the legitimacy of using macroscopic analogies and concepts for the description of molecular properties.[2]

Fuzzy set methods have been developed for a variety of applications, initially mostly in engineering and technology.[3-11] However, many applications in the natural sciences quickly followed.[12-28] The Heisenberg relationship and many other aspects of quantum mechanics can be interpreted in terms of fuzzy sets.[12-16] A straightforward extension of these ideas to some of the elementary concepts of chemistry suggests the following rather natural, fuzzy set representations:

(1) Molecular nuclear configurations within the nuclear configuration space and potential energy hypersurface model of conformational changes and chemical reactions,[17,18] in particular, the distribution and representation of symmetry domains in the nuclear configuration space,[19] as well as reaction mechanisms[18,20]

(2) Molecular symmetry and quasisymmetry, using the syntopy model and related approaches[21-24]

(3) Fuzzy clustering of protein structural classes[25]

(4) Molecular chirality[26] and various other, more general symmetry deficiencies[27]

(5) Electron density and related fuzzy Hausdorff distance problems[28,29]

(6) Various more general molecular shape problems[1,30-40]

In some of the preceding representations, a natural interrelationship between fuzziness and resolution is used, leading to resolution-based chirality, symmetry and similarity measures.[1,27]

Both *global* and *local* shape properties of molecules can be described using a fuzzy set formalism. This approach is suitable for the description of various *functional groups*, the local shape changes induced within various molecular moieties by the rest of the molecule, and some effects of shape and shape changes on chemical reactivity.[40] In particular, the density domain approach to chemical bonding[35,36] provides a quantum-chemical, topological description of functional groups[40] and a consistent framework for a detailed shape characterization of global and local features of molecules.[1,30-40]

More recently, fuzzy electron density modeling of large molecules have been improved to a level comparable to that achieved earlier for small molecules. The *additive fuzzy density fragmentation* (AFDF) scheme of Mezey was described in a general form earlier.[36,37] The simplest version of this scheme, the Mulliken–Mezey AFDF method, is the basis of the molecular electron density lego assembler (MEDLA) technique of Walker and Mezey for generating ab initio quality electron densities for macromolecules.[41-47] The MEDLA method can be applied, virtually without any size limitation, to truly large molecules. This newly available option provided by the MEDLA technique extends the scope of the shape group method[1,30-40,47,48] to ab initio quality electron density shape analysis of proteins and supramolecular structures.

It is now possible to analyze macromolecular electron densities at a resolution far exceeding the resolution of current x-ray diffraction and other experimental and macromolecular computational techniques. The MEDLA method presents a new perspective for the analysis of global and local shape, molecular similarity, and complementarity.

The Mulliken–Mezey AFDF scheme and the more general AFDF schemes[36,37] also serve as the basis for the adjustable density matrix assembler (ADMA) method.[49–52] The ADMA method generates ab initio quality macromolecular *density matrices*, which can be used for the computation of a variety of ab initio quality properties for macromolecules. The ADMA method is also suitable for the calculation of ab initio quality electronic densities, however, additional molecular properties, such as forces and energies, can also be calculated. These options of the ADMA method are expected to be useful in macromolecular conformational analysis, geometry optimizations, and in computational studies of protein folding.

II. A BRIEF REVIEW OF SOME FUZZY SET CONCEPTS RELEVANT TO THE MOLECULAR SHAPE PROBLEM

The theory of fuzzy sets has experienced an extremely rapid development since the original introduction of the subject by Zadeh.[3,4] A very readable, more recent introduction to the fundamentals and some of the more advanced topics of fuzzy set theory has been given from the dual perspectives of systematic theory and applications.[10,11] Some additional applications to the fundamentals of quantum mechanics,[12–16] as well as several chemical applications also were described earlier,[17–29] as well as in previous chapters of this book. Some of the earlier chemical applications involved models where the concept of crisp nuclear geometry in the topological description of reaction mechanisms was replaced with a fuzzy topological approach.[17] Additional applications involve formal fuzzy set approaches to reaction mechanisms and potential energy surfaces,[18–20] syntopy,[21,24] protein classification,[25] chirality,[26] problems of more general symmetry deficiencies,[27] electronic density,[28] and a chemically motivated choice for a generalization of the Hausdorff metric for fuzzy sets.[29]

A detailed review of the basic concepts of fuzzy sets can be found in other chapters of this volume. Here only the specific notations and the fuzzy set concepts most relevant to the molecular shape problem are reviewed, followed by a simple proof for a special fuzzy set generalization of the Hausdorff distance, motivated by the quantum chemical properties of fuzzy electronic densities of molecules.

The notion $\mu_Y(x)$ is used for the *membership function* of elements x of set X in fuzzy set Y.

For fuzzy sets used in this study various set operations are required. If A and B are fuzzy subsets of U, then the *fuzzy intersection*, that is, the result of the operation "A and B" is denoted by $A \cap B$, and is interpreted as a fuzzy subset C of set U, where the corresponding membership function fulfills the condition

$$\mu_C(u) = \min\{\mu_A(u), \mu_B(u)\} \tag{1}$$

for every $u \in U$.

The *fuzzy union*, that is, the result of the operation "A or B" is denoted by $A \cup B$, and is interpreted as a fuzzy subset D of set U, where the membership function fulfills the condition

$$\mu_D(u) = \max\{\mu_A(u), \mu_B(u)\} \tag{2}$$

for every $u \in U$.

A fuzzy generalization of the *complement* is obtained as the result of operation "not A," denoted by A', interpreted as a fuzzy subset A' of set U, where the membership function fulfills the relation

$$\mu_{A'}(u) = 1 - \mu_A(u) \tag{3}$$

for $u \in U$.

In some instances, operations between fuzzy subsets are also required if these subsets have different parent sets. If the operation "A and B" is applied to a fuzzy subset A of set U and a fuzzy subset B of a different set V, then the result $A \cap B$ is a fuzzy relation R_{AB} in the product set $U \otimes V$, where the corresponding membership function is defined as

$$\mu_{R_{AB}}((u,v)) = \min\{\mu_A(u), \mu_B(v)\}. \tag{4}$$

The *α-cut* of a fuzzy subset A of a set X is the crisp set of all those points x of X where the membership function $\mu_A(x)$ is equal to the value α:

$$G_A(\alpha) = \{x : \mu_A(x) = \alpha\}. \tag{5}$$

According to an alternative definition, one may consider a set where $\mu_A(x)$ is greater than or equal to the value α:

$$D_A(\alpha) = \{x : \mu_A(x) \geq \alpha\}. \tag{6}$$

These two choices for α-cuts are analogous to the electron density level sets and the density domains used in the topological analysis of molecular shapes. A third alternative corresponds to the case where the inequality $\geq$ is replaced with the strict inequality $>$ in definition (6).

III. A GENERALIZATION OF THE HAUSDORFF METRIC FOR FUZZY SETS

For the description of shape differences between fuzzy objects, such as molecular electron density clouds, it is useful to generalize the Hausdorff metric for fuzzy sets.[29] The ordinary Hausdorff distance,[53] a formal distance $h(A, B)$ reflecting the differences between two sets A and B of a metric space X is the smallest value r such that each ball of radius r centered at any point of either set contains at least one point of the other set.

For a formal definition of the Hausdorff distance, first we shall review some relevant concepts. We assume that A and B are subsets of a set X, and for points of X a distance function is already defined. For example, if X is the ordinary, three-dimensional Euclidean space E^3 and if the points **a** and **b** of X are represented by their three Cartesian coordinates a_1, a_2, a_3, and b_1, b_2, a_b, respectively, where **a** and **b** can be written as column vectors

$$\mathbf{a} = (a_1, a_2, a_3)', \tag{7}$$

$$\mathbf{b} = (b_1, b_2, b_3)', \tag{8}$$

then a suitable distance function is the Pythagorean distance

$$d(\mathbf{a}, \mathbf{b}) = \left[(a_1 - b_1)^2 + (a_2 - b_2)^2 + (a_3 - b_3)^2 \right]^{1/2}. \tag{9}$$

Note that infinitely many other choices $d'(\mathbf{a}, \mathbf{b})$ for a distance function can be made, as long as these choices fulfill the conditions for a metric:

$$d'(\mathbf{a}, \mathbf{b}) \geq 0, \tag{10}$$

$$d'(\mathbf{a}, \mathbf{b}) = 0 \quad \text{if and only if } \mathbf{a} = \mathbf{b}, \tag{11}$$

$$d'(\mathbf{a}, \mathbf{b}) = d'(\mathbf{b}, \mathbf{a}) \quad (\text{symmetry}), \tag{12}$$

$$d'(\mathbf{a}, \mathbf{b}) \leq d'(\mathbf{a}, \mathbf{b}) + d'(\mathbf{a}, \mathbf{b}) \quad (\text{triangle inequality}). \tag{13}$$

If a metric is given for the underlying set X, say, the Pythagorean distance function is given within the three-dimensional Euclidean space E^3, then the distance between a point **x** of X,

$$\mathbf{x} \in X, \tag{14}$$

and a subset A of X,

$$A \subset X \tag{15}$$

is defined by

$$d(\mathbf{x}, A) = \inf_{\mathbf{a} \in A} \{ d(\mathbf{x}, \mathbf{a}) \}, \tag{16}$$

that is, as the greatest lower bound of distances between points **a** of A and the point **x**. If A is a closed set and the distance d is continuous, then the infimum becomes minimum, that is, this distance is the minimum distance between points **a** of A and the point **x**.

After these preparations, the Hausdorff distance $h(A, B)$ between two subsets A and B of X is defined as

$$h(A, B) = \sup_{\substack{\mathbf{a} \in A \\ \mathbf{b} \in B}} \{d(\mathbf{a}, B), d(\mathbf{b}, A)\}, \qquad (17)$$

that is, as the lowest upper bound of distances between points $\mathbf{a}$ of A and the set B and distances between points $\mathbf{b}$ of B and the set A. If both A and B are closed sets and the actual distance function $d(\mathbf{a}, \mathbf{b})$ is continuous, then the supremum in the definition becomes maximum. For example, the Hausdorff distance between two superimposed molecular contour surfaces (which are closed sets) is the minimum r value such that any point on either contour surface has at least one point of the other contour surface within a distance r.

The Hausdorff distance $h(A, B)$ is a proper metric within any family of compact sets, for example, $h(A, B)$ is zero if and only if the two sets are the same,

$$A = B. \qquad (18)$$

The Hausdorff distance was applied in chemistry in various chirality studies. The Hausdorff distance was used to measure the deviation of a chiral nuclear arrangement from some arbitrary reference arrangement, as proposed by Rassat.[54] Mislow and co-workers used the Hausdorff distance between the object and its optimally overlapping mirror image to provide a chirality measure of the second type.[55,56] Using this Hausdorff distance criterion, Buda and Mislow determined the "most chiral" constrained and unconstrained simplexes in two and three dimensions, that is, the most chiral triangles and tetrahedra.[56]

We need a generalization of the Hausdorff metric for fuzzy sets, suitable for the description of the differences between fuzzy electronic densities where the electron density value $\rho_A(\mathbf{r})$ at each point $\mathbf{r}$ of the space can be regarded as a measure of the point belonging to the molecule A. Naturally, if the electron density contribution from a molecule A is large at a given location $\mathbf{r}$, then this point $\mathbf{r}$ belongs to the molecule to a greater degree than a point $\mathbf{r}'$ where the electron density contribution is negligibly small. Using fuzzy set formalism, this approach can be formulated in terms of membership functions if the electron density is properly scaled, so all numerical values of the scaled densities fall within the unit interval I. This can be achieved by simple linear scaling, for example, by dividing all density values with the maximum density within the molecule A:

$$\mu_{A(\rho)}(\mathbf{r}) : E^3 \to I, \qquad (19)$$

$$\mu_{A(\rho)}(\mathbf{r}) = \rho_A(\mathbf{r}) / \rho_{A, \max}. \qquad (20)$$

A physically more relevant description of the actual space requirement of the molecule A is obtained if the scaling is nonlinear, for example, if the membership function is taken as

$$\mu_{A(\rho,\tau)}(\mathbf{r}) = 1 - \exp\left(-\tau\rho_A^2(\mathbf{r})\right), \tag{21}$$

where τ is a suitable exponential scaling factor. Alternative approaches involve a threshold value above which density the point fully belongs to the molecule, and a gradual decrease of the value of the membership function is obtained only below this threshold. All of the foregoing choices follow a natural chemical expectation, for the membership function emphasizes the importance of high densities; the "grade of belonging" of the high density regions to the molecule A is greater than that of the low density regions.

For all fuzzy sets, including three-dimensional functions of electron density-like continua provided with suitable membership functions, the differences between the corresponding fuzzy sets can be expressed by a metric based on a generalization of the Hausdorff distance. The basic idea is to take the ordinary Hausdorff distances $h(\alpha)$ for the α-cuts of the fuzzy sets for all relevant α values, scale the Hausdorff distance $(h)\alpha$ according to the α value, and from the family of the scaled Hausdorff distances, the supremum determines the fuzzy metric distance $f(A, B)$ between the fuzzy sets A and B. If, in addition, the relative positions of the fuzzy sets A and B are allowed to change, then the infimum of the $f(A, B)$ values obtained for the various positionings determines a fuzzy metric of the dissimilarities of the intrinsic *shapes* of the two fuzzy sets.

First we shall discuss the case of "unscaled" fuzzy Hausdorff-type metric for general fuzzy sets. This metric does not serve our purposes well, since it places equal weights on differences between points with large and very small membership values. In application to molecular electron densities, regions with very small electron densities are rather extensive yet they are of little chemical significance. Placing equal emphasis on these regions and on regions of higher electron density results in a physically unjustified bias. A better representation of physical reality can be obtained if a scaling by the actual density value is applied, as outlined in the previous paragraph.

Nevertheless, an unscaled fuzzy Hausdorff metric provides important insight into the generalization of concepts originally proposed and proved for crisp sets to fuzzy sets, and some features of the proof can be utilized in the proof of the metric properties of the scaled fuzzy Hausdorff-type metric described here. For this purpose, first a proof is presented for the metric properties of an unscaled fuzzy Hausdorff metric, equivalent to the metric proposed by Puri and Ralescu,[6] followed by a proof of the metric properties of an α-scaled fuzzy Hausdorff-type metric, suitable for fuzzy electron densities.

For two fuzzy sets A and B, consider their α-cuts $G_A(\alpha)$ and $G_B(\alpha)$, respectively, for each membership function value α. The ordinary Hausdorff distances for each pair of α-cuts are denoted by

$$h(G_A(\alpha), G_B(\alpha)).$$

A function $g(A, B)$, equivalent to the fuzzy Hausdorff distance suggested by Puri and Ralescu,[6] is defined as

$$g(A, B) = \sup_{\alpha \in [0,1]} \{h(G_A(\alpha), G_B(\alpha))\}. \tag{22}$$

To show that this function $g(A, C)$ is a proper metric, we show that the four general conditions for a metric [relations (10)–(13)] are fulfilled.

(1) *Nonnegativity*:

$$g(A, B) \geq 0. \tag{23}$$

The set $\{h(G_A)(\alpha), G_B(\alpha))\}$ in definition (22) of $g(A, B)$ contains values of ordinary Hausdorff distances that are nonnegative. Consequently, the supremum over this set is also nonnegative.

(2) *Zero distance for identity*:

$$g(A, B) = 0 \quad \text{iff } A = B. \tag{24}$$

If the $g(A, B)$ supremum in definition (22) is zero, this implies that every individual, ordinary Hausdorff distance $h(G_A(\alpha'), G_B(\alpha'))$ of α-cuts in set $\{h(G_A(\alpha), G_B(\alpha))\}$ is zero:

$$h(G_A(\alpha'), G_B(\alpha')) = 0. \tag{25}$$

Hence all pairs of α-cuts for A and B agree for each α value:

$$G_A(\alpha) = G_B(\alpha). \tag{26}$$

Hence, between the two fuzzy sets A and B there exists a one-to-one and onto membership function preserving assignment of points. Consequently, the two fuzzy sets A and B are identical,

$$A = B. \tag{27}$$

On the other hand, if $A = B$, then for each α value

$$G_A(\alpha) = G_B(\alpha). \tag{28}$$

Consequently,

$$g(A, B) = \sup_{\alpha \in [0,1]} \{h(G_A(\alpha), G_B(\alpha))\} = 0. \tag{29}$$

(3) *Symmetry*:

$$g(A, B) = g(B, A). \tag{30}$$

Since each ordinary Hausdorff distance $h(G_A(\alpha'), G_B(\alpha'))$ of α-cuts in set $\{h(G_A(\alpha), G_B(\alpha))\}$ is symmetric with respect to interchange of A and B,.

$$h(G_A(\alpha'), G_B(\alpha')) = h(G_B(\alpha'), G_A(\alpha')), \qquad (31)$$

the supremum $g(A, B)$ in definition (22) of function $g(A, B)$ is also necessarily symmetric: $g(A, B) = g(B, A)$.

(4) *Triangle inequality*:

$$g(A, B) + g(B, C) \geq g(A, C). \qquad (32)$$

Take three fuzzy sets A, B, and C and their α-cuts $G_A(\alpha)$, $G_B(\alpha)$, and $G_C(\alpha)$, respectively, for each membership function value α. Assume that the α-cuts $G_A(\alpha)$, $G_B(\alpha)$, and $G_C(\alpha)$ depend at least piecewise continuously on the α parameter from the unit interval $[0, 1]$, where the intervals of continuity have nonzero lengths and where continuity is understood within the metric topology of the underlying space X. For the three pairs formed from these three fuzzy sets, the ordinary Hausdorff distances $h(G_A(\alpha), G_B(\alpha))$, $h(G_B(\alpha), G_C(\alpha))$, and $h(G_A(\alpha), G_C(\alpha))$ are well defined for each triple of α-cuts.

On the closed set $[0, 1]$, the Hausdorff distance $h(G_A(\alpha), G_B(\alpha))$ is an at least piecewise continuous function of the level set value α. This function either attains its maximum $h(G_A(\alpha'), G_B(\alpha'))$ at some value α' within $[0, 1]$ or it converges to its supremum value

$$\lim_{\alpha \to \alpha'} h(G_A(\alpha), G_B(\alpha)) = \sup_{\alpha \in [0, 1]} \{h(G_A(\alpha), G_B(\alpha))\} \quad (33)$$

as α converges within $[0, 1]$ to some value α' at some discontinuity. Equation (33) also applies to the case when $h(G_A(\alpha), G_B(\alpha))$ attains its maximum at some value α'. Hence, for uniformity, this limit form will be used:

$$g(A, B) = \sup_{\alpha \in [0, 1]} \{h(G_A(\alpha), G_B(\alpha))\}$$

$$= \lim_{\alpha \to \alpha'} h(G_A(\alpha), G_B(\alpha)). \qquad (34)$$

Similarly, there exist values α'' and α''' within $[0, 1]$, such that

$$g(B, C) = \sup_{\alpha \in [0, 1]} \{h(G_B(\alpha), G_C(\alpha))\} = \lim_{\alpha \to \alpha''} h(G_B(\alpha), G_C(\alpha))$$

$$(35)$$

and

$$g(A,C) = \sup_{\alpha \in [0,1]} \{h(G_A(\alpha), G_C(\alpha))\}$$

$$= \lim_{\alpha \to \alpha'''} h(G_A(\alpha), G_C(\alpha)). \qquad (36)$$

As follows from the properties of the supremum,

$$\sup_{\alpha \in [0,1]} \{h(G_A(\alpha), G_B(\alpha))\} = \lim_{\alpha \to \alpha'} h(G_A(\alpha), G_B(\alpha))$$

$$\geq \lim_{\alpha \to \alpha'''} h(G_A(\alpha), G_B(\alpha)) \qquad (37)$$

and

$$\sup_{\alpha \in [0,1]} \{h(G_B(\alpha), G_C(\alpha))\} = \lim_{\alpha \to \alpha''} h(G_B(\alpha), G_C(\alpha))$$

$$\geq \lim_{\alpha \to \alpha'''} h(G_B(\alpha), G_C(\alpha)). \qquad (38)$$

The ordinary Hausdorff distance for each set of α-cuts as $\alpha \to \alpha'''$ fulfills the triangle inequality

$$h(G_A(\alpha), G_B(\alpha)) + h(G_B(\alpha), G_C(\alpha))$$

$$\geq h(G_A(\alpha), G_C(\alpha)). \qquad (39)$$

Consequently,

$$\lim_{\alpha \to \alpha'''} h(G_A(\alpha), G_B(\alpha)) + \lim_{\alpha \to \alpha'''} h(G_B(\alpha), G_C(\alpha))$$

$$\geq \lim_{\alpha \to \alpha'''} h(G_A(\alpha), G_C(\alpha)). \qquad (40)$$

However, according to the inequalities (37) and (38), the two terms on the left-hand side of inequality (40) do not decrease if the limits for $\alpha \to \alpha'''$ are replaced by the corresponding optimum limits of $\alpha \to \alpha'$ and $\alpha \to \alpha''$, respectively:

$$\lim_{\alpha \to \alpha'} h(G_A(\alpha), G_B(\alpha)) + \lim_{\alpha \to \alpha''} h(G_B(\alpha), G_C(\alpha))$$

$$\geq \lim_{\alpha \to \alpha'''} h(G_A(\alpha), G_C(\alpha)). \qquad (41)$$

Combining this result with Eqs. (34), (35), and (36), one obtains

$$g(A, B) + g(B, C) \geq g(A, C). \qquad (42)$$

Note that in some applications it is meaningful to choose a similarity metric for fuzzy sets in such a way that the differences for α-cuts with large α values have more weight than those with small α values. In other words, when the similarities between two fuzzy sets A and B are assessed, it may appear natural to pay more attention to points with large values of the membership functions than to those with small membership function values. In particular, if the membership function is positive, then the 0-cut $G_A(0)$ of the fuzzy set A is the empty set. For an emphasis of the greater relative importance of the large values of the membership function at the "more committed points" of the fuzzy sets A and B, one may consider alternative choices for a function $g(A, B)$. For example, it is useful to scale the ordinary Hausdorff distance in definition (22) by the α value, leading to a new fuzzy, "commitment-weighted" Hausdorff-type metric $f(A, B)$:

$$f(A, B) = \sup_{\alpha \in [0, 1]} \left\{ \alpha h(G_A(\alpha), G_B(\alpha)) \right\}. \tag{43}$$

As can be easily proven by a simple modification of the proof presented here for the unscaled fuzzy Hausdorff metric, this scaled fuzzy Hausdorff distance is also a metric in the space of fuzzy subsets of the underlying set X. A proof is given in subsequent text.

Indeed, $f(A, B)$ fulfills relations (10)–(13) for a metric, as shown in the following paragraphs.

(1) Each element in the set $\{\alpha h(G_A(\alpha), G_B(\alpha))\}$ in the definition (43) of the commitment-weighted fuzzy Hausdorff-type distance $f(A, B)$ is nonnegative; hence, the supremum over this set is also nonnegative. Consequently,

$$f(A, B) \geq 0. \tag{44}$$

(2) If the $f(A, B)$ supremum in definition (43) is zero, this implies that each α-scaled ordinary Hausdorff distance $\alpha h(G_A(\alpha'), G_B(\alpha'))$ of α-cuts in the set $\{\alpha h(G_A(\alpha), G_B(\alpha))\}$ is zero for any $\alpha > 0$,

$$h(G_A(\alpha'), G_B(\alpha')) = 0. \tag{45}$$

Consequently, all pairs of α-cuts for A and B agree for each α value, where $\alpha > 0$:

$$G_A(\alpha) = G_B(\alpha). \tag{46}$$

Since all pairs of these α-cuts coincide, we conclude that between the two fuzzy sets A and B there must exist a one-to-one and onto correspondence of points that also preserves membership function, $\mu_A(x) = \mu_B(x)$, for every point $x \in X$ where this membership function is positive, $\mu_A(x) = \mu_B(x) = \alpha > 0$. In particular, for any point $x' \in X$, one cannot have membership function values $\mu_A(x') = 0$ and $\mu_B(x') = \alpha' > 0$, since then $x' \in G_A(\alpha')$, but $x' \notin G_B(\alpha)$, which violates Eq. (46) for the choice of $\alpha = \alpha'$. Consequently, for all the remaining points x of the underlying space X, $\mu_A(x) = \mu_B(x) = 0$; hence, $\mu_A(x) = 0$ if and only if $\mu_B(x) = 0$ also holds. Consequently, the two fuzzy sets A and B are identical, $A = B$.

Conversely, if $A = B$, then for each α value,

$$G_A(\alpha) = G_B(\alpha) \tag{47}$$

follows, which implies

$$\alpha h(G_A(\alpha), G_B(\alpha)) = 0 \tag{48}$$

for each α value; hence, $A = B$ implies

$$\sup_{\alpha \in [0,1]} \{\alpha h(G_A(\alpha), G_B(\alpha))\} = 0. \tag{49}$$

One obtains

$$f(A, B) = 0 \quad \text{iff} \quad A = B. \tag{50}$$

(3) The ordinary Hausdorff distance $h(G_A(\alpha'), G_B(\alpha'))$ of each α-cut in the set $\{\alpha h(G_A(\alpha), G_B(\alpha))\}$ is symmetric with respect to interchange of sets A and B,

$$h(G_A(\alpha'), G_B(\alpha')) = h(G_B(\alpha'), G_A(\alpha')). \tag{51}$$

Consequently, the supremum $f(A, B)$ in definition (43) is also necessarily symmetric,

$$f(A, B) = f(B, A). \tag{52}$$

(4) The commitment-weighted fuzzy Hausdorff-type distance $f(A, B)$ also satisfies the triangle inequality.

We assume that the α-cuts $G_A(\alpha)$, $G_B(\alpha)$, and $G_C(\alpha)$ of three fuzzy sets A, B, and C, respectively, depend at least piecewise continuously on the α parameter from the unit interval $[0, 1]$, where continuity is understood within the metric topology of the underlying space X. On the closed interval $[0, 1]$, the scaled Hausdorff distance $\alpha h(G_A(\alpha), G_B(\alpha))$ is an at least piecewise continuous function of the level set value α. This function either attains its maximum $h(G_A(\alpha'), G_B(\alpha'))$ at some value α' within $[0, 1]$ or it converges to its supremum value

$$\lim_{\alpha \to \alpha'} \alpha h(G_A(\alpha), G_B(\alpha)) = \sup_{\alpha \in [0,1]} \{\alpha h(G_A(\alpha), G_B(\alpha))\} \tag{53}$$

as α converges within $[0, 1]$ to some value α' at some discontinuity. Equation (53) is also valid when $\alpha h(G_A(\alpha), G_B(\alpha))$ attains its maximum at some value α',

$$f(A,B) = \sup_{\alpha \in [0,1]} \{\alpha h(G_A(\alpha), G_B(\alpha))\}$$

$$= \lim_{\alpha \to \alpha'} \alpha h(G_A(\alpha), G_B(\alpha)). \qquad (54)$$

For the other two pairs of fuzzy sets, there exist values α'' and α''' within the interval $[0,1]$, such that

$$f(B,C) = \sup_{\alpha \in [0,1]} \{\alpha h(G_B(\alpha), G_C(\alpha))\}$$

$$= \lim_{\alpha \to \alpha''} \alpha h(G_B(\alpha), G_C(\alpha)) \qquad (55)$$

and

$$f(A,C) = \sup_{\alpha \in [0,1]} \{\alpha h(G_A(\alpha), G_C(\alpha))\}$$

$$= \lim_{\alpha \to \alpha'''} \alpha h(G_A(\alpha), G_C(\alpha)), \qquad (56)$$

respectively.

The properties of the supremum imply that for limits of convergence to any other α''' value,

$$\sup_{\alpha \in [0,1]} \{\alpha h(G_A(\alpha), G_B(\alpha))\} = \lim_{\alpha \to \alpha'} \alpha h(G_A(\alpha), G_B(\alpha))$$

$$\geq \lim_{\alpha \to \alpha'''} \alpha h(G_A(\alpha), G_B(\alpha)) \quad (57)$$

and

$$\sup_{\alpha \in [0,1]} \{\alpha h(G_B(\alpha), G_C(\alpha))\} = \lim_{\alpha \to \alpha''} \alpha h(G_B(\alpha), G_C(\alpha))$$

$$\geq \lim_{\alpha \to \alpha'''} \alpha h(G_B(\alpha), G_C(\alpha)) \quad (58)$$

hold.

We also know that the α-scaled ordinary Hausdorff distances for each set of α-cuts taken for each α value as $\alpha \to \alpha'''$ fulfill the triangle inequality

$$\alpha h(G_A(\alpha), G_B(\alpha)) + \alpha h(G_B(\alpha), G_C(\alpha))$$

$$\geq \alpha h(G_A(\alpha), G_C(\alpha)), \qquad (59)$$

which implies

$$\lim_{\alpha \to \alpha'''} \alpha h(G_A(\alpha), G_B(\alpha)) + \lim_{\alpha \to \alpha'''} \alpha h(G_B(\alpha), G_C(\alpha))$$

$$\geq \lim_{\alpha \to \alpha'''} \alpha h(G_A(\alpha), G_C(\alpha)). \qquad (60)$$

If in the first and second terms on the left-hand side of inequality (60) the limits given for $\alpha \to \alpha'''$ are replaced by the optimum limits of $\alpha \to \alpha'$ and $\alpha \to \alpha''$, respectively, then the left-hand side of inequality (60) cannot decrease, as implied by inequalities (57) and (58). One obtains

$$\lim_{\alpha \to \alpha'} \alpha h(G_A(\alpha), G_B(\alpha)) + \lim_{\alpha \to \alpha''} \alpha h(G_B(\alpha), G_C(\alpha))$$
$$\geq \lim_{\alpha \to \alpha'''} \alpha h(G_A(\alpha), G_C(\alpha)). \tag{61}$$

A comparison of this result with Eqs. (54), (55), and (56) proves our conjecture:

$$f(A, B) + f(B, C) \geq f(A, C). \tag{62}$$

Note that if each of the $G_A(\alpha)$, $G_B(\alpha)$, and $G_C(\alpha)$ sets is simply connected for any α parameter value from the unit interval $[0, 1]$, that is, if $G_A(\alpha)$, $G_B(\alpha)$, and $G_C(\alpha)$ change continuously within the unit interval $[0, 1]$, then the foregoing proofs can be simplified by replacing suprema with maxima that are realized by specific α', α'', and α''' values, and one can avoid using limits for $\alpha \to \alpha'$, $\alpha \to \alpha''$, and $\alpha \to \alpha'''$.

Both the unscaled and scaled fuzzy Hausdorff-type metrics $g(A, B)$ and $f(A, B)$ serve as the basis for various choices for similarity measures between fuzzy sets, for example:

$$s_g(A, B) = \exp(-[g(A, B)]^2), \tag{63}$$

$$t_g(A, B) = 1/(1 + [g(A, B)]^2), \tag{64}$$

$$z_g(A, B) = 1/(1 + g(A, B)) \tag{65}$$

and

$$s_f(A, B) = \exp(-[f(A, B)]^2), \tag{66}$$

$$t_f(A, B) = 1/(1 + [f(A, B)]^2), \tag{67}$$

$$z_f(A, B) = 1/(1 + f(A, B)). \tag{68}$$

These similarity measures have the value of 1 for identical fuzzy sets, and they give 0 for two fuzzy sets of infinite value for fuzzy generalizations of Hausdorff-type distances.

Both the unscaled fuzzy Hausdorff-type metric $g(A, B)$ and the scaled fuzzy Hausdorff-type metric $f(A, B)$ have been shown in the preceding text to fulfill the conditions for proper metric if the mutual locations of fuzzy sets A, B, and C are fixed. However, in some applications, various translated, rotated, or reflected versions of these fuzzy sets can be regarded as equivalent. For example, the inherent dissimilarities between two fuzzy objects A and B can be measured by the scaled fuzzy Hausdorff-type distance $g(A, B)$ if the relative positions of the objects correspond to maximum superposition. In fact, a new version $g_{op}(A, B)$ of the fuzzy Hausdorff-type metric $g(A, B)$ can be used:

$$g_{op}(A, B) = \inf_{A_v, B_{v'}} \{g(A_v, B_{v'})\} \tag{69}$$

where A_v and $B_{v'}$ are translated and rotated versions of fuzzy sets A and B. If the infimum is taken for a set $\{g(A_v, B_{v'})\}$ containing all such versions, then the $g_{op}(A, B)$ fuzzy Hausdorff-type metric refers to the optimum superposition of fuzzy sets A and B.

For two molecules A and B, represented by fuzzy sets of electron densities, their inherent dissimilarities can be better measured by the scaled fuzzy Hausdorff-type distance $f(A, B)$, where the relative positions of the molecules correspond to maximum superposition. The new variant $f_{op}(A, B)$ of the scaled fuzzy Hausdorff-type metric $f(A, B)$ is defined as

$$f_{op}(A, B) = \inf_{A_v, B_{v'}} \{f(A_v, B_{v'})\}, \tag{70}$$

where, as before, A_v and $B_{v'}$ are translated and rotated versions of fuzzy sets A and B. If the set $\{f(A_v, B_{v'})\}$ contains all such versions A_v and $B_{v'}$, then the $f_{op}(A, B)$ scaled fuzzy Hausdorff-type metric refers to the optimum superposition of fuzzy sets A and B.

Both of the new $g_{op}(A, B)$ and $f_{op}(A, B)$ functions are proper metrics. We shall present a proof only for the chemically more important metric $f_{op}(A, B)$; the proof is entirely analogous for $g_{op}(A, B)$. Evidently, conditions (10), (11), and (12) of metric apply for the versions A_v and $B_{v'}$ realizing the optimum mutual arrangements. For a proof of the triangle inequality, condition (13), one can first consider a fixed version $B_{v'}$ of fuzzy set B and generate the optimum versions A_v and $C_{v''}$ of fuzzy sets A and C, which realize the optimum mutual arrangements with $B_{v'}$:

$$f_{op}(A, B) = f(A_v, B_{v'}) \tag{71}$$

and

$$f_{op}(B, C) = f(B_{v'}, C_{v''}). \tag{72}$$

Such a common version $B_{v'}$ of fuzzy set B must exist, since by taking an appropriate version A_v of fuzzy set A, an optimum mutual positioning can be obtained for any version of $B_{v'}$; the same applies for $C_{v''}$.

For these versions A_v, $B_{v'}$, and $C_{v''}$, the triangle inequality of the scaled fuzzy Hausdorff-type metric $f(A, B)$ [Eq. (62)] applies:

$$f(A_v, B_{v'}) + f(B_{v'}, C_{v''}) \geq f(A_v, C_{v''}). \tag{73}$$

On the left-hand side the two terms are precisely the $f_{op}(A, B)$ and $f_{op}(B, C)$ values; hence, we have

$$f_{op}(A, B) + f_{op}(B, C) \geq f(A_v, C_{v''}). \tag{74}$$

However, the right-hand side is not necessarily the optimum value for the locations of versions of A and C *relative to each other*. If $f(A_v, C_{v''})$ is not the optimum value, then the fuzzy Hausdorff-type distance between versions of A and C can be reduced by changing the relative positions of versions A_v and $C_{v''}$ of fuzzy sets A and C:

$$f(A_v, C_{v''}) \geq f_{op}(A, C). \tag{75}$$

Combining inequalities (74) and (75), one obtains the triangle inequality for the optimum position version $f_{op}(A, B)$ of the scaled fuzzy Hausdorff-type metric:

$$f_{op}(A, B) + f_{op}(B, C) \geq f_{op}(A, C). \tag{76}$$

Note that the versions of fuzzy sets taken into account in the set $\{f(A_v, B_{v'})\}$ of definition (70) can be restricted to translated versions only. In this case the proof follows the same steps as before, and the translation-restricted $f_{op, tr}(A, B)$ scaled fuzzy Hausdorff-type metric is obtained. Alternatively, the allowed rotations can be confined to some angle interval $\Delta \alpha$, leading to another scaled fuzzy Hausdorff-type metric $f_{op, tr, \Delta \alpha}(A, B)$. Furthermore, if in addition to translated and rotated versions, reflected versions are also included among the versions in the set $\{f(A_v, B_{v'})\}$, then one obtains a new version of scaled fuzzy Hausdorff-type metric, $f_{op, \diamond}(A, B)$. For these metrics, the following general relations hold:

$$f_{op, \diamond}(A, B) \leq f_{op}(A, B) \leq f_{op, tr, \Delta \alpha}(A, B) \leq f_{op, tr}(A, B). \tag{77}$$

By contrast to these new, *optimized* metrics involving free or constrained repositioning of sets A and B, the fuzzy Hausdorff-type metrics defined by Eqs. (43) and (63)–(68) do not involve any repositioning of either set A or B and are referred to as *direct* fuzzy Hausdorff-type metrics.

For applications of the scaled fuzzy Hausdorff-type metric $f_{op}(A, B)$ for assessing the similarity of molecules, the $f_{op}(A, B)$ distance can be used as a *dissimilarity measure*.

Consider a set of molecules $A, B, C, \ldots$ with electron densities $\rho_A(\mathbf{r}), \rho_B(\mathbf{r}), \rho_C(\mathbf{r}), \ldots$ and level sets $G_A(a), G_B(a), G_C(a), \ldots$, respectively, for each density threshold value a. By choosing an appropriate definition for fuzzy membership function describing the fuzzy assignment of points $\mathbf{r}$ of the three-dimensional space to each molecule, such as the membership function $\mu_A(\mathbf{r}) = \rho_A(\mathbf{r})/\rho_{A, max}$ of Eq. (20) or $\mu'_A(\mathbf{r}) = 1 - \exp(-\tau \rho_A^2(\mathbf{r}))$ of Eq. (21), the density-scaled fuzzy Hausdorff-type metric $f_{op}(A, B)$ applies and provides various choices for *fuzzy Hausdorff-type similarity measures* between the molecules, for example:

$$s_{f_{\mathrm{op}}}(A, B) = \exp\left(-\left[f_{\mathrm{op}}(A, B)\right]^2\right), \tag{78}$$

$$t_{f_{\mathrm{op}}}(A, B) = 1/\left(1 + \left[f_{\mathrm{op}}(A, B)\right]^2\right), \tag{79}$$

and

$$z_{f_{\mathrm{op}}}(A, B) = 1/\left(1 + f_{\mathrm{op}}(A, B)\right). \tag{80}$$

In special cases, for example, when comparing molecules fitting within a cavity of an enzyme, various oriented and restricted versions $f_{\mathrm{op,tr}}(A, B)$ or $f_{\mathrm{op,tr},\Delta\alpha}(A, B)$ of the density-scaled fuzzy Hausdorff-type metric $f_{\mathrm{op}}(A, B)$ can be used. In turn, these metrics define various similarity measures, for example, those analogous to the measures given in Eqs. (78)–(80):

$$s_{f_{\mathrm{op,tr}}}(A, B) = \exp\left(-\left[f_{\mathrm{op,tr}}(A, B)\right]^2\right), \tag{81}$$

$$t_{f_{\mathrm{op,tr}}}(A, B) = 1/\left(1 + \left[f_{\mathrm{op,tr}}(A, B)\right]^2\right), \tag{82}$$

$$z_{f_{\mathrm{op,tr}}}(A, B) = 1/\left(1 + f_{\mathrm{op,tr}}(A, B)\right), \tag{83}$$

on the one hand, and

$$s_{f_{\mathrm{op,tr},\Delta\alpha}}(A, B) = \exp\left(-\left[f_{\mathrm{op,tr},\Delta\alpha}(A, B)\right]^2\right), \tag{84}$$

$$t_{f_{\mathrm{op,tr},\Delta\alpha}}(A, B) = 1/\left(1 + \left[f_{\mathrm{op,tr},\Delta\alpha}(A, B)\right]^2\right), \tag{85}$$

$$z_{f_{\mathrm{op,tr},\Delta\alpha}}(A, B) = 1/\left(1 + f_{\mathrm{op,tr},\Delta\alpha}(A, B)\right), \tag{86}$$

on the other hand. Analogous fuzzy similarity measures are assigned to the fuzzy metric $f_{\mathrm{op},\diamond}(A, B)$.

These fuzzy similarity measures of molecules are based on electronic density and fuzzy generalizations of Hausdorff-type metrics. These metrics and similarity measures are applicable for any collection of molecules.

IV. FUZZY SYMMETRY DEFICIENCY MEASURES, FUZZY CHIRALITY MEASURES, AND FUZZY SYMMETRY GROUPS BASED ON THE MASS OF FUZZY SETS AND FUZZY HAUSDORFF-TYPE METRICS

A fuzzy set A is said to have the symmetry element R corresponding to the symmetry operator $\mathbf{R}$ if and only if for every point x and for the transformed point

$$\mathbf{R}x = y \tag{87}$$

the fuzzy membership functions fulfill the condition

$$\mu_A(y) = \mu_A(x). \tag{88}$$

The symbols R for a symmetry element and $\mathbf{R}$ for the corresponding symmetry operator are assumed to be fully specified concerning the geometrical placement of R with respect to set A. In some instances, it appears useful to specify a direct reference to a center c in set A, for example, if c is taken as a formal origin with respect to a local coordinate system in which R is specified, or to use direct reference to direction cosines of rotation axes or normals of reflection planes with respect to local coordinate axes.

A fuzzy set A is said to have the *fuzzy symmetry element* $R(\beta)$ corresponding to the symmetry operation $\mathbf{R}$ at the fuzzy level β of the fuzzy Hausdorff-type similarity measure s_g if and only if the fuzzy similarity measure s_g measure $\mathbf{R}A$ and A is greater than or equal to β:

$$s_g(\mathbf{R}A, A) \geq \beta. \tag{89}$$

Various fuzzy subsets B of A may have the fuzzy symmetry element $R(\beta)$ at different fuzzy levels β of the fuzzy Hausdorff-type similarity measure s_g. Here the concept of fuzzy subset is interpreted in the usual way: for the fuzzy subset B of fuzzy set A the condition

$$\mu_B(x) \leq \mu_A(x), \qquad \forall x \tag{90}$$

must hold. Note that entirely analogous conditions apply for unscaled similarity measures t_g and z_g and scaled similarity measures s_f, t_f, and z_f.

Also note that if $\beta = 1$, that is, if similarity becomes indistinguishability, then the fuzzy symmetry element $R(\beta)$ becomes an ordinary symmetry element corresponding to the symmetry operation $\mathbf{R}$.

The maximum fuzzy level $\beta(A, \mathbf{R}, s_g)$ at which the fuzzy symmetry element $R(\beta)$ is present for the fuzzy set A, that is,

$$\beta(A, \mathbf{R}, s_g) = \sup_{\beta \in [0,1]} \{\beta : s_g(\mathbf{R}A, A) \geq \beta\}, \tag{91}$$

is a measure of the "degree" of symmetry aspect R for fuzzy set A, according to fuzzy Hausdorff-type similarity measure s_g. Analogous degrees of symmetry $\beta(A, \mathbf{R}, t_g)$, $\beta(A, \mathbf{R}, z_g)$, $\beta(A, \mathbf{R}, s_f)$, $\beta(A, \mathbf{R}, t_f)$, and $\beta(A, \mathbf{R}, z_f)$ are defined for the unscaled similarity measures t_g and z_g and scaled similarity measures s_f, t_f, and z_f, respectively.

A *fuzzy symmetry operator* $\mathbf{R}(s_g)$ of fuzzy symmetry element $R(\beta')$ present for fuzzy set A at the fuzzy level β' of the fuzzy Hausdorff-type similarity measure s_g is defined by its action on the fuzzy set A:

$$\mathbf{R}(s_g)A = \mathbf{M}_{A,\mathbf{R}}\mathbf{R}A. \tag{92}$$

Here $\mathbf{R}$ is the ordinary symmetry operator corresponding to the fuzzy symmetry element $R(\beta')$.

The role of operator $\mathbf{M}_{A,\mathbf{R}}$ can be described as follows. To take full advantage of the "fuzzy indistinguishability" of sets A and $\mathbf{R}A$ at the fuzzy level β, operator $\mathbf{M}_{A,\mathbf{R}}$ *resets* the values of fuzzy membership functions whenever this fuzzy indistinguishability is implied by the presence of the fuzzy symmetry element $R(\beta)$ at the fuzzy level β.

In general, the elements of a fuzzy set A are fully specified if their membership functions are given, that is, if for the elements x of the underlying space X, the pairs $(x, \mu_A(x))$ are specified. Using this pair notation, the action of operator $\mathbf{M}_{A,\mathbf{R}}$ can be described by the defining equation

$$\mathbf{M}_{A,\mathbf{R}}\big(x, \mu_{\mathbf{R}A}(x)\big) = \big(x, \mu_A(x)\big) \tag{93}$$

if and only if level β fuzzy indistinguishability is implied by the presence of the fuzzy symmetry element $R(\beta)$ at the fuzzy level β.

Note that if a fuzzy set A has the fuzzy symmetry element $R(\beta)$ corresponding to the symmetry operator $\mathbf{R}$ at the fuzzy level β of the fuzzy Hausdorff-type similarity measure s_g, then the application of $\mathbf{R}$ on A generates a set $\mathbf{R}A$ indistinguishable from set A at the fuzzy level β. For the fuzzy set A the application of symmetry operator $\mathbf{R}$ of fuzzy symmetry element $R(\beta')$ present at the fuzzy level β' is completed by a formal recognition of the indistinguishability of set $\mathbf{R}A$ and set A at the given fuzzy level. This additional step, for which the sufficient and necessary condition is the presence of fuzzy symmetry element $R(\beta')$ at the fuzzy level β', involves operator $\mathbf{M}_{A,\mathbf{R}}$ setting the membership functions of elements of the fuzzy set $\mathbf{R}A$ equal to those of fuzzy set A.

Consequently, if the product

$$\mathbf{R}''(s_g) = \mathbf{R}(s_g)\mathbf{R}'(s_g) \tag{94}$$

of the fuzzy symmetry operators $\mathbf{R}(s_g)$ and $\mathbf{R}'(s_g)$ is applied on a fuzzy set A that has each of the corresponding three fuzzy symmetry elements $R(\beta)$, $R'(\beta)$, and $R''(\beta)$ at fuzzy level β, then $\mathbf{R}'(s_g)A$, hence $\mathbf{R}(s_g)\mathbf{R}'(s_g)A$, as well as $\mathbf{R}''(s_g)A$ are indistinguishable from A at the fuzzy level β. When applying any product

$$\mathbf{R}'''(s_g) \cdots \mathbf{R}''(s_g)\mathbf{R}'(s_g)\mathbf{R}(s_g) \tag{95}$$

of symmetry operators, where each subproduct corresponds to a fuzzy symmetry element of A at the fuzzy level β of similarity measure s_g, then, at the fuzzy level β, the fuzziness does not increase in the sequence

$$
\begin{aligned}
&A, \\
&\mathbf{R}(s_g)A, \\
&\mathbf{R}'(s_g)\mathbf{R}(s_g)A, \\
&\mathbf{R}''(s_g)\mathbf{R}'(s_g)\mathbf{R}(s_g)A, \\
&\vdots \\
&\mathbf{R}'''(s_g)\cdots\mathbf{R}''(s_g)\mathbf{R}'(s_g)\mathbf{R}(s_g)A.
\end{aligned}
\tag{96}
$$

A fuzzy set A has the β' *fuzzy symmetry group* $G(s_g, \beta')$ at fuzzy level β' if A has the fuzzy symmetry element $R(\beta')$ at the fuzzy level β' of the fuzzy Hausdorff similarity measure s_g for each symmetry operation $\mathbf{R}$ of the crisp symmetry group G. The β fuzzy symmetry group $G(s_g, \beta)$ at the fuzzy level β that is the *supremum* of the levels β' of all β' fuzzy symmetry groups $G(s_g, \beta')$ of A is the *fuzzy symmetry group* $G(s_g, \beta)$ of the fuzzy set A:

$$
G(s_g, \beta) = G(s_g, \beta(A, G, s_g)),
\tag{97}
$$

where

$$
\beta = \beta(A, G, s_g) = \sup_{\beta'}\{\beta' : s_g(\mathbf{R}A, A) \ge \beta', \forall \mathbf{R} \in G\}.
\tag{98}
$$

Analogous fuzzy symmetry groups $G(t_g, \beta)$, $G(z_g, \beta)$, $G(s_f, \beta)$, $G(t_f, \beta)$, and $G(z_f, \beta)$ of fuzzy set A are defined for the unscaled similarity measures t_g and z_g and scaled similarity measures s_f, t_f, and z_f, respectively.

A measure $\delta(A, \mathbf{R}, s_g)$ of the *symmetry deficiency* of fuzzy set A in symmetry element R according to the fuzzy Hausdorff-type similarity measure s_g can be defined as

$$
\delta(A, \mathbf{R}, s_g) = 1 - \beta(A, \mathbf{R}, s_g).
\tag{99}
$$

Analogous measures of symmetry deficiency, $\delta(A, \mathbf{R}, t_g)$, $\delta(A, \mathbf{R}, z_g)$, $\delta(A, \mathbf{R}, s_f)$, $\delta(A, \mathbf{R}, t_f)$, and $\delta(A, \mathbf{R}, z_f)$, are defined for the unscaled similarity measures t_g and z_g and scaled similarity measures s_f, t_f, and z_f, respectively.

Alternative symmetry deficiency measures of fuzzy sets are defined following the treatment of symmetry deficiency of ordinary subsets of finite n-dimensional Euclidean spaces, introduced earlier.[27] To this end, we shall use certain concepts derived as generalizations of concepts in crisp set theory.

In some instances, we need a measure $m(A)$ that can be regarded as a generalization of the concept of *mass* for fuzzy sets. This measure $m(A)$ of a fuzzy set A, also called the cardinality[8] of A, is defined as the integral

$$m(A) = \int_X \mu_A(x)\, dx, \tag{100}$$

where integration is for the whole domain X of fuzzy set A and where the area element dx (in general, n-dimensional volume element dx) is interpreted in the context of the metric and dimension of the underlying set X. For discrete sets the integration is replaced by summation.

The measure $m(A)$ of a fuzzy set A is analogous to the concept of area $v(Y)$ or, in general, the n-dimensional volume of an ordinary set Y, where the formal "mass-density" of ordinary set Y can be interpreted as the constant membership function $\mu_Y(x) = 1$ for every point x of set Y. For fuzzy sets, however, the concept of volume does not appear particularly useful, and a formal "total mass," taken as the integral of Eq. (100), appears more appropriate. In this context, the role of volume as applied for weighting purposes for ordinary sets can be played by the formal "total mass" $m(A)$ of fuzzy sets,[29] and this function $m(A)$ will be used for scaling various symmetry deficiency measures derived for fuzzy sets.

Consider a family of symmetry elements

$$R = \{R_1, R_2, \ldots, R_m\}. \tag{101}$$

A fuzzy set A is called an R *set* if A has each of the symmetry elements of family R.

A fuzzy set B is called an R-*deficient* set if B has none of the point symmetry elements of family R. However, by analogy with the case of crisp sets, it takes only infinitesimal distortions to lose a given symmetry element. Consequently, unless further restrictions are applied, the "total mass" difference between a fuzzy set of a specified symmetry and another fuzzy set that does not have this symmetry can be infinitesimal. As a result, R-deficient fuzzy sets and fuzzy R sets can be almost identical. Nevertheless, the actual symmetry deficiencies of fuzzy continua, such as formal molecular bodies represented by fuzzy "clouds" of electron densities, can be defined in terms of the deviations from their maximal R subsets and minimal R supersets, defined in subsequent text.

Fuzzy set B' is a *maximal R subset* of fuzzy set A if B' is an R set, $B' \subset A$, and no R set B'' exists such that $B' \subset B''$, $B' \neq B''$, and $B'' \subset A$. Note that the fuzzy, maximal R subset of B' is not necessarily unique for a given fuzzy set A.

A fuzzy set B is a *maximal mass R subset* of fuzzy set A if B is a fuzzy R set, $B \subset A$, and if for all maximal fuzzy R subsets B' of fuzzy set A, $m(B') \leq m(B)$. The fuzzy, maximal mass R subset B is not necessarily unique for a given fuzzy set A; however, the mass $m(B)$ is already a unique number for each fuzzy set A.

A fuzzy set C' is a *minimal R superset* of fuzzy set A if C' is an R set, $A \subset C'$, and if no fuzzy R set C'' exists such that $C'' \subset C'$, $C' \ne C''$, and $A \subset C''$. The fuzzy, minimal R superset C' is not necessarily unique for a given fuzzy set A.

Fuzzy set C is a *minimal mass R superset* of fuzzy set A if C is an R set, $A \subset C$, and if for all minimal fuzzy R supersets C' of fuzzy set A, $m(C) \le m(C')$. A fuzzy minimal mass R superset C is not necessarily unique for a given fuzzy set A; however, the mass $m(C)$ is a unique number for each fuzzy set A.

If the fuzzy set A itself is an R set, then both B and C are unique and $B = C = A$.

A fuzzy set B' is a *maximal R-deficient subset* of fuzzy set A if B' is an R-deficient fuzzy set, $B' \subset A$, and if no R-deficient fuzzy set B'' exists such that $B' \subset B''$, $B' \ne B''$, and $B'' \subset A$. The fuzzy, maximal R-deficient subset B' is not necessarily unique for a given fuzzy set A.

A fuzzy set B is a *maximal mass R-deficient subset* of fuzzy set A if B is an **R**-deficient fuzzy set, $B \subset A$, and if for all maximal **R**-deficient fuzzy subsets B' of fuzzy set A, the relation $m(B') \le m(B)$ holds. Whereas the fuzzy, maximal mass R-deficient subset B is not necessarily unique for a given fuzzy set A, nevertheless, the total mass $m(B)$ is a unique number for each fuzzy set A.

A fuzzy set C' is a *minimal R-deficient superset* of fuzzy set A if C' is an R-deficient fuzzy set, $A \subset C'$, and if no R-deficient set C'' exists such that $C'' \subset C'$, $C' \ne C''$, and $A \subset C''$. The fuzzy, minimal R-deficient superset C' is not necessarily unique for a given fuzzy set A.

A fuzzy set C is a *minimal mass R-deficient superset* of fuzzy set A if C is an R-deficient fuzzy set, $A \subset C$, and if for all minimal R-deficient supersets C' of fuzzy set A, the relation $m(C) \le m(C')$ holds. The fuzzy, minimal mass R-deficient superset C is not necessarily unique for a given fuzzy set A; nevertheless, the total mass $m(C)$ is a unique number for each fuzzy set A.

If fuzzy set A is an R-deficient fuzzy set, then both B and C are unique and $B = C = A$.

If B is a maximal mass fuzzy R subset of fuzzy set A, and C is a minimal mass fuzzy R superset of fuzzy set A, then the relations among these fuzzy sets define measures for symmetry deficiency. The relationships

$$\delta_{R,B}(A) = 1 - m(B)/m(A) \tag{102}$$

and

$$\delta_{R,C}(A) = 1 - m(A)/m(C) \tag{103}$$

define two R-deficiency measures—the *internal R-deficiency measure* $\delta_{R,B}(A)$ and the *external R-deficiency measure* $\delta_{R,C}(A)$, respectively. The average of these two measures defines the *R-deficiency measure*

$$\delta_{R,B}(A) = (\delta_{R,B}(A) + \delta_{R,C}(A))/2. \qquad (104)$$

For any fuzzy R subset B', R superset C', maximal mass fuzzy R subset B, and minimal mass fuzzy R superset C of any fuzzy set A, the relations

$$m(B') \leq m(C') \qquad (105)$$

and

$$m(C) - m(B) \leq m(C') - m(B') \qquad (106)$$

hold.

Chirality, an important shape property of molecules, can be regarded as the lack of certain symmetry elements. Chirality measures are in fact measures of symmetry deficiency.[27] These principles, originally used for crisp sets,[27] also apply for fuzzy sets. Considering the case of three-dimensional chirality, the lacking point symmetry elements are reflection planes σ and rotation-reflections S_{2k} of even indices. Whereas the lacking symmetry elements can be of different nature in different dimensions, nevertheless, all the concepts, definitions, and procedures discussed in this section have straightforward generalizations for any finite dimension n.

For a chiral fuzzy object A, the largest achiral fuzzy object that fits within A, as well as the smallest achiral fuzzy object that contains A, are of special importance. Following the method used for general symmetry deficiencies,[27] one may compare the masses m of these fuzzy objects, and use these comparisons to evaluate the degree of the deviation of the fuzzy object A from achirality. To avoid pathological cases, we shall follow the restriction used for general symmetry deficiencies and consider only fuzzy objects A that have finite, nonzero mass and are "nowhere infinitely thin."

Fuzzy set B' is a *maximal achiral subset* of fuzzy set A if B' is achiral, $B' \subset A$, and if no achiral fuzzy set B'' exists such that $B' \subset B''$, $B' \neq B''$, and $B'' \subset A$. Note that a maximal achiral subset B' is not necessarily unique for a given fuzzy set A.

Fuzzy set B is a *maximal mass achiral subset* of fuzzy set A if B is achiral, $B \subset A$, and if for the fuzzy volumes of all maximal achiral subsets B' of fuzzy set A, the inequality $m(B') \leq m(B)$ applies. Note that, for a given fuzzy set A, such a maximal mass achiral subset B is not necessarily unique. Nevertheless, the fuzzy mass $m(B)$ is a unique number for each fuzzy set A.

A fuzzy set C' is a *minimal achiral superset* of a fuzzy set A if fuzzy set C' is achiral, $A \subset C'$, and if no achiral fuzzy set C'' exists such that $C'' \subset C'$, $C' \neq C''$, and $A \subset C''$. Note that a minimal achiral superset C' is not necessarily unique for a given fuzzy set A.

Fuzzy set C is a *minimal mass achiral superset* of fuzzy set A if the fuzzy set C is achiral, $A \subset C$, and for all minimal achiral supersets C' of fuzzy set A, $m(C) \leq m(C')$. For a given fuzzy set A, a minimal mass achiral superset C is not necessarily unique. Nevertheless, the mass $m(C)$ is a unique number for each fuzzy set A.

If fuzzy set A is achiral, then both the minimal achiral fuzzy superset B and the minimal mass achiral fuzzy superset C are unique and $B = C = A$.

Two fuzzy chirality measures,

$$\chi_B(A) = 1 - m(B)/m(A) \tag{107}$$

and

$$\chi_C(A) = 1 - m(A)/m(C), \tag{108}$$

are defined, where the first measure, $\chi_B(A)$, is a natural, fuzzy set extension of the measure obtained using the maximum overlap criterion between mirror images.[54–64]

The fuzzy Hausdorff-type similarity measures can be used for a direct comparison of a fuzzy set A and its various R sets and R-deficient sets, providing alternative fuzzy symmetry deficiency measures.

Let $s_H(A, B)$ denote any one of the fuzzy Hausdorff-type similarity measures, for example, any one of the examples

$$
\begin{aligned}
s_H(A, B) \in \big\{ & s_g(A, B), t_g(A, B), z_g(A, B), \\
& s_f(A, B), t_f(A, B), z_f(A, B), \\
& s_{f_{op}}(A, B), t_{f_{op}}(A, B), z_{f_{op}}(A, B), \\
& s_{f_{op,tr}}(A, B), t_{f_{op,tr}}(A, B), z_{f_{op,tr}}(A, B), \\
& s_{f_{op,tr,\Delta\alpha}}(A, B), t_{f_{op,tr,\Delta\alpha}}(A, B), z_{f_{op,tr,\Delta\alpha}}(A, B), \\
& s_{f_{op,\diamond}}(A, B), t_{f_{op,\diamond}}(A, B), \text{ and } z_{f_{op,\diamond}}(A, B) \big\},
\end{aligned}
\tag{109}
$$

discussed previously, or any one of the fuzzy similarity measures defined analogously. Let fuzzy set D denote any one of the R subsets B, B' or R supersets C, C' of fuzzy set A:

$$D \in \{B, B', C, C'\}. \tag{110}$$

Here fuzzy set B' is a maximal R subset, fuzzy set B is a maximal mass R subset, fuzzy set C' is a minimal R superset, and fuzzy set C is a minimal mass R superset of fuzzy set A.

A *fuzzy symmetry deficiency measure* $\delta_{H,R,D}(A)$ of fuzzy set A is provided by

$$\delta_{H,R,D}(A) = \inf_{D_\kappa} \{1 - s_H(A, D_\kappa)\}, \tag{111}$$

where the index H refers to the actual choice of the fuzzy Hausdorff-type similarity measures from list (109) and index D refers to the choice of the type set D from the family of R subsets B, B' and R supersets C, C' of fuzzy set A. Since most of these sets are not unique for a given fuzzy set A, the infimum is taken for all such fuzzy sets D_κ.

For example, the simplest of these fuzzy symmetry deficiency measures,

$$\delta_{S_g, R, B'}(A) = \inf_{B'_\kappa} \left\{ 1 - s_g(A, B'_\kappa) \right\}, \tag{112}$$

involves a maximal R subset B' of fuzzy set A, and the similarity measure $s_g(A, B)$ based on the Gaussian transformation of the unscaled fuzzy Hausdorff-type metric $g(A, B)$. In chemical applications involving fuzzy electron density, the scaled version

$$\delta_{S_f, R, B'}(A) = \inf_{B'_\kappa} \left\{ 1 - s_f(A, B'_\kappa) \right\} \tag{113}$$

appears more appropriate, where the similarity measure $s_f(A, B)$ of the Gaussian transformation of the scaled, fuzzy Hausdorff-type metric $f(A, B)$ is involved.

The fuzzy Hausdorff-type similarity measures can also be used for a direct comparison of a fuzzy set A and various extremal achiral fuzzy sets associated with it, providing alternative *fuzzy chirality measures*.

Let fuzzy set D denote any one of the extremal achiral fuzzy sets associated with fuzzy set A:

$$D \in \{B, B', C, C'\}. \tag{114}$$

Here fuzzy set B' is a maximal achiral subset, fuzzy set B is a maximal mass achiral subset, fuzzy set C' is a minimal achiral fuzzy superset, and fuzzy set C is a minimal mass achiral superset of fuzzy set A.

A *fuzzy chirality measure* $\chi_{H,D}(A)$ of fuzzy set A is provided by

$$\chi_{H,D}(A) = \inf_{D_\kappa} \left\{ 1 - s_H(A, D_\kappa) \right\}, \tag{115}$$

where the index H refers to the actual choice of the fuzzy Hausdorff-type similarity measures from list (109) and index D refers to the choice of the type D from the family of the extremal achiral fuzzy sets associated with fuzzy set A. Similarly to the case of R sets, most of these sets are not unique for a given fuzzy set A, and in definition (115), the infimum is taken for all such fuzzy sets D_κ.

The simplest of these fuzzy chirality measures,

$$\chi_{S_g, B'}(A) = \inf_{B'_\kappa} \left\{ 1 - s_g(A, B'_\kappa) \right\}, \tag{116}$$

involves a maximal achiral subset B' of fuzzy set A and the similarity measure $s_g(A, B)$ based on a Gaussian transformation of the unscaled fuzzy Hausdorff-type metric $g(A, B)$. In applications for the evaluation of chirality measures of molecules, where the chirality of a molecular body is manifested by the fuzzy electron density, the scaled version

$$\chi_{S_f, B'}(A) = \inf_{B'_\kappa} \{1 - s_f(A, B'_\kappa)\} \tag{117}$$

places more emphasis on the differences within the high electron density regions. Here the similarity measure $s_f(A, B)$ of the Gaussian transformation of the scaled, fuzzy Hausdorff-type metric $f(A, B)$ provides a choice well justified by the importance of high electron density regions in molecules.

V. ANOTHER FUZZY SYMMETRY APPROACH: SYNTOPY AND SYNTOPY GROUPS

Syntopy and syntopy groups were introduced in an early approach to a fuzzy set representation of approximate symmetry,[21-24] where imperfect symmetry is regarded as fuzzy symmetry. Whereas any symmetry is a discrete property within a metric space, it is natural to consider a fuzzy set approach for a continuous extension of the discrete symmetry concept to quasisymmetric objects, such as some "almost symmetric" molecular structures.[21-24] The syntopy approaches take into account the nonlocalized, quantum-mechanical, fuzzy nature of nuclear arrangements of molecules.

A nuclear configuration K of a molecular conformation, represented by a point K in a metric nuclear configuration space M, either has or does not have a given point symmetry. As the nuclear configuration K changes, the symmetry changes abruptly and discontinuously. By contrast, most other molecular properties vary continuously with the nuclear configuration K. These contrasting features of symmetry and other molecular properties serve as strong motivation for a generalization of symmetry, which allows a more uniform treatment of all molecular properties and the utilization of some of the advantages offered by symmetry for molecular structures that exhibit only approximate symmetry and are only similar to symmetric structures.

The discontinuous symmetry changes and the binary nature of the presence or absence of symmetry elements hinders the application of point group symmetry methods for general molecular structures. In the syntopy approach, based on fuzzy set theory, the discrete concept of point symmetry is replaced by a continuous concept and is applicable to cases of "almost" symmetric or quasisymmetric molecular arrangements. When replacing symmetry with syntopy, some of the advantages of the group theoretical treatment of truly symmetric structures are retained for most molecular arrangements.

In the syntopy group approach,[21-24] the sharply defined families of nuclear arrangements having a specified point symmetry are replaced by fuzzy sets (syntopy sets) of nuclear arrangements having some degree of symmetry resemblance to arrangements of perfect point symmetries. The formalism provides the syntopy sets with a group theoretical characterization, in fact, the syntopy groups retain some of the group theoretical, algebraic relations among the point symmetry operators of arrangements of some exact symmetry. The fuzzy syntopy sets form the syntopy groups. The syntopy groups also have additional, continuous features, as a consequence of the replacement of the binary membership functions of ordinary sets with the continuous membership functions of the fuzzy syntopy sets.

The first syntopy approach[21-23] was based on energy conditions of interconversion paths between the actual nuclear configuration and a configuration with a specified exact symmetry. These energy conditions define fuzzy membership functions for all possible nuclear configurations with respect to syntopy sets, expressing the energy-dependent accessibility of each specified symmetry. For each nuclear configuration K, the membership functions express the "grade of belonging" of configuration K to each possible ideal symmetry. Since the fuzzy membership functions are parametrized by a threshold for the energy cost of converting each nuclear configuration to structures of ideal symmetry, syntopy reflects some of the dynamic properties of symmetry. The syntopy sets define energy-dependent syntopy groups of quasisymmetric, general nuclear arrangements. As can be shown,[21-24] syntopy is a generalization of point symmetry: each energy-dependent syntopy group can be converted into a conventional point symmetry group by varying a continuous parameter. The energy-dependent syntopy groups depend on the electronic state, that is, on the potential energy hypersurface describing the molecular distortions. The energy cost of each geometric interconversion of nuclear configurations is different for each potential energy surface. Consequently, a different syntopy model is obtained for each electronic state.

Another syntopy approach, the fundamental syntopy,[24] is based on a fuzzy set representation of the geometric resemblance of nuclear arrangements to those of exact point symmetries. The fundamental syntopy is a common syntopy model for all electronic states of all possible arrangements of the given collection of atoms, that is, for all potential surfaces associated with the given stoichiometry. In fact, this syntopy is the *underlying syntopy* for the given stoichiometry, also called the *fundamental syntopy*, of the nuclear configuration space M for the given stoichiometry. The fundamental syntopy does not depend on energy conditions associated with individual potential surfaces. The parameters of the fundamental syntopy depend on universal geometrical criteria that define the fuzzy membership functions and the associated syntopy groups.

All possible energy-based syntopies generated by the various potential energy surfaces of the given stoichiometry are interrelated through the fundamental syntopy of the nuclear configuration space M.

A fuzzy set generalization of nuclear point symmetry in terms of these two syntopy models is applicable to all nuclear arrangements. Using appropriate membership functions, syntopy provides a measure of symmetry resemblance of actual, general nuclear configurations to ideal, fully symmetric nuclear configurations.

Within the syntopy models discussed, the concept of point symmetry is generalized from a restricted family of symmetric nuclear arrangements to all possible nuclear arrangements by using fuzzy sets and the symmetry resemblance between actual molecular arrangements to those having some ideal, exact symmetry.

Various generalizations of syntopy are possible by replacing the configuration space distance by some alternative similarity measures. In fact, any of the similarity measures and symmetry deficiency measures may serve as a suitable parameter in the definition of fuzzy syntopy membership functions.

One straightforward extension of the syntopy approach is obtained if the configuration space metric used in the definitions of both the energy-dependent and fundamental syntopies is replaced by the fuzzy Hausdorff-type distance between the fuzzy electronic densities associated with the corresponding nuclear configurations. In these electron-density-based (EDB) syntopy approaches, by using the energy-dependent EDB syntopy groups and the EDB fundamental syntopy groups, the emphasis is placed on the fuzzy molecular bodies instead of the point configurations of nuclei.

VI. A THIRD FUZZY SYMMETRY APPROACH: SYMMORPHY AND FUZZY SYMMORPHY GROUPS BASED ON FUZZY HAUSDORFF-TYPE METRICS

Another description of approximate symmetry, based on symmorphy groups, can also be generalized using the fuzzy Hausdorff-type distances.

In symmetry the metric properties are preserved, whereas in syntopy some of the topological properties are preserved along with the essential algebraic structure of point symmetry groups, where the elements of the syntopy groups are derived from ordinary point symmetry operators.[22-24] The algebraic structure of syntopy groups is closely related to that of the point symmetry groups, and in the limiting case of fixed nuclear configurations, the two groups agree.

Another generalization of symmetry and a fundamentally different algebraic structure is obtained in the symmorphy approach.[65,66] (The symmorphy concept used in this chapter should not be confused with the symmorphic space groups of crystallography, taken as semidirect products or split extensions of group theory.[67]) The introduction of *symmorphy groups*,[65,66] in the sense used here, was based on a family of operations that *preserve the general morphology of objects*.

This family of operators can be regarded as an extension of the family of point symmetry operators. Symmorphy is a particular extension of the point symmetry group concept of finite point sets, such as a collection of atomic nuclei, to the symmorphy group concept of a complete algebraic shape characterization of continua, such as the three-dimensional electron density cloud of a molecule. In fact, this extension can be generalized for fuzzy sets.

First we shall discuss ordinary symmorphy, followed by an extension to fuzzy sets.

The set of symmetry elements of a molecule of a given nuclear configuration K can be regarded as a special selection from the set of all possible reflection planes, all possible proper and improper rotation axes by all possible angles, and all central inversions centered at each point of the three-dimensional space. The selection is based on the molecule, where the criterion is the indistinguishability of the original nuclear arrangement and the one obtained after carrying out the point symmetry operation associated with the selected plane, axis, or point. Note two important aspects:

(1) It is the *molecule* that selects the symmetry elements from the infinite family of potential planes, axes, angles, and points.

(2) All the point symmetry operators **R** so defined are *linear* operators,

$$\mathbf{R}(\alpha A + \beta B) = \alpha \mathbf{R} A + \beta \mathbf{R} B, \tag{118}$$

where A and B represent two molecules (or their sets of nuclei) and α and β can be interpreted as scaling factors.

These two observations serve as the basis for the generalization of symmetry to symmorphy. For a general point set configuration K, it is K that selects a subset of special reflections, rotations, and inversions from the set of infinitely many such operations of the space. The conditions for selection is the indistinguishability of the original and transformed configurations.

In symmorphy, the preceding interpretation of symmetry and the selection principle are extended in two ways:

(1) The family of all possible reflections, proper and improper rotations, and inversions is extended to the family G_{hom} of all homeomorphisms of three-dimensional space (i.e., to all continuous, one-to-one and onto assignments of the points of

three-dimensional space to the points of three-dimensional space with continuous inverse transformations).

(2) The discrete set of the nuclear point distribution is generalized to more general three-dimensional objects, for example, to the continuous molecular charge density functions $\rho(\mathbf{r})$.

Of course, among the transformations in family G_{hom} one finds not only all the distance-preserving symmetry operations, but also all the continuous deformation operations.

The family of all possible homeomorphisms of three-dimensional space is a group G_{hom}. Evidently, any two such homeomorphisms applied consecutively correspond to one such homeomorphism (closure property). The unit element is the identity transformation. Each homeomorphism has an inverse, and the product of homeomorphisms is associative.

The shape of a continuum A can be characterized by a subfamily $G_{\text{sph}}(A)$ of G_{hom}, where the subfamily $G_{\text{sph}}(A)$ contain all those homeomorphisms $\mathbf{S}$ from G_{hom} that bring the transformed object $\mathbf{S}A$ into an arrangement that is indistinguishable from A. Since for such homeomorphisms $\mathbf{S}$ the morphologies of the transformed object $\mathbf{S}A$ and the original object A are indistinguishable, $\mathbf{S}$ is called a *symmorphy transformation* of object A.

The set $G_{\text{sph}}(A)$ of all symmorphy transformations $\mathbf{S}$ of the object A is a subgroup of G_{hom}. Clearly, if neither $\mathbf{S}_1$ and $\mathbf{S}_2$ alters the appearance of the morphology of A, then the transformations $\mathbf{S}_3$, defined as transformation $\mathbf{S}_1$ followed by $\mathbf{S}_2$ and denoted as the formal product

$$\mathbf{S}_3 = \mathbf{S}_2\mathbf{S}_1, \tag{119}$$

does not alter the morphology of object A either; hence, $\mathbf{S}_3$ is also a symmorphy transformation of object A. This establishes the closure property for the family $G_{\text{sph}}(A)$ of symmorphy transformations of A. The unit element of group G_{hom} is also a symmorphy transformation. If $\mathbf{S}$ is a symmorphy transformation of object A, then the inverse $\mathbf{S}^{-1}$ of $\mathbf{S}$ in group G_{hom} is also a symmorphy transformation of object A; hence, $\mathbf{S}^{-1}$ is the inverse of $\mathbf{S}$ in the family $G_{\text{sph}}(A)$ of symmorphy transformation of A. Associativity is also inherited from G_{hom}. Consequently, the family $G_{\text{sph}}(A)$ of all symmorphy transformations $\mathbf{S}$ of object A is a group–a subgroup of G_{hom}.

In the special case of selecting the three-dimensional object as a molecular charge density function $\rho(\mathbf{r})$, the analogies between point symmetry and symmorphy are rather clear. By analogy with the point symmetry of nuclear arrangements, the molecular charge density function $\rho(\mathbf{r})$ can provide a criterion for selecting the symmorphy transformations of $\rho(\mathbf{r})$ from the infinite family G_{hom} of homeomorphisms of the three-dimensional space. A homeomorphism S is a symmorphy transformation of the molecular charge density function $\rho(\mathbf{r})$ *if and only if the original and transformed electronic charge densities are indistinguishable.*

For the molecular charge density function $\rho(\mathbf{r})$, all those homeomorphisms **S** of family G_{hom} are symmorphy transformations for which

$$\rho(\mathbf{Sr}) = \rho(\mathbf{r}) \tag{120}$$

for every point $\mathbf{r}$ of the ordinary, three-dimensional space. In an alternative notation, we may write

$$\mathbf{S}\rho(\mathbf{r}) = \rho(\mathbf{r}). \tag{121}$$

If the molecular charge density function $\rho(\mathbf{r})$ has some symmetry element, then the corresponding symmetry operation **R** is among the symmorphy transformations of $\rho(\mathbf{r})$.

Some simplifications are possible if equivalence classes of symmorphy transformations can be defined where operations **S** from the same class transform the space "occupied" by the object A the same way and differ only in parts of the space where A is not present. Furthermore, using the Brouwer fixed point theorem, a subgroup structure of symmorphy groups $G_{\mathrm{sph}}(A)$ provides a more detailed characterization of molecular shape.[1, 65, 66] These aspects will not be reviewed here.

Fuzzy symmorphy and *fuzzy symmorphy groups* can be defined using the same steps as those in the development of fuzzy symmetry groups based on fuzzy similarity measures of various fuzzy Hausdorff-type metrics, as discussed in Section IV of this chapter. In ordinary symmorphy, those homeomorphisms **S** are selected by the given object A for which **S**A and A are indistinguishable. For crisp sets, indistinguishability is a discrete property; two crisp objects are either distinguishable or not, at least at the given level of resolution.

Within a fuzzy set framework, the discrete aspect of indistinguishability can be replaced by *fuzzy indistinguishability* in a variety of ways. One natural approach involves fuzzy Hausdorff-type distances.

Let $s_h(A, B)$ denote any one of the direct fuzzy Hausdorff-type similarity measures based on the unscaled or scaled fuzzy Hausdorff-type metrics $g(A, B)$, and $f(A, B)$. For example, choose $s_h(A, B)$ as one of the similarity measures defined in Eqs. (63)–(68):

$$s_h(A, B)$$
$$\in \{ s_g(A, B), t_g(A, B), z_g(A, B), s_f(A, B), t_f(A, B), z_f(A, B) \}. \tag{122}$$

We say that fuzzy sets A and B are s_h-indistinguishable at the fuzzy level β if

$$s_h(A, B) \geq \beta. \tag{123}$$

We shall apply the concept of fuzzy indistinguishability to a fuzzy set A and to the image $\mathbf{S}A$ of A obtained by a homeomorphism from the group G_{hom}.

By analogy with fuzzy symmetry elements, a fuzzy set A is said to have the *fuzzy symmorphy element* $S(\beta)$ corresponding to the symmorphy operation $\mathbf{S}$ at the fuzzy level β of the fuzzy Hausdorff-type similarity measure s_g if and only if the fuzzy similarity measure s_g between $\mathbf{S}A$ and A is greater than or equal to β:

$$s_g(\mathbf{S}A, A) \geq \beta. \tag{124}$$

If in relationship (124) the fuzzy similarity measure s_g is replaced by either one of the unscaled similarity measures t_g and z_g or the scaled similarity measures s_f, t_f, and z_f, then the *fuzzy symmorphy element* $S(\beta)$ is specified in terms of the corresponding fuzzy Hausdorff-type similarity measure.

In the limiting case of $\beta = 1$, similarity becomes indistinguishability. In this case, the fuzzy symmorphy element $S(\beta)$ becomes an ordinary symmorphy element corresponding to the symmorphy operation $\mathbf{S}$. The maximum fuzzy level $\beta(A, \mathbf{S}, s_g)$ at which the fuzzy symmorphy element $S(\beta)$ is present for the fuzzy set A is given as

$$\beta(A, \mathbf{S}, s_g) = \sup_{\beta \in [0,1]} \{\beta : s_g(\mathbf{S}A, A) \geq \beta\}, \tag{125}$$

The quantity $\beta(A, \mathbf{S}, s_g)$ is a measure of the "degree" of symmorphy aspect S for fuzzy set A, as specified by the fuzzy Hausdorff-type similarity measure s_g. Analogous quantities, denoted by $\beta(A, \mathbf{S}, t_g)$, $\beta(A, \mathbf{S}, z_g)$, $\beta(A, \mathbf{S}, s_f)$, $\beta(A, \mathbf{S}, t_f)$, and $\beta(A, \mathbf{S}, z_f)$, are defined in terms of the unscaled similarity measures t_g and z_g and the scaled similarity measures s_f, t_f, and z_f, respectively, providing alternative measures for the degree of symmorphy S.

A *fuzzy symmorphy operator* $\mathbf{S}(s_g)$ of fuzzy symmorphy element $S(\beta')$ exhibited by fuzzy set A at the fuzzy level β' (according to the fuzzy Hausdorff-type similarity measure s_g) is defined by its action on the given fuzzy set A:

$$\mathbf{S}(s_g)A = \mathbf{M}_{A,\mathbf{S}}\mathbf{S}A. \tag{126}$$

In the preceding definition, $\mathbf{S}$ is the ordinary symmorphy operator that corresponds to the fuzzy symmorphy element $S(\beta')$ present for fuzzy set A at the fuzzy level β', according to the fuzzy Hausdorff-type similarity measure s_g.

Operator $\mathbf{M}_{A,\mathbf{S}}$ has the same role as operator $\mathbf{M}_{A,\mathbf{R}}$ in the case of fuzzy symmetry operators. To exploit the "fuzzy indistinguishability" of sets A and $\mathbf{S}A$ at the fuzzy level β, operator $\mathbf{M}_{A,\mathbf{S}}$ resets the values of fuzzy membership functions if and only if the fuzzy symmorphy element

$S(\beta)$ is present for fuzzy set A at the fuzzy level β, that is, if and only if there exists a "level β fuzzy indistinguishability" between sets A and SA.

We recall that the elements x of a fuzzy set A are fully specified if the pairs $(x, \mu_A(x))$ are given. The action of operator $\mathbf{M}_{A,S}$ is defined by

$$\mathbf{M}_{A,S}(x, \mu_{SA}(x)) = (x, \mu_A(x)), \tag{127}$$

where the pair notation $(x, \mu_B(x))$ is used for an element x of membership function $\mu_A(x)$ in the fuzzy set B.

By analogy with fuzzy symmetry operations, if a fuzzy set A has the three fuzzy symmorphy elements $S(\beta)$, $S'(\beta)$ and $S''(\beta)$ at fuzzy level β and if the product

$$\mathbf{S}''(s_g) = \mathbf{S}(s_g)\mathbf{S}'(s_g) \tag{128}$$

of the corresponding fuzzy symmorphy operators $\mathbf{S}(s_g)$ and $\mathbf{S}'(s_g)$ is applied on the fuzzy set A, then fuzzy sets $\mathbf{S}'(s_g)A$, $\mathbf{S}(s_g)\mathbf{S}'(s_g)A$, and $\mathbf{S}''(s_g)A$ are indistinguishable from A at the fuzzy level β. This indistinguishability is formally utilized in the step of resetting membership functions using operators $\mathbf{M}_{A,S'}$, $\mathbf{M}_{A,S}$, and $\mathbf{M}_{A,S''}$, as required by definitions (126) and (127). When applying any product of fuzzy symmorphy operators associated with fuzzy symmorphy elements present for molecule A at the fuzzy level β of similarity measure s_g, the fuzziness does not change.

We say that a fuzzy set A has the β' *fuzzy symmorphy group* $G_{\mathrm{sph}}(s_g, \beta')$ *at fuzzy level* β' if A has the fuzzy symmorphy element $R(\beta')$ at the fuzzy level β' of the fuzzy Hausdorff similarity measure s_g for each symmorphy operation $\mathbf{S}$ of the crisp symmorphy group G_{sph}. The β fuzzy symmorphy group $G_{\mathrm{sph}}(s_g, \beta)$ at a level β that is the *supremum* of the levels β' of all β' fuzzy symmorphy groups $G_{\mathrm{sph}}(s_g, \beta')$ of A is the *fuzzy symmorphy group* $G_{\mathrm{sph}}(s_g, \beta)$ of the fuzzy set A:

$$G_{\mathrm{sph}}(s_g, \beta) = G_{\mathrm{sph}}(s_g, \beta(A, G_{\mathrm{sph}}, s_g)), \tag{129}$$

where

$$\beta = \beta(A, G_{\mathrm{sph}}, s_g) = \sup_{\beta'}\left\{\beta' : s_g(SA, A) \geq \beta', \forall S \in G_{\mathrm{sph}}(A)\right\}. \tag{130}$$

By replacing s_g with any one of the unscaled similarity measures t_g and z_g or the scaled similarity measures s_f, t_f, and z_f, the analogous fuzzy symmorphy groups $G_{\mathrm{sph}}(t_g, \beta)$, $G_{\mathrm{sph}}(z_g, \beta)$, $G_{\mathrm{sph}}(s_f, \beta)$, $G_{\mathrm{sph}}(t_f, \beta)$, and $G_{\mathrm{sph}}(z_f, \beta)$ of fuzzy set A are defined, respectively.

An alternative formulation of *fuzzy indistinguishability* can be given using resolution as a criterion. For two objects that appear indistinguishable at low resolution, the level of resolution can be increased gradually until the objects appear different; the lowest level of resolution that allows one to distinguish the objects can be taken as a limit for fuzzy indistinguishability. Indistinguishability at a low level of resolution does not imply indistinguishability at a higher level of resolution, but the converse is not true: indistinguishability at high resolution implies indistinguishability at all lower levels of resolution. This approach is not entirely new; resolution-based similarity measures were introduced previously,[1,27] and indistinguishability can be regarded as an extreme case of similarity. The fuzzy Hausdorff-type similarity measures used to define the maximal fuzzy level β in establishing fuzzy indistinguishability can be replaced with any one of the resolution-based similarity measures (RBSM) introduced earlier,[1,27] and the entire treatment of fuzzy symmetry and fuzzy symmorphy can be adapted for these resolution-based similarity measures.

VII. PROOF OF THE METRIC PROPERTIES OF THE SYMMETRIC SCALING–NESTING DISSIMILARITY MEASURE

The scaling–nesting similarity measures[1] are based on a simple idea. If the shapes of two objects A and B, of equal volume, are compared, then a pair of scaling factors provides a valid measure of their similarity. Consider all rotated and translated versions A_v and $B_{v'}$ of the two sets A and B, and determine the maximal scaling factor $s_{(AB)}$ applied to a version A_v of A that allows the rescaled version A_v of A to fit within B. Similarly, determine the maximum scaling factor $s_{(BA)}$ applied to a version $B_{v'}$ of set B that allows the rescaled version $B_{v'}$ of B to fit within A. These two scaling factors are not, in general, equal, thereby providing asymmetric measures of similarity of the shapes of A and B. It is possible, but not necessary, that

$$s_{(AB)} \neq s_{(BA)}. \tag{131}$$

These asymmetric similarity measures are also called *semisimilarity measures* or one-sided similarity measures. Note, however, that if the shapes of A and B are identical, then both scaling factors $s_{(AB)}$ and $s_{(BA)}$ are equal to 1:

$$s_{(AB)} = s_{(BA)} = 1. \tag{132}$$

In general, by taking the average

$$s_{AB} = [s_{(AB)} + s_{(BA)}]/2, \tag{133}$$

the resulting scaling–nesting similarity measure (SNSM), s_{AB} is symmetric,

$$s_{AB} = s_{BA}, \tag{134}$$

for all sets A and B of equal volume. The symmetric SNSM s_{AB} is the real part of the complex-valued similarity–asymmetric shape tolerance measure $sast_{AB}$:

$$sast_{AB} = [s_{(AB)} + s_{(BA)}]/2 + i[s_{(AB)} - s_{(BA)}]/2. \qquad (135)$$

Whereas the symmetric real part of this measure $sast_{AB}$ describes similarity, the imaginary part describes the asymmetry of shape tolerance. The effect of interchanging A and B corresponds to complex conjugation of $sast_{AB}$.

If the two sets A and B to be compared are of different volumes, then an initial rescaling of these volumes, for example, a rescaling to unit volume, is applied before the determination of the $s_{(AB)}$, $s_{(BA)}$, and s_{AB} values. These similarity measures were originally proposed[1] for constrained and unconstrained shape comparisons and for measuring dissimilarity,

$$d_s(A, B) = 1 - s_{AB}, \qquad (136)$$

in particular, to measure the dissimilarity between a chiral object A and its mirror image $A^\diamond$, providing a chirality measure[1]

$$\chi_{s_{AB}}(A) = d_s(A, A^\diamond) = 1 - s_{AA^\diamond}. \qquad (137)$$

By applying the SNSM similarity measure to mirror images, the quantity $s_{AA^\diamond}$ is a measure of *achirality*, whereas the dissimilarity measure $d_s(A, A^\diamond)$, denoted as $\chi_{s_{AB}}(A)$, is a measure of chirality, where the interrelation (137) between $\chi_{s_{AB}}(A)$ and $s_{AA^\diamond}$ implies that this measure can take values from the unit interval. The measure $\chi_{s_{AB}}(A)$, first proposed[1] as an example of dissimilarity measures of the second kind, is zero for achiral objects and takes positive values for all chiral objects. Objects perceived as having prominent chirality tend to have large $\chi_{s_{AB}}(A)$ values. The SNSM measures have also been applied to more general molecular shape problems.[68] More recently, Klein showed[69] that by a logarithmic transformation of the scaling factors s_{AB}, a metric can be constructed to provide a proper distance-like measure of dissimilarity of shapes.

It can also be shown,[70] however, that the SNSM dissimilarity measure, defined as $d_{s_{AB}} = 1 - s_{AB}$, is a proper metric itself and there is no need for logarithmic transformations. We use the acronym SNDSM for this scaling–nesting dissimilarity metric $d_s(A, B)$. A proof of the metric properties of SNDSM $d_s(A, B)$ is given subsequently.

Take three sets A, B, and C, all of unit volume.

(1) Since

$$0 \leq s_{AB} \leq 1, \qquad (138)$$

the inequality for nonnegativity,

$$d_s(A, B) \geq 0, \tag{139}$$

follows. Note that an upper bound also applies and the inequality

$$d_s(A, B) \leq 1 \tag{140}$$

also holds.

(2) Since $s_{AB} = 1$ iff $A = B$ (then $s_{(AB)} = s_{(BA)} = 1$ also holds),

$$d_s(A, B) = 0 \quad \text{iff } A = B \tag{141}$$

follows.

(3) The SNSM measure s_{AB} is symmetric, $s_{AB} = s_{AB}$. Consequently,

$$d_s(A, B) = 1 - s_{AB} = 1 - s_{BA} = d_s(B, A), \tag{142}$$

that is, $d_s(A, B)$ is also symmetric.

(4) The triangle inequality

$$d_s(A, C) \leq d_s(A, B) + d_s(B, C) \tag{143}$$

is subsequently proven.

A simple inequality follows from the definition of the scaling–nesting semisimilarity measure $s_{(AB)}$. For three objects X, Y, and Z, all of unit volume, $s_{(XY)}$ scales X so a version X_v of it fits within Y, and $s_{(YZ)}$ scales Y so a version $Y_{v'}$ of it fits within Z. Consequently, a scaling factor $s_{(XY)}s_{(YZ)}$ certainly reduces X so that a version X_w of it fits within Z. However, by definition, $s_{(XZ)}$ is the largest scaling factor of X that allows X to fit within Z. Consequently,

$$s_{(XY)}s_{(YZ)} \leq s_{(XZ)}. \tag{144}$$

At first sight, this inequality appears to have little use for our purposes. This is a relationship for *products* of one-sided scaling factors, that is, for the semisimilarity measures $s_{(XY)}$, and the relationship we seek to prove is for *sums* of measures $d_s(A, B)$ derived from the *symmetric* SNSM s_{AB}.

However, this relationship can be combined with another simple inequality that leads to a proof. For any two real numbers a and b of the unit interval

$$0 \leq a, b \leq 1, \tag{145}$$

the following holds:

$$a + b - ab \leq 1. \tag{146}$$

Clearly, for $0 \leq a, b \leq 1$,

$$0 \leq (1 - b)(1 - a) \tag{147}$$

holds, and rearrangement on the right-hand side gives

$$0 \leq 1 - a - b + ab, \tag{148}$$

that is, $a + b - ab \leq 1$, as claimed in inequality (158).

By expressing $d_s(A,C)$, $d_s(A,B)$, and $d_s(B,C)$ in terms of the semisimilarity measures $s_{(AC)}$, $s_{(CA)}$, $s_{(AB)}$, $s_{(BA)}$, $s_{BC)}$, and $s_{(CB)}$, using defining equations (133) and (136), and substituting these expressions into the inequality (143) we seek to prove, we write an equivalent inequality

$$1 - \left(s_{(AC)} + s_{(CA)} \right)/2 \leq 1 - \left(s_{(AB)} + s_{(BA)} \right)/2 + 1 - \left(s_{(BC)} + s_{(CB)} \right)/2. \tag{149}$$

This inequality can be rearranged into

$$s_{(AB)} + s_{(BC)} - s_{(AC)} + s_{(CB)} + s_{(BA)} - s_{(CA)} \leq 2, \tag{150}$$

which must hold if both of the following inequalities hold simultaneously:

$$s_{(AB)} + s_{(BC)} - s_{(AC)} \leq 1, \tag{151}$$

and

$$s_{(CB)} + s_{(BA)} - s_{(CA)} \leq 1. \tag{152}$$

However, according to inequality (144),

$$s_{(AB)}s_{(BC)} \leq s_{(AC)}; \tag{153}$$

hence, for nonnegative $s_{(AB)}$, $s_{(BC)}$, and $s_{(AC)}$, all bounded by $+1$,

$$s_{(AB)} + s_{(BC)} - s_{(AC)} \leq s_{(AB)} + s_{(BC)} - s_{(AB)}s_{(BC)} \tag{154}$$

also holds. Furthermore, since

$$0 \leq s_{(AB)}, s_{(BC)} \leq 1, \tag{155}$$

inequality (146) holds for $s_{(AB)}$ and $s_{(BC)}$, that is,

$$s_{(AB)} + s_{(BC)} - s_{(AB)}s_{(BC)} \leq 1. \tag{156}$$

Comparing this to inequality (154), one obtains

$$s_{(AB)} + s_{(BC)} - s_{(AC)} \leq 1, \tag{157}$$

that is, inequality (151) holds.

A similar argument involving $s_{(CB)}$, $s_{(BA)}$, $s_{(CA)}$, and the product $s_{(CB)}s_{(BA)}$ shows that

$$s_{(CB)} + s_{(BA)} - s_{(CA)} \leq 1, \tag{158}$$

a result identical with inequality (152), is also true simultaneously. Consequently, the scaling−nesting dissimilarity measure $d_s(A,B)$ fulfills the triangle inequality.

The properties (1)–(4) proved in the preceding text imply that the SNDSM $d_s(A, B)$ is a metric. Note that in the foregoing proof there was no restriction concerning the choice of orientation and relative locations of any two sets while attempting to fit one within the other. That is, arbitrarily translated and rotated versions A_v and $B_{v'}$ of each one of sets A and B, scaled or unscaled, can be used. On the other hand, in some applications, some angle restrictions are warranted. Similarly, one may find it physically justified to restrict the range of versions A_v and $B_{v'}$ to those that are derived from sets A and B by some rotation about a fixed axis. These choices are relevant to similarity analysis of molecules in external fields or within enzyme cavities. In general, we denote such a constraint by the abbreviation cnstr. As long as the constraint cnstr is common for all sets involved, the entire proof can be repeated in identical form, and the corresponding constrained SNDSM, denoted by $d_{s_{cnstr}}(A, B)$, is also a metric within the set of all objects obeying the given constraint cnstr.

VIII. CHIRALITY MEASURES AND SYMMETRY DEFICIENCY MEASURES FOR CONTINUA USING THE SNDSM METRIC

If the metric SNDSM $d_s(A, B)$ is applied to mirror images A and $A^\diamond$, then precisely the same chirality measure is obtained as the measure $\chi_{s_{AB}}(A)$ defined earlier.[1] However, the fact that the set of all finite, chiral continua of a given dimension has a simple chirality measure that is compatible with a scaling–nesting *metric* of a formal shape space of all finite continua of the given dimension is a novel aspect that has implications concerning comparisons between general shape changes and interconversions between chiral mirror images.

Besides chirality, more general symmetry deficiencies can be treated within a unified framework using the metric SNDSM $d_s(A, B)$. First we shall describe a general treatment applicable with any metric defined for shapes of finite continua, followed by comments on the special case of SNDSM $d_s(A, B)$.

Choose a point c of set A and regard it as the center for a (possibly only approximate) symmetry element R of A. If a center of mass is defined for set A, then it is advantageous to take c as the center of mass; however, this choice is not required. Determine the smallest positive value n that satisfies the condition

$$\mathbf{R}^n = \mathbf{E}, \tag{159}$$

where $\mathbf{R}^n$ is the nth power of the corresponding symmetry operation $\mathbf{R}$ and $\mathbf{E}$ is the identity operation.

Carry out the operations $\mathbf{R}^i$ ($i = 1, \ldots, n - 1$) associated with this (possibly approximate) symmetry element on set A and generate the union

$A(R, c)$ of all sets $A, \mathbf{R} A, \mathbf{R}^2 A, \ldots, \mathbf{R}^{n-1} A$, that is, construct the set

$$A(R, c) = \bigcup_{i=0, n-1} \mathbf{R}^i A. \tag{160}$$

Clearly, the set $A(R, c)$ has the symmetry element R, since

$$\mathbf{R} A(R, c) = \mathbf{R} \bigcup_{i=0, n-1} \mathbf{R}^i A = \bigcup_{i=0, n-1} \mathbf{R}^{i+1} A = \bigcup_{i=1, n} \mathbf{R}^i A = \bigcup_{i=0, n-1} \mathbf{R}^i A, \tag{161}$$

that is,

$$\mathbf{R} A(R, c) = A(R, c), \tag{162}$$

where the identities

$$\mathbf{R}^n A = \mathbf{E} A = \mathbf{R}^0 A \tag{163}$$

are used.

A comparison of sets A and $A(R, c)$ provides a measure of the symmetry aspect R for set A with respect to center c. In principle, any similarity measure is applicable for this comparison, for example, the Hausdorff metric $h(A, A(R, c))$ provides a valid measure of the symmetry aspect R for set A with respect to center c. If the center of mass is chosen for c, then the simpler notations $A(R)$ and $h(A, A(R))$ are used.

The preceding general treatment of approximate symmetry is especially compatible with the SNDSM metric. The quantity $d_s(A, A(R, c))$ is the corresponding SNDSM "shape distance" between the original set A and the fully R-symmetric set $A(R, c)$ with respect to center c, whereas $d_s(A, A(R))$ is the SNDSM "shape distance" between the original set A and the fully R-symmetric set $A(R)$ with respect to the center of mass of set A. The larger the value for $d_s(A, A(R, c))$ or $d_s(A, A(R))$, the higher the R-symmetry deficiency of set A with respect to the arbitrary center c of the center of mass, respectively. Several related results are discussed in the Appendix.

IX. A FUZZY SCALING–NESTING SIMILARITY MEASURE AND THE FUZZY SCALING–NESTING DISSIMILARITY METRIC

The scaling–nesting semisimilarity measures $s_{(AB)}$ and $s_{(BA)}$, as well as the SNSM s_{AB} and the associated SNDSM metric $d_s(A, B)$, can be generalized easily for fuzzy sets, resulting in the *fuzzy scaling–nesting semisimilarity measures* $\text{fs}_{(AB)}$ and $\text{fs}_{(BA)}$, the symmetric *fuzzy scaling–nesting similarity measure* (FSNSM) fs_{AB}, and a fuzzy version of SNDSM $d_s(A, B)$, a metric $d_{\text{fs}}(A, B)$, referred to as *fuzzy scaling–nesting dissimilarity measure* (FSNDSM).

Consider two fuzzy sets A and B, of equal mass. Consider all rotated and translated versions A_v and $B_{v'}$ of sets A and B. Determine the maximal scaling factor $\text{fs}_{(AB)}$ applied to a version A_v of A that allows the rescaled version $A_{v,\text{scaled}}$ of A_v to fit with B, where the statement "$A_{v,\text{scaled}}$ fitting within B" is interpreted as the fuzzy subset relation

$$\mu_{A_{v,\text{scaled}}}(x) \leq \mu_B(x), \qquad \forall x. \tag{164}$$

Similarly, determine the maximal scaling factor $\text{fs}_{(BA)}$ applied to a version $B_{v'}$ of fuzzy set B that allows the rescaled version $B_{v',\text{scaled}}$ of $B_{v'}$ to fit within A,

$$\mu_{B_{v',\text{scaled}}}(x) \leq \mu_A(x), \qquad \forall x. \tag{165}$$

The fuzzy set scaling factors have very similar properties to those of the ordinary SNSM scaling factors. The two scaling factors $\text{fs}_{(AB)}$ and $\text{fs}_{(BA)}$ are not necessarily equal and they define asymmetric measures of similarity of the shapes of fuzzy sets A and B, where, in fact,

$$\text{fs}_{(AB)} \neq \text{fs}_{(BA)} \tag{166}$$

in most cases. These asymmetric similarity measures are *fuzzy semisimilarity measures* or one-sided fuzzy similarity measures. If the shapes of fuzzy sets A and B are identical, then both scaling factors $\text{fs}_{(AB)}$ and $\text{fs}_{(BA)}$ are equal to 1:

$$\text{fs}_{(AB)} = \text{fs}_{(BA)} = 1. \tag{167}$$

In general, the average

$$\text{fs}_{AB} = \left[\text{fs}_{(AB)} + \text{fs}_{(BA)} \right]/2, \tag{168}$$

the *fuzzy scaling–nesting similarity measure* (FSNSM) fs_{AB} is a symmetric measure of similarity, since the relation

$$\text{fs}_{AB} = \text{fs}_{BA} \tag{169}$$

is valid for all pairs A and B of fuzzy sets of equal mass.

This FSNSM fs_{AB} is the real part of the complex fuzzy similarity–asymmetric shape tolerance measure fsast_{AB}. This measure is defined as $\text{fsast}_{AB} = [\text{fs}_{(AB)} + \text{fs}_{(BA)}]/2 + i[\text{fs}_{(AB)} - \text{fs}_{(BA)}]/2$ and is entirely analogous to the complex similarity–asymmetric shape tolerance measure sast_{AB} given for crisp sets in Eq. (135). The symmetric real part of fsast_{AB} is a measure of similarity, and the antisymmetric imaginary part of fsast_{AB} describes the asymmetry of shape tolerance. If the roles of fuzzy sets A and B are interchanged, then one obtains the complex conjugate $(\text{fsast}_{AB})^*$ of fsast_{AB}.

The fuzzy scaling–nesting semisimilarity measures $\text{fs}_{(AB)}$ satisfy an inequality analogous to inequality (144) of the ordinary scaling–nesting semisimilarity measures $s_{(XY)}$; this property can be verified as follows. For three fuzzy sets A, B, and C, all of unit mass, the factor $\mathrm{fs}_{(AB)}$ scales fuzzy set A so that a version A_v of the scaled A becomes a fuzzy subset of B, and factor $\mathrm{fs}_{(BC)}$ scales fuzzy set B so that a version $B_{v'}$ of the scaled B becomes a fuzzy subset of C. Consequently, the product $\mathrm{fs}_{(AB)}\mathrm{fs}_{(BC)}$ used as a scaling factor certainly reduces A so that a version A_w of the scaled fuzzy set A becomes a fuzzy subset of fuzzy set C. On the other hand, by definition, $\mathrm{fs}_{(AC)}$ is the largest scaling factor of A that allows a version $A_{w'}$ of the scaled fuzzy set A to become a fuzzy subset of fuzzy set C, implying that the inequality

$$\mathrm{fs}_{(AB)}\mathrm{fs}_{(BC)} \leq \mathrm{fs}_{(AC)} \tag{170}$$

holds for fuzzy sets A, B, and C as well.

In terms of FSNSM fs_{AB}, the fuzzy scaling–nesting dissimilarity measure $d_{\mathrm{fs}}(A, B)$ is defined as

$$d_{\mathrm{fs}}(A, B) = 1 - \mathrm{fs}_{AB}. \tag{171}$$

This measure $d_{\mathrm{fs}}(A, B)$ of fuzzy dissimilarity is in fact a metric for fuzzy shapes. With the provision that volume of ordinary sets is replaced with the mass of fuzzy sets, all the steps in the proof of the metric properties of the ordinary SNDSM measure, described in Section VII, can be repeated for FSNDSM $d_{\mathrm{fs}}(A, B)$ in identical form, proving that the fuzzy set version FSNDSM of the SNDSM measure is, indeed, a metric. The fuzzy scaling–nesting dissimilarity measure FSDNSM $d_{\mathrm{fs}}(A, B)$ provides a useful definition for distance between fuzzy sets, interpreted as a metric expressing dissimilarity in a formal space of fuzzy shapes, such as electronic densities of molecules.

X. FUZZY MEASURES OF CHIRALITY AND SYMMETRY DEFICIENCY, FUZZY SYMMETRY GROUPS, AND FUZZY SYMMORPHY GROUPS BASED ON THE FUZZY SCALING–NESTING SIMILARITY MEASURE

The fuzzy measures of symmetry deficiency, fuzzy chirality, fuzzy symmetry, and fuzzy symmetry groups described in Section IV in terms of the fuzzy Hausdorff-type metrics provide new tools for the characterization and comparison of fuzzy objects. However, the computation of the fuzzy Hausdorff-type metric is often cumbersome and alternative formulations may have certain advantages.

The fuzzy scaling–nesting similarity measure fs_{AB} and the fuzzy scaling–nesting dissimilarity metric $d_{\mathrm{fs}}(A, B)$ provide such alternatives. The entire treatment used for the introduction and definition of fuzzy measures of symmetry deficiency, fuzzy chirality, fuzzy symmetry, and fuzzy symmetry groups which involve the fuzzy Hausdorff-type similarity measure s_g, as described in Section IV, can be repeated for FSNSM fs_{AB}. A precise set of definitions and a valid description is obtained by simply replacing the fuzzy Hausdorff-type similarity measure s_g with the fuzzy scaling–nesting similarity measure fs_{AB} in each equation and inequality of Section IV that involves the fuzzy Hausdorff-type similarity measure s_g.

This approach generates intuitively more transparent descriptions of

(1) Fuzzy symmetry elements
(2) Fuzzy measures of symmetry
(3) Fuzzy measures of symmetry deficiency
(4) Fuzzy symmetry operators
(5) Fuzzy symmetry groups
(6) Fuzzy measure of chirality

all based on FSNSM fs_{AB}, also denoted as $fs(A, B)$.

In particular, a fuzzy set A is said to have the *fuzzy symmetry element* $R(\beta)$ corresponding to the symmetry operation $\mathbf{R}$ at the fuzzy level β of the FSNSM $fs_{AB} = fs(A, B)$, if and only if the fuzzy similarity measure fs between $\mathbf{R}A$ and A is greater than or equal to β:

$$fs(\mathbf{R}A, A) \geq \beta. \tag{172}$$

The maximum fuzzy level $\beta(A, \mathbf{R}, fs)$ at which the fuzzy symmetry element $R(\beta)$ is present for the fuzzy set A, that is,

$$\beta(A, \mathbf{R}, fs) = \sup_{\beta \in [0,1]} \{\beta : fs(\mathbf{R}A, A) \geq \beta\}, \tag{173}$$

is a measure of the "degree" of symmetry aspect R for fuzzy set A, according to the FSNSM fs_{AB}.

A measure $\delta(A, \mathbf{R}, fs)$ of the *symmetry deficiency* of fuzzy set A in symmetry element R according to the FSNSM fs_{AB} is defined as

$$\delta(A, \mathbf{R}, fs) = 1 - \beta(A, \mathbf{R}, fs). \tag{174}$$

Assume that $\mathbf{R}$ is the ordinary symmetry operator corresponding to the fuzzy symmetry element $R(\beta')$ present for fuzzy set A according to the FSNSM fs_{AB}. A *fuzzy symmetry operator* $\mathbf{R}(fs)$ of fuzzy symmetry element $R(\beta')$ present for fuzzy set A at the fuzzy level β' of the fuzzy similarity measure fs is defined by its action on the fuzzy set A:

$$\mathbf{R}(fs)A = \mathbf{M}_{A, \mathbf{R}}\mathbf{R}A, \tag{175}$$

where the definition of operator $\mathbf{M}_{A, \mathbf{R}}$ is the same as in Eq. (93) and it involves the actual symmetry operator $\mathbf{R}$.

We say that a fuzzy set A has the β' *fuzzy symmetry group* $G(fs, \beta')$ at *fuzzy level* β' if A has the fuzzy symmetry element $R(\beta')$ at the fuzzy level β' of the FSNSM fs_{AB} for each symmetry operation $\mathbf{R}$ of the crisp symmetry group G.

The β fuzzy symmetry group $G(\text{fs}, \beta)$ at the fuzzy level β that is the *supremum* of the levels β' of all β' fuzzy symmetry groups $G(\text{fs}, \beta')$ of A is the *fuzzy symmetry group* $G(\text{fs}, \beta)$ of the fuzzy set A:

$$G\left(s_g, \beta\right) = G\left(s_g, \beta\left(A, G, s_g\right)\right), \tag{176}$$

where

$$\beta = \beta(A, G, \text{fs}) = \sup_{\beta'} \left\{ \beta' : \text{fs}(\mathbf{R}A, A) \geq \beta', \forall \mathbf{R} \in G \right\}. \tag{177}$$

Alternative fuzzy symmetry deficiency measures are defined in terms of a *maximal R subset B'*, *maximal mass R subset B*, *minimal R superset C'*, and *minimal mass R superset C*, of fuzzy set A, discussed in Section IV. If fuzzy set D denotes any one of the R subsets B, B' or R supersets C, C' of fuzzy set A, $D \in \{B, B', C, C'\}$, then a *fuzzy symmetry deficiency measure* $\delta_{\text{fs}, R, D}(A)$ of fuzzy set A is provided by

$$\delta_{\text{fs}, R, D}(A) = \inf_{D_\kappa} \left\{ 1 - \text{fs}(A, D_\kappa) \right\}, \tag{178}$$

where index D refers to the choice of the type set D from the family of all fuzzy R sets of types B, B', C, and C'.

Most of these R sets are not unique for a given fuzzy set A; hence, the infimum is taken for all such fuzzy sets D_κ.

Fuzzy chirality measures can also be defined in terms of a *maximal achiral subset B'*, *maximal mass achiral subset B*, *minimal achiral subset C'*, and *minimal mass achiral subset C*, of fuzzy set A, discussed in Section IV. If fuzzy set D denotes any one of these fuzzy sets, $D \in \{B, B', C, C'\}$, then a *fuzzy chirality measure* $\chi_{\text{fs}, D}(A)$ of fuzzy set A is provided by

$$\chi_{\text{fs}, D}(A) = \inf_{D_\kappa} \left\{ 1 - \text{fs}(A, D_\kappa) \right\}. \tag{179}$$

Sets B, B', C, and C' are not necessarily unique for a given fuzzy set A; therefore, in definition (196) the infimum is taken for all such fuzzy sets D_κ.

The same treatment can be applied in the derivation of *fuzzy symmorphy groups* based on FSNSM. In Section VI, fuzzy symmorphy and fuzzy symmorphy groups were developed based on the fuzzy Hausdorff-type similarity measure s_g. The steps of the entire derivation can be repeated for FSNSM fs$_{AB}$. A valid description of fuzzy symmorphy is obtained if the fuzzy Hausdorff-type similarity measure s_g is replaced with the fuzzy scaling–nesting similarity measure fs$_{AB}$ in each equation and inequality of Section VI that involves the fuzzy Hausdorff-type similarity measure s_g.

This leads to FSNSM-based descriptions of

(1) Fuzzy symmorphy elements
(2) Fuzzy measures of symmorphy
(3) Fuzzy symmorphy operators
(4) Fuzzy symmorphy groups

each based on FSNSM $\text{fs}_{AB} = \text{fs}(A, B)$.

The fuzzy scaling–nesting similarity measure is simpler to visualize than variants of the Hausdorff metric; hence, FSNSM is advantageous in chemical applications of electron density similarity analysis, where chemical intuition and visualization play a prominent role.

A fuzzy set A is said to have the *fuzzy symmorphy element* $S(\beta)$ corresponding to the symmorphy operation $\mathbf{S}$ at the fuzzy level β of the FSNSM $\text{fs}(A, B)$, if and only if the fuzzy similarity measure fs between $\mathbf{S}A$ and A is greater than or equal to β:

$$\text{fs}(\mathbf{S}A, A) \geq \beta. \tag{180}$$

The maximum fuzzy level $\beta(A, \mathbf{S}, \text{fs})$ at which the fuzzy symmorphy element $S(\beta)$ is present for the fuzzy set A, that is,

$$\beta(A, \mathbf{S}, \text{fs}) = \sup_{\beta \in [0, 1]} \{\beta : \text{fs}(\mathbf{S}A, A) \geq \beta\}, \tag{181}$$

is a measure of the "degree" of symmorphy aspect S for fuzzy set A, according to the FSNSM fs_{AB}.

The measure $\delta(A, \mathbf{S}, \text{fs})$ of the *symmorphy deficiency* of fuzzy set A in symmorphy element S according to the FSNSM fs_{AB} is defined as

$$\delta(A, \mathbf{S}, \text{fs}) = 1 - \beta(A, \mathbf{S}, \text{fs}). \tag{182}$$

Take $\mathbf{S}$, the ordinary symmorphy operator corresponding to the fuzzy symmorphy element $S(\beta')$ present for fuzzy set A according to the FSNSM fs_{AB}. A *fuzzy symmorphy operator* $\mathbf{S}(\text{fs})$ of fuzzy symmorphy element $S(\beta')$ present for fuzzy set A at the fuzzy level β' of the fuzzy similarity measure fs is defined by its action on the fuzzy set A:

$$\mathbf{S}(\text{fs})A = \mathbf{M}_{A,\mathbf{S}}\mathbf{S}A, \tag{183}$$

where the definition of operator $\mathbf{M}_{A,\mathbf{S}}$ is the same as in Eq. (127) and it involves the actual symmorphy operator $\mathbf{S}$.

We say that a fuzzy set A has the β' *fuzzy symmorphy group* $G_{\text{sph}}(\text{fs}, \beta')$ *at fuzzy level* β' if A has the fuzzy symmorphy element $S(\beta')$ at the fuzzy level β' of the FSNSM fs_{AB} for each symmorphy operation $\mathbf{S}$ of the crisp symmorphy group G_{sph}. The β fuzzy symmorphy group $G_{\text{sph}}(\text{fs}, \beta)$ at the fuzzy level β that is the *supremum* of the levels β' of all β' fuzzy symmorphy groups G_{sph}(fs, β') of A is the *fuzzy symmorphy group* G_{sph}(fs, β) of the fuzzy set A:

$$G_{sph}\left(s_g, \beta\right) = G_{sph}\left(s_g, \beta\left(A, G, s_g\right)\right), \tag{184}$$

where

$$\beta = \beta\left(A, G_{sph}, \text{fs}\right) = \sup_{\beta'}\left\{\beta': \text{fs}(SA, A) \geq \beta', \forall S \in G_{sph}\right\}. \tag{185}$$

The fuzzy symmorphy approach provides an alternative tool for shape analysis of fuzzy objects. In the chemical context, the deformations of objects surrounding a given molecule can be thought of as a deformation of the space. For example, the deformations of enzymes during a partial folding or unfolding process of a protein may correspond to a fuzzy symmorphy that leaves the actual shape of the enzyme's cavity and the shape of a molecule within the cavity "indistinguishable at some fuzzy level β" from their original fuzzy shapes. In such problems, the fuzzy symmorphy described here appears to provide the right tools for shape analysis.

XI. THE CENTER OF MASS OF A FUZZY SET, THE CENTER OF MOLECULAR ELECTRON DENSITY, AND FUZZY CENTRAL MEASURES OF SYMMETRY DEFICIENCY

In many applications of continuum mechanics the center of mass of objects has special significance. It appears advantageous to use an analogous concept[29] in fuzzy set theory, taking the value of membership function $\mu_A(x)$ at each point x of a fuzzy set A in the role of mass density and the integral defined in Eq. (100) as the total mass of the fuzzy set A. This is easily accomplished if the underlying set X can be interpreted as a Euclidean space with a well-defined Cartesian coordinate system. In this case, the center of mass $c(A)$ of fuzzy set A is defined as

$$c(A) = \int_X \mu_A(x)\, x\, dx \left/ \left[\int_X \mu_A(x)\, dx\right]\right. = \int_X \mu_A(x)\, x\, dx / m(A), \tag{186}$$

where integration is for the whole domain X of fuzzy set A and where x is regarded as a formal position vector of the corresponding element x, with components interpreted in the context of the metric and dimension of the underlying space X.

Whereas the molecular center of mass is of importance in both dynamics and spectroscopy, a formal center of the electronic density distribution has direct significance in shape characterization. A suitable definition of this latter center may differ from the molecular center of mass. The fuzzy set model of electron densities is represented by the corresponding unscaled and transformed versions of the membership functions $\mu_{A_{(\rho)}}(\mathbf{r})$ and $\mu_{A_{(\rho,\tau)}}(\mathbf{r})$, respectively, as defined by Eqs. (19)–(21). Using the concept of the center of mass of fuzzy sets, the unscaled and transformed versions of the electron density center of a molecule A can be taken as

$$c\left(A_{(\rho)}\right) = \int_X \mu_{A_{(\rho)}}(x)\,x\,dx \Big/ \left[\int_X \mu_{A_{(\rho)}}(x)\,dx\right] \qquad (187)$$

and

$$c\left(A_{(\rho,\tau)}\right) = \int_X \mu_{A_{(\rho,\tau)}}(x)\,x\,dx \Big/ \left[\int_X \mu_{A_{(\rho,\tau)}}(x)\,dx\right], \qquad (188)$$

respectively.

When fuzzy set shape analysis methods requiring formal reference points are used for molecules, the centers $c(A_{(\rho)})$ and $c(A_{(\rho,\tau)})$ of the electronic density appear as a natural choice for reference.

Using the center of mass concept of fuzzy sets, the symmetry deficiency measure of finite continua, described in Section VIII, can be generalized for fuzzy sets. The chirality and more general symmetry deficiencies of fuzzy sets can be treated within a unified framework using the fuzzy metric FSNDSM $d_{fs}(A, B)$.

Although the treatment subsequently described is applicable for any reference point c of X for any fuzzy set A, the choice of reference point as the center of mass $c(A)$ of fuzzy set A is usually advantageous in chemical applications, where the electronic density center $c(A_{(\rho)})$ or $c(A_{(\rho,\tau)})$ of molecule A can be selected.

We regard point c as the center for a (possibly only approximate) symmetry element R of A. Following the treatment for "crisp" continua, described in Section VIII, we take the symmetry operation $\mathbf{R}$ associated with this (possibly approximate) symmetry element R, and carry out the operations $\mathbf{R}^i$ $(i = 1,\ldots,n-1)$ on set A, resulting in the fuzzy sets $A, \mathbf{R}A, \mathbf{R}^2A,\ldots,\mathbf{R}^{n-1}A$, where the value n satisfies the conditions

$$n > 1 \qquad (189)$$

and

$$\mathbf{R}^n = \mathbf{E}, \qquad (190)$$

where $\mathbf{E}$ is the identity operation.

The fuzzy union $A(R,c)$ of all the fuzzy sets $A, \mathbf{R}A, \mathbf{R}^2A,\ldots,\mathbf{R}^{n-1}A$,

$$A(R,c) = \bigcup_{i=0,n-1} \mathbf{R}^iA, \qquad (191)$$

has the symmetry element R, since

$$\mathbf{R}A(R,c) = \mathbf{R} \bigcup_{i=0,n-1} \mathbf{R}^iA = \bigcup_{i=0,n-1} \mathbf{R}^{i+1}A = \bigcup_{i=1,n} \mathbf{R}^iA = \bigcup_{i=0,n-1} \mathbf{R}^iA.$$

(192)

Consequently,

$$\mathbf{R}A(R,c) = A(R,c).$$

(193)

If the center of mass of the fuzzy set A is chosen as reference point c, then the notation $A(R,c)$ is replaced by the simpler notation $A(R)$.

The dissimilarity of fuzzy sets A and $A(R,c)$ provides a measure of the symmetry aspect R for set A with respect to center c. A large measure of dissimilarity implies a higher degree of symmetry deficiency of fuzzy set A, with respect to symmetry represented by element R. This symmetry deficiency can be described using either one of the fuzzy set dissimilarity metrics. For example, if the fuzzy metric FSNDSM $d_{fs}(A, B)$ is used, then one obtains the fuzzy symmetry deficiency measure $\delta_{fs,R,c}(A)$:

$$\delta_{fs,R,c}(A) = d_{fs}(A, A(R,c)).$$

(194)

Alternative fuzzy symmetry deficiency measures $\delta_{H,R,c}(A)$ of fuzzy set A with respect to reference point c and symmetry represented by element R are defined as

$$\delta_{H,R,c}(A) = 1 - s_H(A, A(R,c)),$$

(195)

where s_H is one of the fuzzy Hausdorff-type similarity measures from list (109). The simplest of these fuzzy symmetry deficiency measures is

$$\delta_{s_g,R,c}(A) = 1 - s_g(A, A(R,c)),$$

(196)

using the similarity measure $s_g(A, B)$ based on a Gaussian transformation of the unscaled fuzzy Hausdorff-type metric $g(A, B)$. For fuzzy electron densities, the similarity measure $s_f(A, B)$ based on the Gaussian transformation of the scaled fuzzy Hausdorff-type metric $f(A, B)$ is a natural choice:

$$\delta_{s_f,R,c}(A) = 1 - s_f(A, A(R,c)).$$

(197)

XII. THE FUZZY AVERAGE OF CRISP SETS, THE FUZZY AVERAGE OF FUZZY SETS, THE CRISP AVERAGE OF CRISP SETS, THE CRISP AVERAGE OF FUZZY SETS, AND RELATED FUZZY SYMMETRY MEASURES

Consider a family F of crisp or fuzzy subsets,

$$F = \{F_1, F_2, \ldots, F_m\},$$

(198)

of an underlying set X. The average of these sets $F_1, F_2, \ldots, F_m$ can be interpreted as a fuzzy subset F_{fav}, the fuzzy average of family F, defined[29] by its fuzzy membership function $\mu_{F_{fav}}(x)$ as

$$\mu_{F_{fav}}(x) = \left[\sum_{k=1,m} \mu_{F_k}(x) \right] \Big/ m. \tag{199}$$

The fuzzy average F_{fav} of a family F of crisp sets $F_1, F_2, \ldots, F_m$ is itself a crisp set if and only if

$$F_1 = F_2 = \cdots = F_m, \tag{200}$$

and in this case

$$F_{fav} = F_1 = F_2 = \cdots = F_m \tag{201}$$

also holds.

In all other cases, the fuzzy average F_{fav} of a family F of crisp sets is a genuine fuzzy set, since the average in definition (199) is over the necessarily binary membership function values for crisp sets, and if not all of the equations in Eq. (200) hold, for example, if

$$F_k \neq F_{k'}, \tag{202}$$

then for some x element the membership function values $\mu_{F_k}(x)$ and $\mu_{F_{k'}}(x)$ must differ,

$$\mu_{F_k}(x) \neq \mu_{F_{k'}}(x), \tag{203}$$

resulting in an average value $\mu_{F_{fav}}(x)$ that must fall within the open unit interval

$$0 < \mu_{F_{fav}}(x) < 1. \tag{204}$$

The fuzzy average F_{fav} of a family F of fuzzy sets $F_1, F_2, \ldots, F_m$ is necessarily a fuzzy set.[29]

The crisp average F_{crav} (or simply F_{av}) of a family F of crisp or fuzzy or mixed sets $F_1, F_2, \ldots, F_m$ is defined in terms of the crisp α-cut $D_{F_{fav}}(\alpha)$ of the fuzzy average F_{fav} of the family F, where the unique α value is selected so that the mass of $D_{F_{fav}}(\alpha)$ is equal that of the fuzzy average F_{fav}:

$$F_{crav} = F_{av} = D_{F_{fav}}(\alpha), \tag{205}$$

where for α,

$$m\big(D_{F_{fav}}(\alpha)\big) = m(F_{fav}). \tag{206}$$

The fuzzy and crisp average concepts are the basis of further measures of approximate symmetry of a crisp or fuzzy set A. Consider a point c as the center for a (possibly only approximate) symmetry element R of A.

Take the symmetry operation $\mathbf{R}$ associated with this (possibly approximate) symmetry element R, and generate the sets $A, \mathbf{R}A, \mathbf{R}^2 A, \ldots, \mathbf{R}^{n-1}A$, by carrying out the corresponding operations $\mathbf{R}^i$ ($i = 1, \ldots, n-1$) on set A, where, as in the previous discussions, for the value n the conditions $n > 1$ and $\mathbf{R}^n = \mathbf{E}$ hold, with $\mathbf{E}$ the identity operation.

If family F is chosen as

$$F = \{A, \mathbf{R}A, \mathbf{R}^2 A, \ldots, \mathbf{R}^{n-1}A\}, \tag{207}$$

then both the fuzzy average $F_{f\text{av}}$ and the crisp average F_{av} of a family F are R sets,

$$\mathbf{R}F_{f\text{av}} = F_{f\text{av}} \tag{208}$$

and

$$\mathbf{R}F_{\text{av}} = F_{\text{av}}, \tag{209}$$

and any one of the fuzzy Hausdorff-type similarity measures $s_H(A, B)$,

$$s_H(A, B)$$

$$\in \{s_g(A, B), t_g(A, B), z_g(A, B), s_f(A, B), t_f(A, B), z_f(A, B)\}, \tag{210}$$

provides a valid measure of approximate symmetry of set A, with respect to center c and the (possibly approximate) symmetry element R:

$$s_{H, R, c, F_{f\text{av}}}(A) = s_H(A, F_{f\text{av}}) \tag{211}$$

and

$$s_{H, R, c, F_{\text{av}}}(A) = s_H(A, F_{\text{av}}). \tag{212}$$

Similarly, the fuzzy scaling–nesting similarity measures serve as the basis of measures of approximate symmetry, with respect to the fuzzy average.

Valid measures of approximate symmetry are given for a set A, with respect to some symmetry element R and some center c, using the fuzzy scaling–nesting dissimilarity metric FSNDSM $d_{\text{fs}}(A, B)$ or the scaling–nesting similarity measure SNSM fs_{AB}, expressed between A and $F_{f\text{av}}$, on the one hand, and between A and F_{av}, on the other hand:

$$s_{\text{fs}, R, c, F_{f\text{av}}}(A) = s_{\text{fs}}(A, F_{f\text{av}}) = 1 - \text{fs}_{AF_{f\text{av}}} \tag{213}$$

and

$$s_{\text{fs}, R, c, F_{\text{av}}}(A) = s_{\text{fs}}(A, F_{\text{av}}) = 1 - \text{fs}_{AF_{\text{av}}}. \tag{214}$$

XIII. TWO GENERALIZATIONS OF THE ZPA FOLDING–UNFOLDING CONTINUOUS SYMMETRY MEASURES FOR CONTINUA USING THE SNDSM METRIC AND THE HAUSDORFF METRIC

The intuitive, visual detection of symmetry often follows comparisons between rotated and reflected images of objects.[71] Analogies with this process of symmetry detection can be formulated in a rigorous framework using a formal "folding–unfolding" technique. This idea is the basis of important numerical measures of approximate symmetry, introduced by Zabrodsky, Peleg, and Avnir,[72–75] who named this family of approximate symmetry measures continuous symmetry measures. To distinguish these measures from other measures having some continuous aspects, here we shall refer to the measures introduced by Zabrodsky, Peleg, and Avnir as ZPA folding-unfolding continuous symmetry measures or, for short, ZPA measures.

Whereas the ZPA approach was originally developed for finite point sets, such as a family of nuclear locations in three dimensions, the original authors pointed out techniques that generalize the ZPA approach to continuum sets. For a finite, discrete point set D, successive powers of a potential symmetry operator $\mathbf{R}$ are assigned and applied to elements of an ordered sequence of points of set D, essentially "folding up" the point set D into a set that usually has a much smaller diameter than the original set D. The average location of points of this folded set is determined, followed by the application of the inverses of each of the previously used powers of the symmetry operator to this average, effectively "unfolding" the folded point set into a set having the exact symmetry corresponding to the given symmetry operator $\mathbf{R}$. Either the size of the folded set or the deviation between the original set and the final "unfolded" set can be used as a measure of deviation from the exact symmetry.

For crisp continuum sets and fuzzy sets, the crisp and fuzzy versions of the SNSM metric, as well as the crisp and fuzzy versions of the Hausdorff metric, provide generalizations of the ZPA approach. The generalization we shall describe requires a computational technique applicable for generating a "crisp average" A_{crav} of a family

$$F_A = \{A_1, A_2, \ldots, A_m\} \tag{215}$$

of m crisp closed and bounded subsets A_i of an n-dimensional Euclidean space X.

To this end, we shall identify each point x of a set A_i by suitable hyperpolar coordinates of one radial coordinate and $n - 1$ angle coordinates, where the origin is attached to a specified point c of set A_i. Here n is the dimension of the underlying space X. In the most common, three-dimensional case, which is important in chemistry, the usual polar coordinates r, ϕ, and θ can be used, with respect to the center c of each set A_i and with reference to Cartesian coordinate axes defined parallel to axes of a Cartesian coordinate system given in three-dimensional space.

For a given set A_i, its convex hull C_i is unique. In the three-dimensional case, for each choice of the (ϕ, θ) pair, a unique line segment $q_i(\phi, \theta)$ is specified that connects a selected reference point (e.g., the center of A_i)

$$c_i \in C_i \tag{216}$$

with the unique point $y_i(\phi, \theta)$ of the following properties:

(1) $y_i(\phi, \theta) \in C_i$
(2) $y_i(\phi, \theta)$ has polar angle coordinate (ϕ, θ)
(3) $y_i(\phi, \theta)$ has the longest distance from center c_i

$$y_i(\phi, \theta) \in C_i: d(c_i, y_i(\phi, \theta))$$
$$= \sup\{d(c_i, x): x = x(\phi, \theta), x \in C_i\}. \tag{217}$$

It is possible that along this line segment $q_i(\phi, \theta)$ some points do not belong to set A_i. Nevertheless, we shall use a formal path parameterization along the line $q_i(\phi, \theta)$ that varies only within set A_i and ignores the possible gaps or any missing subsegments at the ends of the line segment.

Consider a parameterization for the formal path $q_i(u, (\phi, \theta))$, taken for each fixed pair (ϕ, θ) as a mapping from the unit interval $I = [0, 1]$ to underlying space X, where the image of this mapping is the line segment $q_i(\phi, \theta)$ and the parametrization by u is proportional to the arc length along the path $q_i(u, (\phi, \theta))$:

$$q_i(u, (\phi, \theta)): I = [0, 1] \to q_i(\phi, \theta) \subset X; \tag{218}$$

$$q_i(0, (\phi, \theta)) = c_i \tag{219}$$

and

$$q_i(1, (\phi, \theta)) = y_i(\phi, \theta). \tag{220}$$

Using the membership function $\mu_{A_i}(x)$ of crisp set A_i, we define a new parameter $t(w, (\phi, \theta))$:

$$t_i(w, (\phi, \theta)) = \int_{[0, w]} \mu_{A_i}(q_i(z, (\phi, \theta))) \, dz, \tag{221}$$

where integration is for the subset $[0, w]$ of the unit interval $I = [0, 1]$. Informally, the new parameter $t_i(w, (\phi, \theta))$ "leaves out the gaps" along line $q_i(\phi, \theta)$, where a subinterval of $q_i(\phi, \theta)$ lies outside of set A_i. Scaling $t_i(w, (\phi, \theta))$ by the effective length $t_i(1, (\phi, \theta))$ of line segment $q_i(\phi, \theta)$

within set A_i, a function $v_i(w,(\phi,\theta))$ is defined thus:

$$v_i(w,(\phi,\theta)) = t_i(w,(\phi,\theta))/t_i(1,(\phi,\theta)). \tag{222}$$

A possibly discontinuous path $p_i(u,(\phi,\theta))$ along segment $q_i(\phi,\theta)$ is defined, involving only those points of $q_i(\phi,\theta)$ which fall within set A_i:

$$p_i(u,(\phi,\theta)): I = [0,1] \to q_i(\phi,\theta) \cap A_i \subset X, \tag{223}$$

$$p_i(u,(\phi,\theta)) = q(v_i^{-1}(u,(\phi,\theta))). \tag{224}$$

Using this formalism, any point

$$x \in q_i(\phi,\theta) \cap A_i \tag{225}$$

can be specified uniquely by a set of three polar coordinates (u,ϕ,θ) and the serial index i of set A_i:

$$x = x(i,u,\phi,\theta). \tag{226}$$

As a consequence of the definition of parameter u, for every choice of values (i,u,ϕ,θ), there exists precisely one point $x(i,u,\phi,\theta)$ that must be an element of set A_i. Conversely, each point of each set A_i can be given in the form $x = x(i,u,\phi,\theta)$. Consequently, this approach provides a unified system of parametrization for points of each member of the family $A_1, A_2, \ldots, A_m$ of sets.

For a given choice of local origin points $c_i \in A_i$ and local Cartesian coordinate axes used to define polar angles ϕ,θ for each set A_i, the crisp average $A_{\text{crav}}(F_A)$ of sets $A_1, A_2, \ldots, A_m$ is defined by

$$A_{\text{crav}}(F_A) = \bigcup_{u,\phi,\theta} (1/m) \sum_{i=1,m} x(i,u,\phi,\theta), \tag{227}$$

where in the summation we consider the vector representation of each point $x(i,u,\phi,\theta)$.

The concept of crisp average of sets is the basis of a simple generalization of the ZPA approach from finite, discrete point sets to continua, described as follows.

Consider a set A and a (possibly approximate) symmetry element R, where the associated symmetry operator $\mathbf{R}$ leaves at least one point of the convex hull C of set A invariant. We assume that a reference point $c \in C$, a fixed point of $\mathbf{R}$, and a local Cartesian coordinate system of origin c are specified, where the coordinate axes are oriented according to the usual conventions with respect to the symmetry operator $\mathbf{R}$. For example, if R is a C_7 rotation axis, then the z axis of the local Cartesian system is chosen to coincide with this C_7 axis, whereas if R is a reflection plane, then the z axis may be chosen perpendicular to this plane.

Take the powers $\mathbf{R}^0 = \mathbf{E}, \mathbf{R}, \mathbf{R}^2, \ldots, \mathbf{R}^{m-1}$ of the symmetry operator $\mathbf{R}$, where m is the smallest positive integer that satisfies the condition $\mathbf{R}^m = \mathbf{E}$

of Eq. (159). Partition the Euclidean space X into m segments $X_0, X_1, \ldots, X_{m-1}$, where

$$X = \bigcup_{i=0, m-1} X_i \qquad (228)$$

and

$$\mathbf{R} X_i = X_{(i+1) \bmod m}. \qquad (229)$$

For example, if one considers the three-dimensional case and R is a reflection plane, then $m = 2$ and the two half spaces with boundary plane the reflection plane R fulfill the conditions; if R is a C_k rotation axis or an S_k axis, then $m = k$ and each segment can be taken as a wedge of the edge C_k or the S_k axis and of wedge angle $2\pi/k$; if R is a point of inversion i, then $m = 2$ and the two half spaces with boundary plane the (x, y) plane fulfill the conditions.

We shall use the notation P for the specification of the actual convention used in the positioning of R with respect to crisp continuum set A and for the partitioning $X_0, X_1, \ldots, X_{m-1}$ of the space X.

The ZPA folding–unfolding operation can be generalized for the crisp continuum set A, symmetry operator $\mathbf{R}$, and the given partitioning P of X, using the concept of crisp average of sets as follows. By taking segments $A_0, A_1, A_2, \ldots, A_{m-1}$ of set A using the definition

$$A_i = A \cap X_i, \qquad (230)$$

the following sets are generated:

$$A_0, \mathbf{R}^{m-1} A_1, \mathbf{R}^{m-2} A_2, \ldots, \mathbf{R}^{m-j} A_j, \ldots, \mathbf{R} A_{m-1}.$$

Using the notation

$$B_j = \mathbf{R}^{m-j} A_j, \qquad (231)$$

the B_j sets are the "folded" versions of subsets A_j of set A, according to the various powers of the (possibly only approximate) symmetry operator $\mathbf{R}$, where the "folding" terminology of Zabrodsky, Peleg, and Avnir is applied to the case of continuum set A.

The folded set A_{fold}, defined as the crisp average of three B_j sets, can be generated using the technique described in the first part of this section:

$$A_{\mathrm{fold}} = A_{\mathrm{crav}}(F_A) = \bigcup_{u, \phi, \theta} (1/m) \sum_{i=0, m-1} x(i, u, \phi, \theta). \qquad (232)$$

The "unfolding" of this crisp average $A_{crav}(F_A)$ using various powers of symmetry operator $\mathbf{R}$ leads to the following sets: $A_{fold}, \mathbf{R}^{1-m}A_{fold}, \mathbf{R}^{2-m}A_{fold}, \ldots, \mathbf{R}^{j-m}A_{fold}, \ldots, \mathbf{R}^{-1}A_{fold}$.

Take the union of all these sets as the folded–unfolded set $A_{f, uf, R, P}$ of set A according to symmetry element R and partitioning P:

$$A_{f, uf, R, P} = \bigcup_{i = 0, m-1} \mathbf{R}^{-i}A_{fold}. \tag{233}$$

The folded–unfolded set $A_{f, uf, R, P}$ of set A is evidently an R set, that is, $A_{f, uf, R, P}$ has exact R symmetry.

The dissimilarity of A and $A_{f, uf, R, P}$ provides a symmetry deficiency measure analogous to the ZPA continuous symmetry measure of discrete point sets. As a dissimilarity measure, both the SNDSM metric and the Hausdorff metric, or any other dissimilarity measure suitable for continua, are applicable.

For the SNDSM metric, the symmetry deficiency is taken as $d_{sZPA}(A, A_{f, uf, R, P})$, describing the SNDSM "shape distance" between the crisp continuum set A and the fully R-symmetric "folded–unfolded" set $A_{f, uf, R, P}$ of set A. This measure of symmetry deficiency is dependent on the positioning P of R with respect to set A and on the choice of an associated partitioning of A. An SNDSM measure for the symmetry deficiency of set A is obtained that is independent of positioning and partitioning if the infimum of $d_{sZPA}(A, A_{f, uf, R, P})$ is generated over all positionings and partitionings fulfilling the conditions laid out in their respective definitions:

$$d_{sZPA}(A, A_{f, uf, R}) = \inf_P \{ d_{sZPA}(A, A_{f, uf, R, P}) \}. \tag{234}$$

Using the Hausdorff metric for the dissimilarity of sets A and $A_{f, uf, R, P}$, one obtains another generalization of the ZPA continuous symmetry measure of discrete point sets to crisp continuum sets, leading to a new symmetry deficiency measure $h_{ZPA}(A, A_{f, uf, R, P})$ that is a valid measure of the symmetry aspect R for crisp set A.

The actual value of this Hausdorff measure of symmetry deficiency also depends on the positioning P of R with respect to A as well as on the choice of an associated partitioning of A.

Another measure for the symmetry deficiency of set A, independent of positioning and partitioning, is given by the infimum of $h_{ZPA}(A, A_{f, uf, R, P})$ taken over all the allowed positionings and partitionings. This measure, $h_{ZPA}(A, A_{f, uf, R})$, is defined as

$$h_{ZPA}(A, A_{f, uf, R}) = \inf_P \{ h_{ZPA}(A, A_{f, uf, R, P}) \}. \tag{235}$$

These symmetry deficiency measures and others employing different general dissimilarity measures exploit the advantages of the elegance of the ZPA approach. By providing an extension of the methodology to crisp continuum sets, the folding–unfolding approach becomes applicable for continuous molecular isodensity surfaces (MIDCOs).

XIV. FUZZY SET GENERALIZATIONS OF ZPA FOLDING–UNFOLDING CONTINUOUS SYMMETRY MEASURES BASED ON THE FUZZY FSNDSM METRIC AND FUZZY HAUSDORFF-TYPE METRICS

The fuzzy average F_{fav} of a family of crisp or fuzzy sets $F_1, F_2, \ldots, F_m$ has a much simpler construction[29] than the average of crisp sets used in Section XIII and that also simplifies the extension of ZPA-type folding–unfolding continuous symmetry measures to both crisp continua and fuzzy sets, using dissimilarity metrics designed for fuzzy sets and used as the actual measures of symmetry deficiency. Here we shall follow the technique[29] based on the fuzzy membership function $\mu_{F_{\text{fav}}}(x)$ of the fuzzy average F_{fav}, as defined earlier by Eq. (199).

Consider a crisp or fuzzy subset A of the Euclidean space X, a (possibly approximate) symmetry element R, and the associated symmetry operator $\mathbf{R}$. A fixed point of $\mathbf{R}$ is chosen as a reference point $c \in X$, and a local Cartesian coordinate system of origin c is specified, with coordinate axes oriented according to the usual conventions with respect to the symmetry operator $\mathbf{R}$, as described for crisp sets in Section XIII.

Choose m as the smallest positive integer that satisfies the condition $\mathbf{R}^m = \mathbf{E}$ of Eq. (171) and take the powers $\mathbf{R}^0 = \mathbf{E}, \mathbf{R}, \mathbf{R}^2, \ldots, \mathbf{R}^{m-1}$ of the symmetry operator $\mathbf{R}$. Following the technique described in Section XIII, partition the Euclidean space X into m segments $X_0, X_1, \ldots, X_{m-1}$, where the union of these segments generates the space X,

$$X = \bigcup_{j=0, m-1} X_j, \tag{236}$$

and where segments of subsequent indices are related to one another by the symmetry operation

$$\mathbf{R} X_j = X_{(j+1) \bmod m}. \tag{237}$$

The interpretation of these segments and the notation P for the convention used for the positioning of R with respect to set A and for the partitioning $X_0, X_1, \ldots, X_{m-1}$ of the space X are the same as those used for crisp sets, described in Section XIII.

The segment A_j of the crisp or fuzzy set A is defined as

$$A_j = A \cap X_j, \tag{238}$$

and the "folded" version of the jth segment A_j of the crisp or fuzzy set A, according to the $(m-j)$th power $\mathbf{R}^{m-j}$ of the (possibly only approximate) symmetry operator $\mathbf{R}$ is denoted by

$$B_j = \mathbf{R}^{m-j}A_j. \tag{239}$$

For the family S_A of segments $A_0, A_1, A_2, \ldots, A_{m-1}$, the corresponding folded sets $A_0, \mathbf{R}^{m-1}A_1, \mathbf{R}^{m-2}A_2, \ldots, \mathbf{R}^{m-j}A_j, \ldots, \mathbf{R}A_{m-1}$ are obtained, that is, the family S_B of sets $B_0, B_1, B_2, \ldots, B_{m-1}$ is generated.

The fuzzy folded set $A_{f\text{fold}}$ is the fuzzy average $S_{B\text{fav}}$ of these B_j sets, defined by the fuzzy membership function $\mu_{S_{B\text{fav}}}(x)$ as follows:

$$\mu_{A_{f\text{fold}}}(x) = \mu_{S_{B\text{fav}}}(x) = \left[\sum_{k=0,m-1} \mu_{B_k}(x) \right] \bigg/ m. \tag{240}$$

The "unfolding" of the fuzzy folded set $A_{f\text{fold}}$ is obtained using the appropriate inverse powers of symmetry operator $\mathbf{R}$ generating the following sets: $A_{f\text{fold}}, \mathbf{R}^{1-m}A_{f\text{fold}}, \mathbf{R}^{2-m}A_{f\text{fold}}, \ldots, \mathbf{R}^{j-m}A_{f\text{fold}}, \ldots, \mathbf{R}^{-1}A_{f\text{fold}}$. The fuzzy union of all these sets is the folded–unfolded fuzzy set $A_{\text{ff,uf},R,P}$ of crisp or fuzzy set A, generated according to symmetry element R and the actual partitioning P:

$$A_{\text{ff,uf},R,P} = \bigcup_{j=0,m-1} \mathbf{R}^{-j}A_{f\text{fold}}. \tag{241}$$

The folded–unfolded set $A_{\text{ff,uf},R,P}$ of crisp or fuzzy set A is a fuzzy R set, by construction.

Fuzzy dissimilarity measures, such as the fuzzy FSNDSM metric $\text{fs}(A, B)$, and any one of the fuzzy Hausdorff-type dissimilarity metrics, for example, $f(A, B)$, can be applied to the pair of set A and the folded–unfolded set $A_{\text{ff,uf},R,P}$. These fuzzy dissimilarity measures generate fuzzy symmetry deficiency measures analogous to the ZPA continuous symmetry measure of discrete point sets.

If the FSNDSM metric $\text{fs}(A, B)$ is selected, then the corresponding symmetry deficiency measure is defined as $d_{\text{fs}}(A, A_{\text{ff,uf},R,P})$. The measure $d_{\text{fs}}(A, A_{\text{ff,uf},R,P})$ describes the fuzzy FSNDSM "shape distance" between the crisp or fuzzy set A and the fully R-symmetric, fuzzy, folded–unfolded set $A_{\text{ff,uf},R,P}$ of the original set A. By analogy with the case of crisp sets discussed in Section XIII, the $d_{\text{fs}}(A, A_{\text{ff,uf},R,P})$ symmetry deficiency measure is P-dependent, that is, it depends on the positioning P of R with respect to A and on the choice of the corresponding partitioning of the underlying Euclidean space X and that of the original crisp or fuzzy set A.

By taking the infimum for all the allowed choices of P, a symmetry deficiency measure of crisp or fuzzy set A is obtained that is independent of positioning and partitioning. The corresponding $d_{\text{fs}}(A, A_{\text{ff,uf},R})$

measure is defined as the infimum of $d_{fs}(A, A_{ff, uf, R, P})$ generated over all the allowed positionings and partitionings:

$$d_{fs}(A, A_{ff, uf, R}) = \inf_{P}\{d_{fs}(A, A_{ff, uf, R, P})\}. \qquad (242)$$

Using any one of the versions of the fuzzy Hausdorff-type metrics for the dissimilarity of sets A and $A_{ff, uf, R, P}$, for example, the "commitment weighted" fuzzy Hausdorff-type dissimilarity metric $f(A, B)$, one obtains another generalization of the ZPA continuous symmetry measure of discrete point sets to crisp or fuzzy sets. The corresponding symmetry deficiency measure $f(A, A_{ff, uf, R, P})$ provides a measure for the symmetry aspect R for crisp or fuzzy set A, with reference to the given positioning P of R with respect to A and to the choice of the associated partitioning of A.

For a symmetry deficiency measure of set A, independent of positioning and partitioning, the infimum of $f(A, A_{ff, uf, R, P})$ can be taken over all the allowed positionings and partitionings P. This measure, $f(A, A_{ff, uf, R})$, is defined as

$$f(A, A_{ff, uf, R}) = \inf_{P}\{f(A, A_{ff, uf, R, P})\}. \qquad (243)$$

Following the principles of the ZPA approach, these symmetry deficiency measures are generalizations of the folding–unfolding approach, equally applicable to crisp continuum sets and fuzzy sets, for example, to entire electron density distributions of molecules and various molecular fragments representing fuzzy functional groups.

XV. THE CHIRAL RACEMIZATION PATH PROBLEM IN n-DIMENSIONS AND MISLOW'S LABEL PARADOX

Interconversion paths between chiral mirror images, often referred to as racemization paths, have fascinated chemists for some time.[76] An interesting problem has been studied by Mislow, Buda, and Poggi-Corradini,[56,77,78] who investigated the conditions for the existence of fully chiral paths interconverting chiral mirror images, that is, rearrangements along which no achiral arrangements occur, yet the initial and final arrangements are chiral mirror images of each other. They developed an elegant approach for the special case of unlabeled three-dimensional tetrahedra, and proved the existence of fully chiral paths interconverting chiral mirror images of tetrahedra.[56,77] More recently, Mislow also demonstrated[78] that for fully labeled, three-dimensional, chiral tetrahedra, no path that interconverts chiral mirror images can preserve chirality. In an independent study,[79,80] a general proof was presented for all n-chiral simplices of distinguishable vertices in any finite dimension n, stating that no chirality-preserving path exists that interconverts mirror images of n-chiral simplices.

In three dimensions, chiral tetrahedra with unlabeled vertices are interconvertible to their mirror images along paths that preserve chirality. Usually, the equivalence of labels on the vertices of a chiral polyhedron increases the chances that a motion converts the chiral polyhedron into an achiral one. Conversely, distinct labels of vertices usually increase the chances of obtaining a chiral polyhedron. Based on these general trends, it is somewhat unexpected that chiral tetrahedra with equivalent vertex labels can get converted into their mirror images without encountering achiral arrangements, whereas chiral tetrahedra with distinct vertex labels must become achiral when converted into their mirror images.

These counterintuitive properties of "racemization paths" of three-dimensional labeled and unlabeled chiral tetrahedra, noted by Mislow,[78] are referred to as Mislow's label paradox. More recently, it has been shown[80,81] that Mislow's label paradox is general for n-chiral simplices in all finite dimensions n, and sufficient and necessary partial vertex labeling conditions have been given[80,81] for chirality-preserving interconversion paths of mirror images of chiral n-dimensional simplexes.

Each point a_k of a point set $a_1, a_2, a_3, \ldots, a_{m-1}, a_m$, of m points can be regarded to carry a distinguishing label, for example, according to the convention used in earlier works,[18,79-81] the serial index k may serve in an additional role as a label. Alternatively, if labeling is to be emphasized, one may write $a_k(k)$ for point a_k. The distinguishing label may differ from the serial index; if two points a_i and a_j are regarded equivalent, this may be expressed by specifying a symmetry operation or a permutation operation that interrelates them, or their equivalence can be emphasized by displaying the same label for both, for example, $a_i(k)$ and $a_j(k)$.

Two relevant theorems were proven in an earlier study on n-dimensional transformations preserving or abandoning n-chirality.[79] Here only the main results (Theorems 3 and 10 of the study[79]) are presented, using the explicit labeling notation $a_i(k)$; for the proofs, see the original reference.[79]

(1) No n-chiral simplex

$$S = \{a_1, a_2, a_3, \ldots, a_n, a_{n+1}\} \qquad (244)$$

of $n + 1$ points $a_1(1), a_2(2), a_3(3), \ldots, a_n(n), a_{n+1}(n+1)$ of distinct labels can be deformed continuously within E^n into its mirror image

$$S^\diamond = \{a_1^\diamond, a_2^\diamond, a_3^\diamond, \ldots, a_n^\diamond, a_{n+1}^\diamond\} \qquad (245)$$

without passing through an n-achiral intermediate.

(2) Any n-chiral point set

$$S = \{a_1, a_2, a_3, \ldots, a_n, a_{n+1}, \ldots, a_{m-1}, a_m\} \qquad (246)$$

of $m \geq n + 2$ points

$$a_1(k_1), a_2(k_2), a_3(k_3), \ldots, a_n(k_n), a_{n+1}(k_{n+1}), \ldots,$$
$$a_{m-1}(k_{m-1}), a_m(k_m),$$

can be deformed continuously within E^n, $n \geq 2$, into its mirror image

$$S^{\diamond} = \{a_1^{\diamond}, a_2^{\diamond}, a_3^{\diamond}, \ldots, a_n^{\diamond}, a_{n+1}^{\diamond}, \ldots, a_{m-1}^{\diamond} a_m^{\diamond}\} \qquad (247)$$

without passing through any n-chiral arrangement, where

$$k_i = k_j \qquad (248)$$

is possible; that is, not necessarily all points $a_i(k_i)$ have distinct labels as some may be equivalent and interrelated by symmetry or by permutational assignment.

A minimal relaxation of the unique labeling condition of statement (1) is sufficient for the existence of a chirality-preserving racemization path. Following the general proof given for the n-dimensional case,[80, 81] this is verified subsequently.

(3) Any n-chiral simplex

$$S = \{a_1, a_2, a_3, \ldots, a_n, a_{n+1}\} \qquad (249)$$

of $n + 1$ points $a_1(1), a_2(2), a_3(3), \ldots, a_n(n), a_{n+1}(n)$, where the first $n - 1$ points have distinct labels and where points $a_n(n)$ and $a_{n+1}(n)$ carry the same label and are regarded equivalent, can be deformed continuously within E^n into its mirror image

$$S^{\diamond} = \{a_1^{\diamond}, a_2^{\diamond}, a_3^{\diamond}, \ldots, a_n^{\diamond}, a_{n+1}^{\diamond}\} \qquad (250)$$

without passing through an n-achiral intermediate.

Proof of statement (3). Point set S represents an n-chiral simplex in E^n; consequently, any subset of k vertices of S defines a unique $(k-1)$-dimensional hyperplane that contains these k points. Take the $(n-2)$-dimensional, unique hyperplane Q that contains the first $n-1$, uniquely labeled vertices, and take the unique $(n-1)$-dimensional hyperplane A that contains the first n vertices of S. Hyperplane A is a maximum achiral subset of S. Furthermore,

$$A \supset Q. \qquad (251)$$

Using hyperplane Q as an $(n-2)$-dimensional rotation axis in the n-dimensional space E^n, rotate plane A along axis Q into the unique new position A', where point a_{n+1} falls within the rotated hyperplane A'. The angle of this rotation is denoted by α. Rotate the original hyperplane A along the axis Q by $\alpha/2$ in the same sense as in the previous rotation. The resulting new hyperplane A'' contains the first $n-1$ vertices of S, but A'' contains neither a_n nor a_{n+1}. Hyperplane A'' must contain some point with different distances to points a_n and a_{n+1}; otherwise, S could not be n-chiral, since then A'' could serve as a reflection hyperplane that reflects S onto itself, with a_n and a_{n+1} assigned to each other. Consider A'' as a reflection plane and take the corresponding reflected image

$$S^\diamond = \left\{ a_1^\diamond, a_2^\diamond, a_3^\diamond, \ldots, a_{n-1}^\diamond a_n^\diamond, a_{n+1}^\diamond \right\} \tag{252}$$

of point set S. Since A'' reflects Q onto itself,

$$a_1^\diamond = a_i, \qquad i = 1, \ldots, n-1, \tag{253}$$

holds for the first $n-1$ uniquely labeled points, whereas the points

$$a_n^\diamond \in A' \tag{254}$$

and

$$a_{n+1}^\diamond \in A \tag{255}$$

cannot belong to the vertex set S.

Select the unique point $q_n \in Q$ with the shortest distance to point $a_n \in A$ and the unique point $q_{n+1} \in Q$ with the shortest distance to point $a_{n+1} \in A'$. Denote the respective distances by d_n and d_{n+1}; neither d_n nor d_{n+1} can be zero, that is, $d_n > 0$ and $d_{n+1} > 0$, since point set S is n-chiral. Points q_n and q_{n+1} are also the unique points of set Q with the shortest distances, d_n and d_{n+1}, from points $a_n^\diamond \in A'$ and $a_{n+1}^\diamond \in A$, respectively, since hyperplanes A and A' are related by reflection plane A''.

Two unit vector $\mathbf{v}_n$ and $\mathbf{v}_{n+1}$ are defined, pointing from q_n and a_n and from q_{n+1} to a_{n+1}, respectively. Using distances d_n and d_{n+1}, two points are selected,

$$b_n = q_n + d_{n+1}\mathbf{v}_n \tag{256}$$

and

$$b_{n+1} = q_{n+1} + wd_n\mathbf{v}_{n+1}, \tag{257}$$

where the weight factor w is taken as $w = 2$ if $d_n = d_{n+1}$ and $w = 1$ otherwise.

As follows from these choices,

$$b_n \in A \tag{258}$$

and

$$b_{n+1} \in A'. \tag{259}$$

Points b_n, a_{n+1}, and $a_{n+1}^{\diamond}$ are at the same d_{n+1} distance from set Q, whereas points b_{n+1}, a_n, and $a_n^{\diamond}$, are at the same d_n distance from set Q, except in the special case of $w = 2$, where b_{n+1} has the distance $2d_n$ from hyperplane Q.

A geometrical path p_n is constructed within hyperplane A from the two straight line segments $[a_n, b_n]$ and $[b_n, a_{n+1}^{\diamond}]$, where line segment $[a_n, b_n]$ is the degenerate, single point segment if $d_n = d_{n+1}$. Another geometrical path p_{n+1} is constructed within the other hyperplane A', using the two straight line segments $[a_{n+1}, b_{n+1}]$ and $[b_{n+1}, a_n^{\diamond}]$.

Continuous parametrizations of these paths are given as

$$p_n(u): [0,1] \to A, \qquad \text{where} \quad p_n(0) = a_n, \, p_n(1) = a_{n+1}^{\diamond}, \qquad (260)$$

and

$$p_{n+1}(u): [0,1] \to A', \quad \text{where} \quad p_{n+1}(0) = a_{n+1}, \, p_{n+1}(1) = a_n^{\diamond}, \quad (261)$$

respectively, where for both $p_n(u)$ and $p_{n+1}(u)$, the parameter u is proportional to the arc length in space E^n.

Using hyperplane A'' as a reflection plane, the reflected image of the point $p_n(0) = a_n$ is $p_{n+1}(1) = a_n^{\diamond}$ and the reflected image of point $p_n(1) = a_{n+1}^{\diamond}$ is $p_{n+1}(0) = a_{n+1}$. Note that for each pair, the A''-reflected mirror images belong to *different* values of the parameter u. Also note that no other point of either path is an A''-reflected mirror image of any point of the other path.

Deform the point set S into its mirror image $S^{\diamond}$ by a concerted motion of simultaneously moving points a_n and a_{n+1} along paths p_n and p_{n+1} to their new locations $a_n^{\diamond}$ and $a_{n+1}^{\diamond}$, respectively, by varying u as a common parameter for both paths $p_n(u)$ and $p_{n+1}(u)$. The choice of weight factor w ensures that in the course of the actual motions of points a_n and a_{n+1} of equivalent labels, at no intermediate locations along their paths will the two moving points a_n and a_{n+1} become A''-reflected mirror images of each other. The first $n-1$ points of S have unique labels; consequently, any symmetry operation that could imply achirality for any superset of S containing the first $n-1$ points with unique labels must leave these first $n-1$ points invariant. Hence, no such potential symmetry operator can interrelate any point from the set of the first $n-1$ vertices with any one of the remaining two points a_n and a_{n+1}. Consequently, the only chance of achirality for an intermediate arrangement along an interconversion path must involve a mirror plane leaving the first $n-1$ points invariant and interconverting some translated versions of points a_n and a_{n+1}. However, we have shown that there exists a concerted motion of points a_n and a_{n+1} that avoids these two points becoming mirror images by any reflection plane containing hyperplane Q. In this concerted motion, the n-chirality of the point set is preserved throughout the entire deformation of the n-dimensional simplex S into its mirror image $S^{\diamond}$. Q.E.D.

This result is the basis of the n-dimensional generalization of Mislow's label paradox. The key observation is that after completing the deformation of S, in the mirror image $S^\diamond$ the roles of the two equivalent points have been switched:

$$p_n(0) = a_n, \tag{262}$$

$$p_n(1) = a_{n+1}^\diamond, \tag{263}$$

whereas

$$p_{n+1}(0) = a_{n+1} \tag{264}$$

and

$$p_{n+1}(1) = a_n^\diamond. \tag{265}$$

This role switch is possible only for points of equivalent labels; we know from the general, n-dimensional result (1) that for any chiral arrangement of $n+1$ nonequivalent points, no achirality-avoiding interconversion path is possible. Whereas equivalence of points increases the probability of achiral arrangements, it also allows role switches, and in the case of simplices, the possibility of a role switch is more significant. In fact, a single role switch is sufficient and there is no need to modify the unique labels of the first $n-1$ points.

A minimal relaxation of the unique labeling constraint, that is, having precisely two labels equivalent and preserving uniqueness for all other labels, is *sufficient* for the existence of chirality-preserving interconversion paths between mirror images of any n-chiral simplex. Since unique labeling excludes the possibility of chirality-preserving interconversion paths, the minimal relaxation of the unique labeling constraint is also a *necessary* condition. Hence, a sufficient and necessary labeling condition is obtained for the general, n-dimensional case, providing a generalization and an explanation of Mislow's label paradox for any finite dimension n, $n \geq 3$.

XVI. SOME DEVELOPMENTS IN THE COMPUTATION OF PROPERTIES OF FUZZY ELECTRON DENSITIES

The computation of fuzzy electron density fragments and the construction of ab initio quality macromolecular electron densities is a novel approach to the quantum-chemical study and detailed modeling of large molecules. The *additive fuzzy density fragmentation* (AFDF) *scheme* of Mezey, described in a general form elsewhere,[36,37] is employed in the molecular electron density lego assembler (MEDLA) technique of Walker and Mezey.[41–46] The simplest version of the AFDF approach—the Mulliken–Mezey AFDF scheme—is suitable for the generation of ab initio quality MEDLA electron densities for virtually any macromolecule, protein, or supramolecular structure.

An example of the MEDLA electron density of bovine insulin protein[44] is shown in Fig. 1, where the fuzzy body of the electronic density cloud is

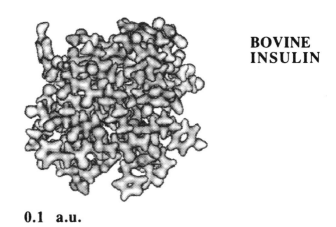

BOVINE
INSULIN

0.1 a.u.

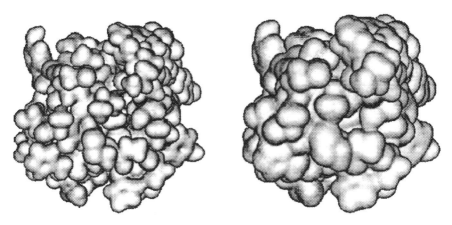

0.01 a.u.   0.001 a.u.

FIGURE 1   The fuzzy body of the electron density of a bovine insulin molecule is represented by three molecular isodensity contour surfaces (MIDCOs), for the density thresholds of 0.1, 0.01, and 0.001 a.u. (atomic unit), respectively, as computed using the MEDLA method. Bovine insulin was among the proteins selected for the first ab initio quality electron density computations for macromolecules.[44]

represented by three molecular isodensity contours (MIDCOs). The density contour surface of high electron density threshold [0.1 atomic unit (a.u)], shown on the top of the figure, fits within the contour surface of threshold value 0.01 a.u (bottom left), which, in turn, fits within the contour surface of threshold 0.001 a.u. (bottom right).

Since the introduction of the MEDLA method, three more recent developments have extended the applications of the Mulliken–Mezey and the more general AFDF schemes. These developments, all utilizing various representations of additive fuzzy subsets of molecular electron densities, are the

(1) Adjustable local density assembler (ALDA) method.[51,52,82]
(2) Adjustable density matrix assembler (ADMA) method.[49-52]
(3) Additive fuzzy density fragment force (AFDFF) method.[39,83]

Each of these methods is based on the AFDF approach. Within the framework of the conventional Hartree–Fock–Roothaan–Hall self-consistent field linear combination of atomic orbitals (LCAO) ab initio representation of molecular wave functions built from molecular orbitals (MOs), the AFDF principle can be formulated using fragment density matrices. For a complete molecule M of some nuclear configuration K, using an atomic orbital (AO) basis of a set of n AOs $\varphi_i(\mathbf{r})$ ($i = 1, 2, \ldots, n$) the $n \times n$-dimensional density matrix $\mathbf{P}$ can be determined using the coefficients of AOs in the occupied MOs. The electronic density $\rho(\mathbf{r})$ of the molecule M, a function of the three-dimensional position variable $\mathbf{r}$, can be written as

$$\rho(\mathbf{r}) = \sum_{i=1}^{n} \sum_{j=1}^{n} P_{ij} \varphi_i(\mathbf{r}) \varphi_j(\mathbf{r}). \tag{266}$$

Within the AFDF scheme, the family of nuclei of molecule M is divided into m mutually exclusive subfamilies, denoted by $f_1, f_2, \ldots, f_k, \ldots, f_m$, providing the atomic orbital reference locations for m density fragments $F_1, F_2, \ldots, F_k, \ldots, F_m$, corresponding to m fragment density functions $\rho^1(\mathbf{r}), \rho^2(\mathbf{r}), \ldots, \rho^k(\mathbf{r}), \ldots, \rho^m(\mathbf{r})$, respectively.

The $\rho^k(\mathbf{r})$ additive, fuzzy fragment density functions are defined in terms of the complete AO set of molecule M and a family of fragment density matrices $\mathbf{P}^1, \mathbf{P}^2, \ldots, \mathbf{P}^k, \ldots, \mathbf{P}^m$.

Although more advanced AFDF methods have also been introduced,[36,37] in most applications to date[41-46] the simplest, Mulliken–Mezey AFDF scheme has been used. This scheme can be regarded as analogous to Mulliken's population analysis technique[84,85] without integration. This AFDF technique (described in the following text) is suitable for the generation of ab initio quality electron densities for virtually any macromolecule, protein, and supramolecular structure.

The simplest, Mulliken–Mezey additive fragment density matrix, AFDM $\mathbf{P}^k$ of dimensions $n \times n$, is defined as follows:

$$
P_{ij}^k = \begin{cases}
P_{ij}, & \text{if both } \varphi_i(\mathbf{r}) \text{ and } \varphi_j(\mathbf{r}) \text{ are AOs centered on} \\
& \text{some nuclei of the } k\text{th fragment,} \\
0.5P_{ij}, & \text{if precisely one of } \varphi_i(\mathbf{r}) \text{ and } \varphi_j(\mathbf{r}) \text{ is centered on} \quad (267) \\
& \text{a nucleus of fragment } k, \\
0, & \text{otherwise.}
\end{cases}
$$

Note that both the density matrix $\mathbf{P}$ of the complete molecule M and the additive fragment density matrix AFDM $\mathbf{P}^k$ of the kth fragment involve the same, full set of AOs. Based on the fragment density matrix $\mathbf{P}^k$, the electron density $\rho^k(\mathbf{r})$ of Mezey's kth additive fuzzy density fragment is defined as

$$
\rho^k(\mathbf{r}) = \sum_{i=1}^{n} \sum_{j=1}^{n} P_{ij}^k \varphi_i(\mathbf{r}) \varphi_j(\mathbf{r}). \tag{268}
$$

Since the mutually exclusive nuclear families $f_1, f_2, \ldots, f_k, \ldots, f_m$ contain all the nuclei of molecule M, the sum of the fragment density matrices $\mathbf{P}^k$ is equal to the density matrix $\mathbf{P}$ of complete molecule M:

$$
P_{ij} = \sum_{k=1}^{m} P_{ij}^k \tag{269}
$$

and

$$
\mathbf{P} = \sum_{k=1}^{m} \mathbf{P}^k. \tag{270}
$$

Since the electron density depends linearly on the elements of the density matrices, exact additivity of the fragment matrix elements P_{ij}^k and that of the fragment density matrices $\mathbf{P}^k$ imply exact additivity of the fragment densities $\rho^k(\mathbf{r})$, that is, the sum of the fuzzy fragment densities $\rho^k(\mathbf{r})$ is exactly equal to the electron density $\rho(\mathbf{r})$ of the complete molecule M:

$$
\rho(\mathbf{r}) = \sum_{k=1}^{m} \rho^k(\mathbf{r}). \tag{271}
$$

Within any given ab initio Hartree–Fock LCAO method, the Mulliken–Mezey AFDF scheme is, indeed, exactly additive. This and the more advanced AFDF schemes[36,37] can be used to build ab initio quality electron densities for large molecules, based on precalculated fuzzy electron density fragments, where the local surroundings and interactions involving each fragment within a large enough neighborhood are accurately represented and incorporated in the density representation of the fragment. To build ab initio quality electron densities for a large "target" molecule from "custommade" fragments, these fragment densities are first generated from ab initio Hartree–Fock LCAO electron density calculations for smaller "parent" molecules $M_1, M_2, \ldots, M_k, \ldots, M_m$, each artificially distorted to match the local nuclear arrangement and surroundings in the target molecule.

The Mulliken–Mezey AFDF scheme (MM-AFDF) has been used to generate a numerical electron density fragment database[41] containing many "surroundings-dependent" versions of electron densities for various molecular fragments. The molecular electron density lego assembler (MEDLA) method of Walker and Mezey[41–46] is based on the MM-AFDF scheme, using a numerical electron density fragment databank—the MEDLA databank.[41]

According to detailed test calculations and using the standard 6-31G** basis set results as the benchmark, the MEDLA electron densities are more accurate than the standard 3-21G ab initio results and are virtually indistinguishable from the results of standard 6-31G** ab initio electron density calculations.[42–46]

The AFDF approach can be applied within an alternative framework for building macromolecular electron densities that do not rely on a numerical electron density database. Within the *adjustable local density assembler* (ALDA) method,[51,52,82] the AFDF approach is used to generate fragment density matrices AFDM. These fragment density matrices are obtained from standard ab initio calculations for appropriately distorted parent molecules $M_1, M_2, \ldots, M_k, \ldots, M_m$, each reproducing the local nuclear arrangement and surroundings of the fragment in the target molecule. The fragment density matrix elements for each AFDM $\mathbf{P}^k$, augmented with basis set and nuclear geometry information, are stored in an ALDA database. At this stage, no fragment densities are calculated. An ALDA database is much smaller than a MEDLA database and generates electron densities of comparable accuracy. Note that in a MEDLA database, the storage of numerical electron density values at several million grid points is required for each fragment.

Within the ALDA method,[51,52, 82] the actual fragment density contributions to the "target" macromolecule are computed only when the macromolecular electron density is assembled. This step requires the evaluation of the sum of Eq. (268) for each fragment AFDF $\rho^k(\mathbf{r})$, followed by the calculation of the sum in Eq. (271) for the target macromolecule M. Note that this process is quadratic in the number of AOs of the fragments; however, it is still linear in the number of fragments of the macromolecule. Since the fragment size is bounded by the size of the parent molecules, limited by the feasibility of standard ab initio calculations, the essential dependence of the ALDA method is *linear* in the macromolecular size.

The fragment densities are added up at *any* desired set of grid points **r**, as implied by Eq. (271), resulting in the ALDA electron density $\rho(\mathbf{r})$ of the target macromolecule M. The fragment densities in Eq. (271) converge to zero rapidly with the distance from the nearest nucleus of each nuclear family f_k. In practice, a cutoff is employed for the density contributions, implying a distance cutoff, analogous to the cutoff used for the MEDLA fragments. If for a given macromolecule M the same set of grid points and the same set of parent molecules $M_1, M_2, \ldots, M_k, \ldots, M_m$, are used in ALDA and MEDLA calculations, then the ALDA and MEDLA methods give exactly equivalent results. The ALDA method is slower than the MEDLA method; however, it avoids the high memory demand of the MEDLA database.

On the other hand, the ALDA method can easily generate macromolecular electron densities using any numerical grid, for example, one that involves a dense grid near the nuclei and lower grid density at the low electron density, peripheral regions of the macromolecule. This feature offers some advantages in various property calculations, such as the evaluation of electrostatic potentials.

Another advantage is the possibility of a simple, approximate readjustment of the electronic density for minor nuclear geometry variations.[51,52,82] Most of the electronic density follows any minor rearrangement of the nuclei, a feature well described within the LCAO approach, where the AOs are centered on the nuclei; hence, they also "move" if the nuclei move. If the nuclear rearrangement is small, then an acceptable approximation of the new electronic density can be obtained by using the original fragment density matrices AFDM $\mathbf{P}^k$ and the given set of AOs centered at the new nuclear locations. Whereas this is only a rather crude approximation, it reproduces the major part of the new electronic density as long as the nuclear rearrangement is very small.[51,52,82] Following this simple treatment, some of the fundamental conditions for density matrices are no longer guaranteed; for example, the idempotency condition for the density matrix of a reassembled parent molecule may be violated. However, for small geometry distortions, these violations are significant only if electron densities of high quality are needed. If high accuracy is required, then one is either forced to generate the actual fragment density matrix for the exact nuclear geometry required, by simply taking a new, distorted parent molecule, or one may apply some correction algorithm that reduces the error of the approximate electron density at the distorted nuclear geometry.

Löwdin's symmetric orthogonalization method[86–89] is an often employed technique for the generation of orthonormal molecular basis sets. Since within most LCAO MO methods, density matrices are determined by AO basis set coefficients, and idempotency of density matrices is a property easily controlled on an orthonormal basis, Löwdin's transformation is especially suitable for density matrix transformations. Similar methods are used in quantum crystallography,[90] motivated by studies on molecular wave functions and N-representable density matrices[91-119] that can be fitted to crystallographic diffraction data, to generate "experimental" density matrices for molecules based on experimental electronic densities.

Within the ALDA technique, if the change of the nuclear geometry K is small, it is often unwarranted to carry out an actual modification of the fragment density matrices.[51,52,82] However, if such modification is required, then a simple correction of the density matrix for the same set of AOs and for a slightly different nuclear geometry K' is based on the restoration of idempotency, using fractional powers of the overlap matrix **S**, employed in Löwdin's symmetric orthogonalization method.[86-89] This idempotency correction of the ALDA density matrices for distorted nuclear geometries can be viewed as "orthonormalization–deorthonormalization," employing two Löwdin-type transformations for two different nuclear geometries. Since the method involves transformed overlap matrices, the technique is referred to as the SALDA method.[110] Idempotency of the density matrix **P** at the new nuclear configuration K' can be restored by the application of Löwdin's symmetric $\mathbf{S}(K)^{1/2}$ transformation to the molecular density matrix $\mathbf{P}(K)$ obtained for the original nuclear configuration K, followed by a symmetric inverse transformation using $\mathbf{S}(K')^{-1/2}$, where $\mathbf{S}(K)$ and $\mathbf{S}(K')$ are the AO overlap matrices for nuclear configurations K and K', respectively.

The idempotency condition of the molecular density matrix $\mathbf{P}(K)$ for nuclear configuration K can be written as the relation

$$\mathbf{P}(K) * \mathbf{P}(K) = \mathbf{P}(K), \tag{272}$$

where the multiplication $*$ is interpreted in terms of a matrix product involving the overlap matrix $\mathbf{S}(K)$:

$$\mathbf{P}(K)\mathbf{S}(K)\mathbf{P}(K) = \mathbf{P}(K). \tag{273}$$

By applying Löwdin's transformation[86-89] to density matrix $\mathbf{P}(K)$, multiplication of Eq. (273) from both left and right by $\mathbf{S}(K)^{1/2}$ gives

$$\mathbf{S}(K)^{1/2}\mathbf{P}(K)\mathbf{S}(K)^{1/2}\mathbf{S}(K)^{1/2}\mathbf{P}(K)\mathbf{S}(K)^{1/2} = \mathbf{S}(K)^{1/2}\mathbf{P}(K)\mathbf{S}(K)^{1/2}, \tag{274}$$

implying that the product $\mathbf{S}(K)^{1/2}\mathbf{P}(K)\mathbf{S}(K)^{1/2}$ is idempotent with respect to ordinary matrix multiplication.

By generating the inverse of Löwdin's transformation using the appropriate power $\mathbf{S}(K')^{-1/2}$ of the new overlap matrix $\mathbf{S}(K')$ at the new nuclear configuration K', an improved, and *idempotent* approximation $\mathbf{P}(K',[K])$ to the density matrix $\mathbf{P}(K')$ at the new nuclear configuration K'

is obtained:

$$\mathbf{P}(K',[K]) = \mathbf{S}(K')^{-1/2}\mathbf{S}(K)^{1/2}\mathbf{P}(K)\mathbf{S}(K)^{1/2}\mathbf{S}(K')^{-1/2}. \quad (275)$$

Exact idempotency follows from

$$\mathbf{P}(K',[K]) * \mathbf{P}(K',[K])$$

$$= \mathbf{P}(K',[K])\mathbf{S}(K')\mathbf{P}(K',[K])$$

$$= \mathbf{S}(K')^{-1/2}\mathbf{S}(K)^{1/2}\mathbf{P}(K)\mathbf{S}(K)^{1/2}\mathbf{S}(K')^{-1/2}\mathbf{S}(K')\mathbf{S}(K')^{-1/2}$$

$$\times \mathbf{S}(K)^{1/2}\mathbf{P}(K)\mathbf{S}(K)^{1/2}\mathbf{S}(K')^{-1/2}$$

$$= \mathbf{S}(K')^{-1/2}\mathbf{S}(K)^{1/2}\mathbf{P}(K)\mathbf{S}(K)^{1/2}\mathbf{S}(K)^{1/2}\mathbf{P}(K)\mathbf{S}(K)^{1/2}\mathbf{S}(K')^{-1/2}$$

$$= \mathbf{S}(K')^{-1/2}\mathbf{S}(K)^{1/2}\mathbf{P}(K)\mathbf{S}(K)\mathbf{P}(K)\mathbf{S}(K)^{1/2}\mathbf{S}(K')^{-1/2}$$

$$= \mathbf{S}(K')^{-1/2}\mathbf{S}(K)^{1/2}\mathbf{P}(K)\mathbf{S}(K)^{1/2}\mathbf{S}(K')^{-1/2}$$

$$= \mathbf{P}(K',[K]), \quad (276)$$

where the identities

$$\mathbf{S}(K')^{-1/2}\mathbf{S}(K')\mathbf{S}(K')^{-1/2} = \mathbf{I} \quad (277)$$

and

$$\mathbf{P}(K)\mathbf{S}(K)^{1/2}\mathbf{S}(K)^{1/2}\mathbf{P}(K) = \mathbf{P}(K)\mathbf{S}(K)\mathbf{P}(K) = \mathbf{P}(K) \quad (278)$$

have been used.

Within the framework of the SALDA method,[110] restoring idempotency of density matrices at displaced nuclear geometries involves the parent molecules, where for small geometry variations a new, approximate, but exactly idempotent density matrix, as well as the associated, improved fragment density matrices can be computed using the above method.

The same Löwdin-type orthonormalization–deorthonormalization method can also be applied for the restoration of the idempotency of geometry-dependent macromolecular density matrices within the context of the recently introduced ADMA macromolecular density matrix technique, reviewed in the following text.

The Mulliken–Mezey AFDF scheme and the more general AFDF schemes[36,37] also serve as the basis for the adjustable density matrix assembler (ADMA) method.[49−52] The ADMA method generates ab initio quality macromolecular *density matrices*, which can be used for the computation of electron density and a variety of other ab initio quality properties for macromolecules, such as forces and energies. These options of the ADMA method are expected to provide new directions in macromolecular conformational analysis and macromolecular geometry optimizations. Two areas of special importance are ab initio level computational studies of protein folding and spatial rearrangements of supramolecular structures.

If macromolecular density matrices are available, several new applications of quantum chemistry become possible. The ADMA technique provides the bridge between mainstream quantum chemistry and the additive fuzzy fragmentation approach. Most of the routine quantum chemical computational techniques, such as expectation value computations for various properties, become applicable to macromolecules if a macromolecular density matrix can be computed.

The ADMA technique is based on additive fragment density matrices (Afdms), defined within a consistent framework of AO representations for the fragments and the large target molecule.

If the AFDF scheme is augmented with a compatibility condition for AO bases and for the fragment neighborhoods for the target macromolecule as well as for the series of small parent molecules, then a direct, *algebraic* application of the AFDF principle can be used for the construction of ab initio quality approximate density matrices for given target macromolecules. The simplest of the AFDF schemes—the Mulliken–Mezey scheme—is also the simplest to adapt for the direct, algebraic generation of density matrices for target macromolecules.

The macromolecular density matrix constructed from the fragment density matrices within the ADMA framework represents the same level of accuracy as the electron densities obtained with the MEDLA and ALDA methods. The effects of interactions between local fragment representations are determined to the same level of accuracy within the ADMA, the MEDLA, and the ALDA approaches. The ADMA direct density matrix technique allows small readjustments of nuclear geometries, in a manner similar to the ALDA technique; however, within the ADMA framework, the geometry readjustment can be carried out directly on the macromolecule.

Following the earlier, more detailed introductions to the ADMA method,[51,52] a short description of the special feature of the ADMA technique is given subsequently.

Appropriately defined, *mutually compatible*, additive fragment density matrices (MC-AFDM) $\mathbf{P}^k$ obtained from small parent molecules are combined to form the ADMA macromolecular density matrix $\mathbf{P}$, where "mutual compatibility" implies that

(a) The mutual orientations of AO basis sets are constrained and
(b) The fragment choices must fulfill a target–parent fragmentation compatibility condition.

The ADMA method relies on a fragment density matrix database where the fragment density matrices must fulfill condition (b). Condition (a) can be satisfied using a suitable transformation of the fragment density matrices to physically equivalent fragment density matrices defined with respect to a properly oriented AO basis set. Since the final, macromolecular density matrix must refer to a single AO basis set, all the fragment density matrices ready for assembly must refer to local coordinate systems with coordinate axes oriented the same way as the axes of the macromolecular coordinate system.

The kth fragment density matrix $\mathbf{P}^k(\varphi)$ is obtained from an ab initio calculation for the parent molecule M_k within a local coordinate system. Vector $\varphi^{(k)}(\mathbf{r})$ represents the set of AOs of the parent molecule M_k, with reference to the local coordinate system. In a local coordinate system with axes aligned with those of the macromolecular coordinate system, vector $\psi^{(k)}(\mathbf{r})$ represents the same sequence of AOs at the same nuclear centers. These two representations are related by the orthogonal matrix transformation $\mathbf{T}^{(k)}$,

$$\psi^{(k)}(\mathbf{r}) = \mathbf{T}^{(k)}\varphi^{(k)}(\mathbf{r}), \tag{279}$$

where $\mathbf{T}^{(k)}$ is a block-diagonal matrix, containing the one-dimensional identity matrix for each s-orbital, an ordinary three-dimensional rotation matrix for each set of p-orbitals, the standard five-dimensional transformation matrix for each set of five orthonormalized d-orbitals, etc.

The density matrix $\mathbf{P}^k$ for the fragment k, with reference to the common coordinate axis orientation, is obtained by the similarity transformation

$$\mathbf{P}^k = \mathbf{T}^{(k)}\mathbf{P}^k(\varphi)\mathbf{T}'^{(k)}. \tag{280}$$

This fragment density matrix $\mathbf{P}^k$ is used to build the ADMA macromolecular density matrix $\mathbf{P}$.

The main component of target–parent fragmentation compatibility is the following condition: if the nuclei of the target molecule M are classified into m families, then each parent molecule M_k may contain only complete nuclear families $f_{k'}$ from the target molecule M.

A given parent molecule M_k either contains the complete nuclear family $f_{k'}$ as part of the surroundings for the actual nuclear set f_k of the fragment density matrix $\mathbf{P}^k$ or M_k does not contain any part of nuclear family $f_{k'}$. In addition, some peripheral H nuclei (or possibly other nuclei) may be needed to construct a viable parent molecule M_k to tie off dangling bonds. These extra nuclei are at large distances from the actual nuclear set f_k of the fragment density matrix $\mathbf{P}^k$.

Usually, the centrally located nuclear family f_k of parent molecule M_k is surrounded by a sufficiently thick layer of "coordination shell" of nuclear families $f_{k'}$, and the additional H (or other) nuclei are positioned far from the nuclei of the actual family f_k. Consequently, the density matrix elements between the orbitals of these additional atoms and the orbitals centered on nuclei of central family f_k are negligibly small. This allows a distinction between density matrix elements, and only those density matrix elements P_{ij}^k of each parent molecule M_k are involved in the construction of the final, ADMA macromolecular density matrix $\mathbf{P}$, which fulfill the following two conditions:

(a) P_{ij}^k of $\mathbf{P}^k$ has nonzero contribution as defined by the selection condition of the actual AFDF scheme (e.g., the Mulliken–Mezey scheme).

(b) P_{ij}^k of $\mathbf{P}^k$ does not involve the peripheral extra H (or other) nuclei of the parent molecules.

For any target macromolecule M, nuclear families f_k and the corresponding choices of parent molecules M_k fulfilling these conditions can always be selected.

Following the original description, the application of the ADMA method involves several steps:

(1) Identification of nuclear families $f_1, f_2, \ldots, f_k, \ldots, f_m$ of the target macromolecule M. The respective numbers of AOs in these families are denoted by $n_1, n_2, \ldots, n_k, \ldots, n_m$. By definition, take

$$n_0 = 0. \tag{281}$$

Define an index

$$f(k, i) = i + \sum_{v=0}^{k-1} n_v. \tag{282}$$

This index $f(k, i)$ is the serial index assigned to basis orbital $\varphi_i(\mathbf{r})$ of nuclear family f_k within the macromolecular density matrix $\mathbf{P}$.

(2) Construct nuclear arrangements for parent molecules $M_1, M_2, \ldots, M_k, \ldots, M_m$, where each parent molecule M_k contains the central nuclear family f_k and additional nuclear families $f_{k'}$ surrounding family f_k in a formal "coordination shell," and possibly, additional, peripheral nuclei attached to dangling bonds linked to some nuclei in families $f_{k'}$. Each nuclear family f_k occurs precisely once in any parent molecule as the central family. However, a given family f_k may occur in several parent molecules $M_{k'}$ as part of their coordination shell. Around the corresponding central nuclear families $f_{k'}$, these coordination shells in parent molecules $M_{k'}$ reproduce the local interactions within the large target molecule M.

(3) A consistent labeling of nuclear families within the target macromolecule and in the various parent molecules $M_1, M_2, \ldots, M_k, \ldots, M_m$ is introduced. For every pair (k, k') of indices, where $k, k' = 1, 2, \ldots, m$, a quantity $c_{k'k}$ is defined:

$$c_{k'k} = \begin{cases} 1, & \text{if nuclear family } f_{k'} \text{ contributes to parent molecule } M_k, \\ 0, & \text{otherwise,} \end{cases} \tag{283}$$

where

$$c_{0k} = 0, \quad \text{for each } k = 1, 2, \ldots, m. \quad (284)$$

(4) Ab initio computations are carried out for each parent molecule M_k, and the fragment density matrices $\mathbf{P}^k(\varphi)$ are determined, unless a suitable fragment density matrix is already available in the ADMA databank.

(5) Using the appropriate transformation matrix $\mathbf{T}^{(k)}$, the AO basis of each fragment density matrix $\mathbf{P}^k(\varphi)$ is transformed to an AO basis defined with respect to coordinate axes parellel to those of the common, macromolecular coordinate system, resulting in the fragment density matrix $\mathbf{P}^k$.

(6) If needed, permutations of the AO functions are carried out for each fragment density matrix $\mathbf{P}^k$ to generate a block structure for $\mathbf{P}^k$ where the following statements hold:

(a) The blocks correspond to sets of AOs associated to specific nuclear families $f_{k'}$.

(b) Within each fragment density matrix $\mathbf{P}^k$, all blocks corresponding to each nuclear family $f_{k'}$ follow the same ordering of the AOs.

(c) In each fragment density matrix $\mathbf{P}^k$, the block-row and block-column indices follow a monotonically increasing subsequence of the sequence $f_1, f_2, \ldots, f_k, \ldots, f_m$ of the target macromolecule M.

(d) For each fragment density matrix $\mathbf{P}^k$, the AOs centered on the additional nuclei which "tie off" the peripheral "dangling bonds" in parent molecule M_k are listed at the end of the sequence of AOs.

Evidently, this step (6) is required only if the conditions (a)–(d) are not met.

(7) Each fragment density matrix $\mathbf{P}^k$ is expanded to a fragment density matrix $\mathbf{P}^k(M)$ of the target macromolecule M by inserting blank rows and blank columns of blocks corresponding to AO sets of all nuclear families $f_{k''}$ *not contributing* to the fragment density matrix $\mathbf{P}^k$. (In practice, a simple index transformation is sufficient to identify the x, y index pair of element P_{xy} of the macromolecular density matrix $\mathbf{P}$, where a given element P_{ij}^k of a fragment density matrix $\mathbf{P}^k$ contributes.) The actual serial index of orbital $\varphi_i(\mathbf{r})$ of nuclear family $f_{k'}$ in the fragment density matrix $\mathbf{P}^k$ is given as

$$t(k, k', i) = c_{k'k} \left[i + \sum_{v=0}^{k'-1} c_{vk} n_k \right]. \quad (285)$$

The transformed row and column indices of each fragment density matrix $\mathbf{P}^k$ are determined and the corresponding matrix element is added to the appropriate matrix element of $\mathbf{P}$, following a simple rule: If the condition

$$t(k,k',i)t(k,k'',j) \neq 0 \qquad (286)$$

holds, then set

$$P_{f(k',i),f(k'',j)} = P_{f(k',i),f(k'',j)} + P^k_{t(k,k',i),t(k,k'',j)}. \qquad (287)$$

This procedure is carried out for each nonzero element of each fragment density matrix $\mathbf{P}^k$, resulting in the corresponding ADMA density matrix $\mathbf{P}$ of the target macromolecule.

The preceding adaptation of the ADMA algorithm follows the intuitive model of combining blocks of compatible fragment density matrices. An alternative adaptation, more suitable for direct computer implementation, has also been proposed.[49-51] In the following paragraphs, this second adaptation is outlined. This alternative description of the ADMA algorithm follows more closely the actual computational process.

Each AO $\varphi(\mathbf{r})$ of the final macromolecular AO basis is characterized by three indices and, depending on the context, any of these indices can be used. In the notation $\varphi_{b,k'}(\mathbf{r})$ the indices indicate that this is the bth AO within the set

$$\left\{ \varphi_{a,k'}(\mathbf{r}) \right\}_{a=1}^{n_{k'}} \qquad (288)$$

of AOs used for nuclear family $f_{k'}$. In the notation $\varphi_j^k(\mathbf{r})$ for the same AO, this orbital is viewed as the jth AO within the basis set

$$\left\{ \varphi_i^k(\mathbf{r}) \right\}_{i=1}^{n_{p^k}} \qquad (289)$$

used for the kth fragment density matrix $\mathbf{P}^k$. The number n_{p^k} of AOs used for fragment density matrix $\mathbf{P}^k$ is given by

$$n_{p^k} = \sum_{k'=1}^{m} c_{k'k} n_{k'}. \qquad (290)$$

The same AO can also be designated as $\varphi_y(\mathbf{r})$ if its serial index is y within the basis set

$$\left\{ \varphi_x(\mathbf{r}) \right\}_{x=1}^{n} \qquad (291)$$

of the macromolecular density matrix $\mathbf{P}$. If $\varphi_{a,k'}(\mathbf{r}) = \varphi_x(\mathbf{r})$, then the index x is related to indices a and k' by

$$x = x(k',a,f) = a + \sum_{b=1}^{k'-1} n_b. \qquad (292)$$

Here the symbol f in the argument of index function $x(k', a, f)$ indicates that indices k' and a originate from a nuclear family. If $\varphi_i^k(\mathbf{r}) = \varphi_x(\mathbf{r})$, then index x can be determined from indices i and k for each fragment density matrix $\mathbf{P}^k$ as follows. Define

$$a'_k(k'', i) = i + \sum_{b=1}^{k''} n_b c_{bk}, \tag{293}$$

$$k' = k'(i, k) = \min\{k'' : a'_k(k'', i) \le 0\}, \tag{294}$$

and

$$a_k(i) = a'_k(k', i) + n_{k'}, \tag{295}$$

for all nuclear families $f_{k''}$ for which

$$c_{k''k} \ne 0. \tag{296}$$

The index x assigned to indices i and k is given by

$$x = x(k, i, P) = x(k', a_k(i), f), \tag{297}$$

where the index functions $x(k', a, f)$ defined in Eq. (292) and k' given in Eq. (294) are taken and where P in the argument of the index function $x(k, i, P)$ indicates that indices k and i refer to a fragment density matrix.

Initially, matrix $\mathbf{P}$ is set to zero, followed by testing each element P_{ij}^k of each fragment density matrix $\mathbf{P}^k$ in arbitrary order. By identifying each nonzero element P_{ij}^k of each fragment density matrix $\mathbf{P}^k$ and using the foregoing index assignments, the macromolecular density matrix $\mathbf{P}$ is constructed by setting

$$P_{x(k,i,P), y(k,j,P)} = P_{x(k,i,P), y(k,j,P)} + P_{ij}^k, \qquad \forall P_{ij}^k \ne 0. \tag{298}$$

If the target macromolecule M and the parent molecules $M_1, M_2, \ldots, M_k, \ldots, M_m$ fulfill the compatibility conditions, then either of the two adaptations of the ADMA algorithm generates a macromolecular density matrix $\mathbf{P}$ from the corresponding fragment density matrices $\mathbf{P}^1, \mathbf{P}^2, \ldots, \mathbf{P}^k, \ldots, \mathbf{P}^m$.

The ADMA algorithm for the generation of the macromolecular density matrix $\mathbf{P}$, defined in terms of the macromolecular AO basis set $\{\varphi_x(\mathbf{r})\}_{x=1,\ldots,n}$, provides an ab initio quality quantum-chemical description of macromolecule M. In particular, the electronic density obtained by the standard approach [Eq. (266)] is equivalent to that obtained using the MEDLA approach. The interactions between AOs of the central nuclear family f_k and the AOs located at the additional nuclei beyond the coordination shell in each parent molecule M_k are ignored within the ADMA scheme. If these matrix elements belong to AO pairs separated by distances which are beyond the distance cutoff of the MEDLA numerical density fragments stored in the MEDLA database, then the corresponding matrix elements are neglected. The ADMA technique reproduces the ab initio quality electron densities obtained with the MEDLA method. If rigorous charge conservation is needed, then the density matrix **P** can be scaled using the same method described for the MEDLA technique.[45]

Using standard density matrix methodology, various other molecular properties can also be calculated.

If one takes a finite upper bound for the computer time requirement for each fragment density matrix, as well as for its share of nonzero index contribution to the final, macromolecular density matrix, then one can show that the computer time requirement of the ADMA computation grows linearly with the number of fragment density matrices.

If nuclear geometry variations are small, they can be treated the same way as in the ALDA method. In particular, the orthonormalization–deorthonormalization method described in the context of the SALDA method[110] can be adapted for the ADMA technique. Employing two Löwdin-type transformations using relatively inexpensive macromolecular overlap matrices $S(K)$ and $S(K')$ for two, slightly different nuclear geometries K and K', a satisfactory approximation $P(K',[K])$ of the density matrix $P(K')$ can be obtained in terms of density matrix $P(K)$. Motivated by the involvement of transformed overlap matrices, this technique is referred to as the SADMA method.[111]

The SADMA option for the accommodation of small geometry variations is advantageous in studies of small macromolecular distortions, of protein folding processes, and potentially in the structure refinement process of x-ray structure determination.

The ADMA and SADMA methods[49–52,111] generate ab initio quality density matrices **P** for large molecules M, while avoiding the computation of macromolecular wave functions. At the Hartree–Fock level, the first-order density matrix **P** fully determines all higher-order density matrices. Within the Hartree–Fock framework, expectation values for one-electron and two-electron operators can be computed using the first-order and second-order density matrices. Consequently, the ADMA and SADMA methods[49–52,111] provide new possibilities for adapting quantum-chemical techniques for macromolecules.

If electronic densities, in particular, if density matrices of large molecules are available, then an important variant of the Hellmann–Feynman theorem[112–114] can be used for the computation of forces acting upon various nuclei, to be used for macromolecular geometry optimization.

Consider the Schrödinger equation for the energy eigenvalue E of a normalizable wave function ψ:

$$\hat{H}\psi = E\psi, \qquad (299)$$

where σ is a real parameter of the molecular Hamiltonian $\hat{H}$. According to the Hellmann–Feynman theorem, the derivative of the energy, with respect to a parameter σ satisfying some constraints, is the expectation value of the derivative of the Hamiltonian:

$$\partial E / \partial \sigma = \langle \psi | \partial \hat{H} / \partial \sigma | \psi \rangle / \langle \psi | \psi \rangle. \tag{300}$$

Hurley proposed a simple, sufficient condition for the applicability of the Hellmann–Feynman theorem.[115,116] If within a variational framework, the family of trial functions is invariant to changes in parameter σ, then the Hellmann–Feynman theorem is satisfied by the optimum trial function. In variational approaches involving Lagrange multipliers, for example, in the Hartree–Fock and multiconfigurational self-consistent field methods, Hurley's condition is fulfilled.[117]

In many approximate methods, the error of calculated Hellmann–Feynman forces is significant. Following the introduction of the force method of direct, analytic differentiation of Hartree–Fock and related approximate energies by Pulay and by Pulay and Meyer,[118–125] the Hellmann–Feynman theorem is rarely used in computational applications. Note, however, that the Hellmann–Feynman theorem still plays a prominent role in studying various special problems.[126–133]

The electrostatic Hellmann–Feynman theorem is a special form of the general Hellmann–Feynman theorem. This form of the theorem can be expressed in terms of electronic density, and no explicit form of the electronic wave function is needed. The electrostatic Hellmann–Feynman theorem is of special significance in view of new developments in the construction of macromolecular electron densities and density matrices without using wave functions.[39]

An important application of the electrostatic Hellmann–Feynman theorem within the AFDF framework is the basis of a novel, macromolecular geometry optimization technique.[39] Assume that the electronic density $\rho(\mathbf{r})$ of a molecule of N nuclei and k electrons is available. The components of the position vector $\mathbf{R}_a$ of nucleus a and those of position vector $\mathbf{r}_i$ of electron i are denoted by X, Y, and Z:

$$\mathbf{R}'_a = (X_a, Y_a, Z_a) \tag{301}$$

and

$$\mathbf{r}'_i = (X_i, Y_i, Z_i), \tag{302}$$

respectively. If the Dirac delta formalism is used, then the electronic density $\rho(\mathbf{r})$ can be expressed in terms of the electronic wave function ψ as

$$\rho(\mathbf{r}) = \left\langle \psi \left| \sum_{i=1}^{k} \delta(\mathbf{r} - \mathbf{r}_i) \right| \psi \right\rangle \Big/ \langle \psi | \psi \rangle. \tag{303}$$

Within the Born–Oppenheimer approximation, the molecular Hamiltonian is given as a sum of the electronic Hamiltonian $\hat{H}_e$ and the nuclear Hamiltonian $\hat{H}_n$,

$$\hat{H} = \hat{H}_e + \hat{H}_n, \tag{304}$$

where

$$\hat{H}_e = \hat{T}_e + \hat{V}_e \tag{305}$$

is a sum of the electronic kinetic energy operator $\hat{T}_e$ and the electronic potential energy operator $\hat{V}_e$, and within the clamped nuclei model,

$$\hat{H}_n = V_{nn} \tag{306}$$

is the nuclear repulsion energy.

The electron–electron, electron–nucleus, and nucleus–nucleus distances r_{ij}, r_{ia}, and r_{ab}, respectively, are defined in terms of the $\mathbf{R}_a$ and $\mathbf{r}_i$ position vectors.

The V_{nn} nuclear repulsion term is given as

$$V_{nn} = \sum_{a < b}^{N} z_a z_b r_{ab}^{-1}, \tag{307}$$

and the electronic kinetic and potential energy operators are

$$\hat{T}_e = -1/2 \sum_{i=1}^{k} \Delta_i \tag{308}$$

and

$$\hat{V}_e = \sum_{i<j}^{k} r_{ij}^{-1} - \sum_{i=1}^{k} \sum_{a=1}^{N} z_a r_{ia}^{-1}, \tag{309}$$

respectively. Using these expressions, the molecular Hamiltonian $\hat{H}$ is written as

$$\hat{H} = -1/2 \sum_{i=1}^{k} \Delta_i + \sum_{i<j}^{k} r_{ij}^{-1} - \sum_{i=1}^{k} \sum_{a=1}^{N} z_a r_{ia}^{-1} \sum_{a<b}^{N} z_a z_b r_{ab}^{-1}. \tag{310}$$

Differentiation of the Hamiltonian $\hat{H}$ of Eq. (310) according to components of nuclear position vector $\mathbf{R}_a$ gives

$$-\partial\hat{H}/\partial\mathbf{R}_a = -\sum_{i=1}^{k} z_a(\mathbf{R}_a - \mathbf{r}_i)/|\mathbf{R}_a - \mathbf{r}_i|^3$$

$$+ \sum_{a<b}^{N} z_a z_b(\mathbf{R}_a - \mathbf{R}_b)/|\mathbf{R}_a - \mathbf{R}_b|^3, \tag{311}$$

defining a formal force operator

$$\mathbf{F}_a = -\partial \hat{H}/\partial \mathbf{R}_a, \tag{312}$$

representing the combined forces due to all the electrons and all the other nuclei, acting upon nucleus a.

On the other hand, if the components of nuclear position vector $\mathbf{R}_a$ are taken as the parameter σ of the Hellmann–Feynman theorem, then, according to the theorem,

$$-\partial E/\partial \mathbf{R}_a = \langle \psi | \mathbf{F}_a | \psi \rangle / \langle \psi | \psi \rangle. \tag{313}$$

By writing the expectation value of operator $\mathbf{F}_a$ for the normalized wave function ψ as $\langle \mathbf{F}_a \rangle$, Eq. (311) takes the form

$$-\partial E/\partial \mathbf{R}_a = \langle \mathbf{F}_a \rangle. \tag{314}$$

The steepest descent direction, that is, the negative gradient of the Born–Oppenheimer potential energy hypersurface[18] can be determined from the forces $-\partial E/\partial \mathbf{R}_a$ acting on nuclei a of coordinates X_a, Y_a, and Z_a. In some applications, the equivalent representation as the Hellmann–Feynman force $\langle \mathbf{F}_a \rangle$ is more advantageous.

In the course of differentiation of the Hamiltonian of Eq. (310) by components of $\mathbf{R}_a$, all the electronic kinetic energy terms and all the terms describing electron–electron interactions are eliminated. Consequently, the force operator $\mathbf{F}_a$ is a one-electron operator and the expectation value $\langle \mathbf{F}_a \rangle$ can be written as

$$\langle \mathbf{F}_a \rangle = -z_a \int \rho(\mathbf{r})(\mathbf{R}_a - \mathbf{r})|\mathbf{R}_a - \mathbf{r}|^{-3} \, d\mathbf{r}$$

$$+ z_a \sum_{a \neq b}^{N} z_b(\mathbf{R}_a - \mathbf{R}_b)|\mathbf{R}_a - \mathbf{R}_b|^{-3}. \tag{315}$$

This expression involves no reference to a wave function, and electronic information is provided by the electronic density $\rho(\mathbf{r})$, described in the Dirac delta formalism as given in Eq. (303). The computation of the expectation value $\langle \mathbf{F}_a \rangle$ of the force operator requires only a simple sum of a classical contribution from the electronic charge density and the internuclear repulsion.

This approach is advantageous if MEDLA, ALDA, or ADMA macromolecular electronic densities are available. By carrying out a simple three-dimensional integration in the first term and a trivial summation in the second term of Eq. (315), the electrostatic Hellmann–Feynman theorem can be used for the computation of forces acting on the nuclei of macromolecules.

Within the MEDLA and ALDA methods, the application of the electrostatic Hellmann–Feynmann theorem involves numerical integration. If the electron density of the target macromolecule is generated numerically on a three-dimensional grid, then special care is required for the integration near the nuclei. A grid that is satisfactory in regions of the space where the density varies little is not necessarily suitable for numerical integration near the nuclei where the electronic density shows large variations.

This numerical problem of integration can be avoided using the ADMA technique. Within the ADMA method, the integration in Eq. (361) can be performed using the analytical expressions of macromolecular density matrices and AOs. As an option of the ADMA algorithm,[49] the calculated ADMA Hellmann–Feynman forces can be used for macromolecular geometry optimization and macromolecular conformational analysis.

APPENDIX

Some special properties of R sets of a crisp or fuzzy set A are discussed herein.

Take the family of sets $A, \mathbf{R}A, \mathbf{R}^2A, \ldots, \mathbf{R}^{n-1}A$, where n is subject to condition (159), and take the R set

$$A(R, c) = \bigcup_{i=0, n-1} \mathbf{R}^i A \qquad (316)$$

as given in Eq. (160). This construction is of special importance, as shown by the following theorem.

THEOREM. *Each minimal R superset $C'(A)$ of A can be constructed as a union $\bigcup_{i=0, n-1} \mathbf{R}^i A$ with respect to some version of symmetry operator $\mathbf{R}$.*

Proof. Since $C'(A)$ is as an R superset of A, there must exist at least one center c and a set of defining parameters for a version of a symmetry element of type R that qualifies $C'(A)$ as an R set. Take the symmetry operation $\mathbf{R}$ that corresponds to this version R. Then, for the union of the family $A, \mathbf{R}A, \mathbf{R}^2A, \ldots, \mathbf{R}^{n-1}A$ of sets with respect to this symmetry operator $\mathbf{R}$,

$$\bigcup_{i=0, n-1} \mathbf{R}^i A = C'(A), \qquad (317)$$

that is, the minimal R superset $C'(A)$ of A can be reconstructed in the form as given in Eq. (160). Q.E.D.

A very similar approach is applied to maximum R subsets $B'(A)$ of a crisp or fuzzy set A.

A formal center c is selected for a (possibly only approximate) symmetry element R of set A. It is advantageous to take c as the center of mass of A if a center of mass is defined for set A. The smallest positive value n satisfying the condition

$$\mathbf{R}^n = \mathbf{E} \tag{318}$$

is determined, where $\mathbf{R}^n$ is the nth power of the symmetry operation $\mathbf{R}$ corresponding to R, and $\mathbf{E}$ is the identity operation, similarly to the case described by Eq. (159).

The symmetry operations $\mathbf{R}^i$ ($i = 1, \ldots, n-1$) are applied to set A, resulting in the sets $A, \mathbf{R}A, \mathbf{R}^2A, \ldots, \mathbf{R}^{n-1}A$. The intersection of all these sets is denoted by

$$A(R, c, \text{int}) = \bigcap_{i=0, n-1} \mathbf{R}^i A. \tag{319}$$

The set $A(R, c, \text{int})$ must have the symmetry element R, since

$$\mathbf{R}\,A(R, c, \text{int}) = \mathbf{R} \bigcap_{i=0, n-1} \mathbf{R}^i A = \bigcap_{i=0, n-1} \mathbf{R}^{i+1}A$$
$$= \bigcap_{i=1, n} R^i A = \bigcap_{i=0, n-1} \mathbf{R}^i A, \tag{320}$$

where in the last two intersections, the identities

$$\mathbf{R}^n A = \mathbf{E}A = \mathbf{R}^0 A \tag{321}$$

apply. Consequently,

$$\mathbf{R}\,A(R, c, \text{int}) = A(R, c, \text{int}). \tag{322}$$

A measure of the symmetry aspect R for set A is obtained with respect to center c if one compares sets A and $A(R, c, \text{int})$. Using the Hausdorff metric

$$h(A, A(R, c, \text{int})), \tag{323}$$

a valid measure of the symmetry aspect R is obtained for set A.

If the center of mass is chosen as reference c, then one may use the simpler notations $A(R, \text{int})$ and $h(A, A(R, \text{int}))$.

If the SNDSM metric is used, then the quantity $d_s(A, A(R, c, \text{int}))$ is a "shape distance" between set A and the R-symmetric subset $A(R, c, \text{int})$ with respect to center c. If the center of mass is the reference, then the simpler notation $d_s(A, A(R, \text{int}))$ can be used.

An interesting property of maximal R subsets is established by a theorem analogous to the theorem for minimal R supersets.

THEOREM. *Each maximal R subset $B'(A)$ of A can be generated as an intersection $\bigcap_{i=0, n-1} \mathbf{R}^i A$ with respect to some version of the symmetry operator $\mathbf{R}$.*

Proof. Since $B'(A)$ is an R subset of A, there must exist at least one center c and a set of defining parameters P for a version of a symmetry element of type R that qualifies $B'(A)$ as an R set. Choose the symmetry operation $\mathbf{R}$ that corresponds to this version R. For the intersection of the family $A, \mathbf{R}A, \mathbf{R}^2A, \ldots, \mathbf{R}^{n-1}A$ of sets generated using this symmetry operator $\mathbf{R}$,

$$\bigcap_{i=0, n-1} \mathbf{R}^iA = B'(A), \tag{324}$$

that is, the maximal R subset $B'(A)$ of A can be written in the form as given in Eq. (319). Q.E.D.

As follows from its construction, the fuzzy average of crisp or fuzzy sets $A, \mathbf{R}A, \mathbf{R}^2A, \ldots, \mathbf{R}^{n-1}A$ does have the corresponding fuzzy R symmetry.

ACKNOWLEDGMENTS

Special thanks are due to Dr. G. M. Maggiora for stimulating discussions and for pointing out important references. The original research work leading to the methods and results described in this report was supported by both strategic and operating research grants from the Natural Sciences and Engineering Research Council of Canada.

REFERENCES

1. P. G. Mezey, *Shape in Chemistry: An Introduction to Molecular Shape and Topology.* VCH, New York, 1993.
2. R. G. Woolley, *Adv. Phys.* **25**, 27 (1976).
3. L. A. Zadeh, *Inform. Control* **8**, 338 (1965).
4. L. A. Zadeh, *J. Math. Anal. Appl.* **23**, 421 (1968).
5. A. Kaufmann, *Introduction à la Théorie des Sous-Ensembles Flous.* Masson, Paris, 1973.
6. M. L. Puri and D. A. Ralescu, *J. Math. Anal. Appl.* **114**, 409 (1986).
7. D. Dubois and H. Prade, "Evaluation of Fuzzy Sets and Fuzzy Integration: A Synthetic Discussion," in *Abstracts of the Second Joint IFSA-EF and EURO-WG Workshop,* pp. 131–135, 1986.
8. H. Bandemer and W. Näther, *Fuzzy Data Analysis.* Kluwer Academic, Dordrecht, 1992.
9. E. Sanchez and M. M. Gupta (eds.), *Fuzzy Information, Knowledge Representation and Decision Analysis.* Pergamon, Elmsford, NY, 1983.
10. Z. Wang and G. J. Klir, *Fuzzy Measure Theory.* Plenum, New York, 1992.
11. G. J. Klir and B. Yuan, *Fuzzy Sets and Fuzzy Logic, Theory and Applications.* Prentice-Hall, Englewood Cliffs, NJ, 1995.
12. E. Prugovecki, *Found. Phys.* **4**, 9 (1974).
13. E. Prugovecki, *Found. Phys.* **5**, 557 (1975).
14. E. Prugovecki, *J. Phys. A* **9**, 1851 (1976).
15. S. T. Ali and H. D. Doebner, *J. Math. Phys.* **17**, 1705 (1976).
16. S. T. Ali and E. Prugovecke, *J. Math. Phys.* **18**, 219 (1977).
17. P. G. Mezey, "Topological Model of Reaction Mechanism," in *Structure and Dynamics of Molecular Systems* (R. Daudel, J.-P. Korb, J.-P. Lemaistre, and J. Maruani, eds.), Vol. 1, pp. 57–70. Reidel, Dordrecht, 1985.

18. P. G. Mezey, *Potential Energy Hypersurfaces*. Elsevier, Amsterdam, 1987.
19. P. G. Mezey, *Potential Hypersurfaces*. pp. 187, 364. Elsevier, Amsterdam, 1987.
20. P. G. Mezey, "From Geometrical Molecules to Topological Molecules: A Quantum Mechanical View," *in Molecular Organization and Engineering* (J. Maruani, ed.), Vol. 2, p. 61. Kluwer Academic, Dordrecht, 1988.
21. J. Maruani and P. G. Mezey, *C. R. Acad. Sci. Paris Sér. II* **305**, 1051 (1987); **306**, 1141 (1987).
22. P. G. Mezey and J. Maruani, *Mol. Phys.* **69**, 97 (1990).
23. J. Maruani and P. G. Mezey, *J. Chim. Phys.* **87**, 1025 (1990).
24. P. G. Mezey and J. Maruani, *Int. J. Quantum Chem.* **45**, 177 (1993).
25. C.-T. Zhang, K.-C. Chou, and G. M. Maggiora, *Protein Eng.* **8**, 425 (1995).
26. P. G. Mezey, "A Global Approach to Molecular Chirality," *in New Developments in Molecular Chirality* (P. G. Mezey, ed.). Kluwer Academic, Dordrecht, 1991.
27. P. G. Mezey, *J. Math. Chem.* **11**, 27 (1992).
28. P. G. Mezey, *Advances in Molecular Similarity* **1**, (1996).
29. P. G. Mezey, unpublished.
30. P. G. Mezey, *Int. J. Quantum Chem., Quantum Biol. Symp.* **12**, 113 (1986).
31. P. G. Mezey, *Int. J. Quantum Chem., Quantum Biol. Symp.* **14**, 127 (1987).
32. P. G. Mezey, *J. Comput. Chem.* **8**, 462 (1987).
33. P. G. Mezey, "Molecular Surfaces," *in Reviews in Computational Chemistry* (K. B. Lipkowitz and D. B. Boyd, eds.), Vol. 1, pp. 265–294. VCH, New York, 1990.
34. P. G. Mezey, *J. Chem. Inf. Comput. Sci.* **32**, 650 (1992).
35. P. G. Mezey, *Canad. J. Chem.* **72**, 928 (1994).
36. P. G. Mezey, "Density Domain Bonding Topology and Molecular Similarity Measures," *in Topics in Current Chemistry*: *Molecular Similarity* (K. Sen, ed.), Vol. 173, pp. 63–83. Springer-Verlag, Berlin, 1995.
37. P. G. Mezey, "Methods of Molecular Shape-Similarity Analysis and Topological Shape Design," *in Molecular Similarity in Drug Design* (P. M. Dean, ed.), pp. 241–268. Chapman & Hall, London, 1995.
38. P. G. Mezey, "Molecular Similarity Measures for Assessing Reactivity," *in Molecular Similarity and Reactivity*: *From Quantum Chemical to Phenomenological Approaches* (R. Carbó, ed.), pp. 57–76. Kluwer Academic, Dordrecht, 1995.
39. P. G. Mezey, "Descriptors of Molecular Shape in 3D," *in From Chemical Topology to Three-Dimensional Geometry* (A. T. Balaban, ed.). Plenum, in press.
40. P. G. Mezey, "Functional Groups in Quantum Chemistry," *in Advances in Quantum Chemistry* (P.-O. Löwdin, ed.), Vol. 27, pp. 163–222. Academic Press, New York, 1996.
41. P. D. Walker and P. G. Mezey, *Program MEDLA* 93, Mathematical Chemistry Research Unit, University of Saskatchewan, Canada, 1993.
42. P. D. Walker and P. G. Mezey, *J. Am. Chem. Soc.* **115**, 12423 (1993).
43. P. D. Walker and P. G. Mezey, *J. Am. Chem. Soc.* **116**, 12022 (1994).
44. P. D. Walker and P. G. Mezey, *Canad. J. Chem.* **72**, 2531 (1994).
45. P. D. Walker and P. G. Mezey, *J. Math. Chem.* **17**, 203 (1995).
46. P. D. Walker and P. G. Mezey, *J. Comput. Chem.* **16**, 1238 (1995).
47. P. D. Walker, G. M. Maggiora, M. A. Johnson, J. D. Petke, and P. G. Mezey, *J. Chem. Inf. Comput. Sci.* **35**, 568 (1995).
48. P. D. Walker, G. M. Maggoira, M. A. Johnson, J. D. Petke, and P. G. Mezey, *J. Comput. Chem.* **16**, 1474 (1995).
49. P. G. Mezey, *Program ADMA 95*, Mathematical Chemistry Research Unit, University of Saskatchewan, Saskatoon, Canada, 1995.
50. P. G. Mezey, *Structural Chem.* **6**, 261 (1995).
51. P. G. Mezey, *J. Math. Chem.* **18**, 141 (1995).
52. P. G. Mezey, "Local Shape Analysis of Macromolecular Electron Densities," *in Computational Chemistry: Reviews and Current Trends* (J. Leszczynski, ed.). World Scientific, Singapore, 1996.

53. F. Hausdorff, *Set Theory* (J. R. Auman, transl.), pp. 166–168. Chelsea, New York, 1957.
54. A. Rassat, *C. R. Acad. Sci. Paris Sér. II* **299**, 53 (1984).
55. A. B. Buda, T. Auf der Heyde, and K. Mislow, *Angew. Chem., Int. Ed. Engl.* **31**, 989 (1992).
56. A. B. Buda and K. Mislow, *J. Am. Chem. Soc.* **114**, 6006 (1992).
57. K. Mislow, personal communication.
58. A. I. Kitaigorodskii, *Organic Chemical Crystallography*, Chap. 4. Consultants Bureau, New York, 1961.
59. G. Gilat, *Chem. Phys. Lett.* **121**, 9 (1985).
60. G. Gilat and L. S. Schulman, *Chem. Phys. Lett.* **121**, 13 (1985).
61. G. Gilat, *J. Phys. A* **22**, L545 (1989).
62. A. B. Buda, T. P. E. Auf der Heyde, and K. Mislow, *J. Math. Chem.* **6**, 243 (1991).
63. T. P. E. Auf der Heyde, A. B. Buda, and K. Mislow, *J. Math. Chem.* **6**, 255 (1991).
64. A. B. Buda and K. Mislow, *Elem. Math.* **46**, 65 (1991).
65. P. G. Mezey, "Topology of Molecular Shape and Chirality," *in New Theoretical Concepts for Understanding Organic Reactions* (J. Bertran and I. G. Csizmadia, eds.), pp. 77–99. NATO ASI Series, Kluwer Academic, Dordrecht, 1989.
66. P. G. Mezey, "Three-Dimensional Topological Aspects of Molecular Similarity," *in Concepts and Application of Molecular Similarity* (M. A. Johnson and G. M. Maggiora, eds.), pp. 321–368. Wiley, New York, 1990.
67. M. Senechal, *Crystalline Symmetries, An Informal Introduction.* pp. 70. Hilger, Bristol, 1990.
68. P. G. Mezey, *Int. J. Quantum Chem.* **51**, 255 (1994).
69. D. J. Klein, *J. Math. Chem.*, **18**, 321 (1995).
70. P. G. Mezey, *Int. J. Quantum Chem.*, in press.
71. I. Hargittai and M. Hargittai, *Symmetry through the Eyes of a Chemist.* VCH, Weinheim, 1986.
72. H. Zabrodsky, S. Peleg, and D. Avnir, *J. Am. Chem. Soc.* **114**, 7843 (1992).
73. H. Zabrodsky, S. Peleg, and D. Avnir, *J. Am. Chem. Soc.* **115**, 8278 (1993).
74. H. Zabrodsky and D. Avnir, *Adv. Mol. Struct. Res.* **1**, 1 (1995).
75. H. Zabrodsky and D. Avnir, *J. Am. Chem. Soc.* **117**, 462 (1995).
76. K. Mislow and R. Bolstad, *J. Am. Chem. Soc.* **77**, 6712 (1955).
77. K. Mislow and P. Poggi-Corradini, *J. Math. Chem.* **13**, 209 (1993).
78. K. Mislow, paper presented at the 6th International Conference on Mathematical Chemistry, Pitlochry, Scotland, UK, July 10–14, 1995.
79. P. G. Mezey, *J. Math. Chem.* **17**, 185 (1995).
80. P. G. Mezey, paper presented at the 6th International Conference on Mathematical Chemistry, Pitlochry, Scotland, UK, July 10–14, 1995.
81. P. G. Mezey, *J. Math. Chem.* **18**, 133 (1996).
82. P. G. Mezey, *Program ALDA 95*, Mathematical Chemistry Research Unit, University of Saskatchewan, Saskatoon, Canada, 1995.
83. P. G. Mezey, *Int. J. Quantum Chem.*, in press.
84. R. S. Mulliken, *J. Chem. Phys.* **23**, 1833, 1841, 2338, 2343 (1955).
85. R. S. Mulliken. *J. Chem. Phys.* **36**, 3428 (1962).
86. P.-O. Löwdin, *J. Chem. Phys.* **18**, 365 (1950).
87. P.-O. Löwdin, *Adv. Phys.* **5**, 1 (1956).
88. P.-O. Löwdin, *Adv. Quantum Chem.* **5**, 185 (1970).
89. F. L. Pilar, *Elementary Quantum Chemistry.* McGraw-Hill, New York, 1968.
90. L. Massa, L. Huang, and J. Karle, *Int. J. Quantum Chem.*, Quant. Chem. Symp. **29**, 371 (1995).
91. P.-O. Löwdin, *Phys. Rev.* **97**, 1474 (1955).
92. R. McWeeney, *Rev. Mod. Phys.* **32**, 335 (1960).
93. A. J. Coleman, *Rev. Mod. Phys.* **35**, 668 (1963).
94. W. L. Clinton, A. J. Galli, and L. J. Massa, *Phys. Rev.* **177**, 7 (1969).

95. W. L. Clinton, A. J. Galli, G. A. Henderson, G. B. Lamers, L. J. Massa, and J. Zarur, *Phys. Rev.* **177**, 27 (1969).
96. W. L. Clinton and L. J. Massa, *Int. J. Quantum Chem.* **6**, 519 (1972).
97. W. L. Clinton and L. J. Massa, *Phys. Rev. Lett.* **29**, 1363 (1972).
98. W. L. Clinton, C. Frishberg, L. J. Massa, and P. A. Oldfield, *Int. J. Quantum Chem. Quantum Chem. Symp.* **7**, 505 (1973).
99. G. A. Henderson and R. K. Zimmermann, *J. Chem. Phys.* **65**, 619 (1976).
100. V. G. Tsirel'son, V. E. Zavodnik, E. B. Fonichev, R. P. Ozerov, and I. S. Kuznetsolirez, *Kristallografiya* **25**, 735 (1980).
101. C. Frishberg and L. J. Massa, *Phys. Rev. B* **24**, 7018 (1981).
102. C. Frishberg and L. G. Massa, *Acta Crystallogr. Sect. A* **38**, 93 (1982).
103. L. J. Massa, M. Goldberg, C. Frishberg, R. F. Boehme, and S. J. LaPlaca, *Phys. Rev. Lett.* **55**, 622 (1985).
104. C. Frishberg, *Int. J. Quantum Chem.* **30**, 1 (1986).
105. L. Cohn, C. Frishberg, C. Lee, and L. J. Massa, *Int. J. Quantum Chem., Quantum Chem. Symp.* **19**, 525 (1986).
106. L. J. Massa, *Chem. Scr.* **26**, 469 (1986).
107. R. F. Boehme and S. J. LaPlaca, *Phys. Rev. Lett.* **59**, 985 (1987).
108. K. Tanaka, *Acta Crystallogr. Sect. A* **44**, 1002 (1988).
109. Y. Y. Aleksandrov, V. G. Tsirel'son, I. M. Resnik, and R. P. Ozerov, *Phys. Status Solidi B* **155**, 201 (1989).
110. P. G. Mezey, *Program SALDA 95*, Mathematical Chemistry Research Unit, University of Saskatchewan, Saskatoon, Canada, 1995.
111. P. G. Mezey, *Program SADMA 95*, Mathematical Chemistry Research Unit, University of Saskatchewan, Saskatoon, Canada, 1995.
112. H. Hellmann, *Einführung in die Quantenchemie*, Sect. 54. Deuticke and Co., Leipzig, 1937.
113. R. P. Feynman, *Phys. Rev.* **56**, 340 (1939).
114. S. T. Epstein, "The Hellmann–Feynman Theorem," in *The Force Concept in Chemistry* (B. M. Deb, ed.). Van Nostrand-Reinhold, New York, 1981.
115. A. C. Hurley, *Proc. Roy. Soc. London Ser. A* **226**, 170 (1954).
116. A. C. Hurley, *Proc. Roy. Soc. London Ser. A* **226**, 179 (1954).
117. S. T. Epstein, *Theor. Chim. Acta* **55**, 251 (1980).
118. P. Pulay, *Mol. Phys.* **17**, 197 (1969).
119. P. Pulay, *Mol. Phys.* **18**, 473 (1970).
120. P. Pulay, *Mol. Phys.* **21**, 329 (1971).
121. P. Pulay and W. Meyer, *J. Mol. Spectrosc.* **40**, 59 (1971).
122. P. Pulay and W. Meyer, *J. Chem. Phys.* **57**, 3337 (1972).
123. P. Pulay, "Direct Use of Gradients for Investigating Molecular Energy Surfaces," in *Applications of Electronic Structure Theory* (H. F. Schaefer, ed.). Plenum, New York, 1977.
124. P. Pulay, "Calculation of Forces by Non-Hellmann–Feynman Methods," in *The Force Concept in Chemistry* (B. M. Deb, ed.). Van Nostrand-Reinhold, New York, 1981.
125. P. Pulay, TEXAS (ab initio Hartree–Fock gradient program), *Theor. Chim. Acta* **50**, 299 (1979).
126. R. G. Parr, *J. Chem. Phys.* **40**, 3726 (1964).
127. J. Goodisman, *J. Chem. Phys.* **44**, 2085 (1966).
128. J. Goodisman, *J. Chem. Phys.* **45**, 4689 (1966).
129. J. Goodisman, *J. Chem. Phys.* **47**, 334 (1967).
130. L. Zulicke and H. J. Spangenberg, *Theor. Chim. Acta* **5**, 139 (1966).
131. M. T. Marron, "Energies, Energy Differences and Mechanisms of Internal Motions," in *The Force Concept in Chemistry* (B. M. Deb, ed.). Van Nostrand-Reinhold, New York, 1981.
132. C. Trindle and K. K. George, *Int. J. Quantum Chem.* **10**, 21 (1976).
133. P. G. Mezey, *Mol. Phys.* **47**, 121 (1982).

6

Linguistic Variables in the Molecular Recognition Problem

JÜRGEN BRICKMANN
Institute of Physical Chemistry and
  Darmstadt Center for Scientific Computing
Technical University of Darmstadt
D-64287 Darmstadt, Germany

I. INTRODUCTION

The specific recognition of a molecule within a molecular scenario plays an important role in many chemical processes. For instance, it forms the basis for highly specific reactions in biochemistry and catalysis. A large variety of different factors (energetic, entropic, kinetic, etc.) come into play in a conceptional model approach when we attempt to describe such recognition in a precise way.[1] From a thermodynamic point of view, the specificity of a receptor can be measured by a subgroup A of molecules it recognizes (at a given level of affinity defined by a certain ΔG value) from among a larger ensemble B of molecules that in principle have to be considered. This type of recognition may often be described in terms of the lock-and-key image first introduced by Fischer[2] in 1894.

According to the formalism of information theory, the specificity of a certain class A from a set B of molecules with respect to a given receptor can be measured by the expression[1]

$$S_A = \log\left(\dim(B)/\dim(A)\right), \tag{1}$$

where the dimensions $\dim(A)$ and $\dim(B)$ are simply the number of different molecules in the sets A and B, respectively. High specificity results in high values of S_A. It is obvious that this value depends to a large extent on the way set B is defined. If, for example, an antibody selectively binds 2 of 20 different steroids, one has $S_A = \log_2(20/2) = 3.32$. If, however, 512 potential binding partners are considered, $S_A = \log_2 256 = 8$ results. As long as set B can be well defined as in the case of the steriods, the numerical value of S_A will be useful for the characterization of the receptor selectivity with respect to this class of molecules. In many cases this is, however, not possible. Two simple examples serve to demonstrate this fact. If one asks "How selective is the sweetness receptor with respect to a known sweetener?," one may easily determine the dimension of A, but there is no simple selection criterion available for the B set because the known sweeteners (sucrose, sucralose, saccharine, acesulfame, etc.) belong to quite different classes of molecules. How might one determine the molecules in the reference set B?

This chapter deals with the classification problem, i.e., with the question of how to define the set A and the reference set B. The problem is strongly related to the question of molecular similarity or molecular complementarity in the region of a receptor (if this is known). We are thus looking at molecules from the point of view of a "molecular inspector" and trying to discover which molecules belong to a certain class of possible "keys" to fit some given "lock." The search becomes even more complicated when the lock cannot be specified. In this case one is looking for complementarity between arbitrary regions of one molecule and regions of a second molecule without any knowledge of the patterns being sought.

There is no doubt that the most effective search procedure—for those instances in which it can be applied—is still the "eyeball" technique used by human "searchers".[3] It is easy to see by inspection that a regularly shaped object (the key) "probably fits" into a rigid surface of complementary shape (the lock).

The second section of this chapter deals with the question of how to transform the molecular scenario into a representation for which the eyeball technique can still be used. We shall consider ways in which the molecular scenario can be transformed into one in which human pattern recognition abilities can be successfully used. New instruments of man–machine communication in molecular science will be described and the concept of molecular surfaces will be introduced. These surfaces are envisioned as the interface between different molecules or between a molecule and its solvent. This section also deals with some visualization techniques and the mapping of patterns onto the molecular surfaces; it also demonstrates that this type of molecular representation is well suited to the application of the eyeball technique.

It is well known that the eyeball technique has a variety of limitations. These are significant in all those cases when there is no way to transform the scenario into a representation where human senses are able to recognize data or features. Another limitation is related to the large number of objects within a search. If one has to check all the molecules stored in a structural database (10^4–10^5 molecular structures) to find those molecules which, in principle, can be considered as possible keys for a given receptor (set *B*; see foregoing text), the eyeball technique will no longer be applicable, simply for pragmatic reasons. Such a search can be done only by using the increasing power of modern computational technology.

How can strategies based on human recognition be used for the development of algorithms which can be applied in molecular recognition processes, at least in a preselective manner? The third section of this chapter deals with this question. We will demonstrate that fuzzy set theory offers some promise for a solution. Fuzzy set theory has already been successfully applied in different areas of pattern recognition and at different stages of the recognition process.[3] We shall demonstrate in particular that the concept of linguistic variables can prove very useful in molecular similarity search processes and in the study of complementarity.

The fourth section concerns the question of quantification of shape complementarity. Fuzzy measures, which can be used to quantify the fitting of molecular surface patches, are employed. Some conclusions are drawn in the final section.

II. TRANSFORMATION OF MOLECULAR SCENARIOS TO A THREE-DIMENSIONAL WORLD

Human abilities for pattern recognition can be successfully applied in the field of molecular recognition only when there are procedures and tools available that allow a transformation of the molecular scenario into one which can be manipulated with visual control. This can be realized by exploitation of the concept of molecular surfaces with the aid of modern graphical workstation technologies. We now consider the relevant concepts in some detail.

A. Molecular Surfaces

All intermolecular interactions can be adequately described, at least in principle, by multidimensional scalar and vector fields representing the energetics of a molecular system as functions of both intermolecular distances and orientations as well as intramolecular structure data. The visualization of these fields, however, has to be based on a three-dimensional picture or a two-dimensional projection because human pattern recognition ability is strongly related to the two- and three-dimensional world. Consequently, the multidimensional field has to be reduced to a two- or three-dimensional representation. In molecular science this can be done in many different ways.

We shall not describe all the possibilities of molecular visualizations here, but restrict our discussion to molecular surfaces and the mapping of

a

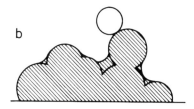
b

FIGURE 1   Hard sphere model of a molecular surface (a) and contact surface (b). The contact surface is generated by rolling a test particle (sphere) over the hard sphere model.

molecular properties onto these surfaces. We may suppose that a molecule encounters the surface of another molecule as a smooth object. Such a surface can be generated as an isosurface of the electron density function of the molecule or simply by rolling a hard sphere model particle over the hard sphere model surface (see Fig. 1).

The hard sphere model, first introduced by Connolly,[4,5] forms a kind of reference standard for molecular surface generation in many molecular modeling packages such as the MOLCAD program,[6] which was developed by the Darmstadt group. The contact surface representation gives the chemist some insight into molecular shape as it would be seen from a particle of given size. Modern workstation technology allows for real-time manipulation (translation, rotation, scaling, stereo projection, etc.), i.e., the three-dimensional world can be directly experienced. Surfaces generated with the same test particle (e.g., a water molecule with an effective sphere radius of $r = 1.4$ Å) can be qualitatively and quantitatively compared. Moreover, the contact surface generated with a water probe is well suited for our discussion of shape fitting (for example, the fitting together of two proteins).

Formal molecular surfaces have become important tools for the interpretation of molecular properties, interactions and processes.[7-9] A detailed review is to be found in Mezey's monographs.[4]

The molecular surface concept is not only useful for a representation of the bulkiness and the shape of molecules. These surfaces can also be used as screens for the visualization of many properties by means of color coding techniques. Color coding is a popular means of displaying scalar information on a surface.[10,11] Every three-dimensional scalar or vector field that may be generated on the basis of the position of atomic or molecular fragments can be visualized by color coding on a given surface.

The activity of a drug is very often related to the complementarity of the molecule and the receptor. Such complementarity is in many cases defined in quite a vague manner. This may be demonstrated by a well-known example. It has been shown by Becket[12] that the $(-)$ form (or l form) of adrenaline fits much better into a hypothetical receptor than the $(+)$ isomer (or d form). The author based his interpretation of the drug receptor interaction on the existence of three different interaction types: a flat (hydrophobic) binding site, a hydrogen bond acceptor site, and a cation–anion interaction without any reference to the microscopic details of the interaction. His model was thus formulated in a completely linguistic manner. The linguistic interactions in this model are not primarily related to specific structure elements of the adrenaline molecule, but to the reaction field between the drug and the receptor. Molecules with completely different structures may generate very similar reaction fields. This is demonstrated with several different sweeteners in Fig. 2, where the local hydrophobicity (see subsequent text) has been mapped onto the molecular surface of different sweeteners.

What are the properties that can be mapped onto the molecular surface to generate patterns that can be used for similarity analysis? In the following sections some traditional possibilities as well as new concepts are described.

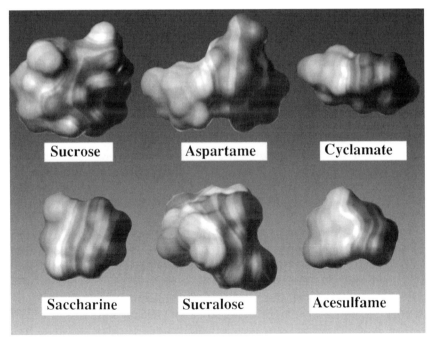

FIGURE 2 Local hydrophobicity on the surface of different sweeteners. Light grey represents hydrophilic behavior, whereas dark grey represents hydrophobic behavior.

B. Electrostatic Maps

Thermodynamically, molecular recognition is dominantly controlled by free energy changes as defined by

$$\Delta G = \Delta H - T \Delta S, \qquad \Delta A = \Delta U - T \Delta S, \qquad (2)$$

where U, H, A, G, and S are the conventional notations for internal energy, enthalpy, free energy, Gibbs free energy, and entropy, respectively. The energetics can in many cases adequately be described on the basis of electrostatic interaction. Typically the electrostatic potential of a unit charge is mapped[13] onto the surface to generate patterns that allow us to identify electrostatic complementarity. Hydrogen bonds are basically influenced by electrostatic interactions as well, although sometimes it is reasonable to map proton donor and acceptor functionality separately.[6]

C. Local Hydrophobicity

Local hydrophobicity[14] plays an important role in molecular recognition processes. It is generally accepted that the hydrophobic interaction between two molecules is related to both energetic and entropic contributions, but there is still no *simple* physical model available for hydrophobicity and hydrophobic interactions. However, there have been several attempts to define relative hydrophobicity values on the basis of empirical findings.

An empirical method for the localization, quantification, and analysis of the relative hydrophobicity of a molecule or a molecular fragment was reported[14] by the author's group. This approach is based on two assumptions, namely:

1. The overall hydrophobicity of a molecule can be obtained as a superposition of fragment contributions. This hydrophobicity may be measured, for example, by the logarithm of the partition coefficient in an octanol/water system, where $\log(P) = -RT\,\Delta G_{\text{transfer}}$ and the transfer free energy $\Delta G_{\text{transfer}}$ refers to one mole of substance going from one solvent to the other.

2. The latter free energy can be represented as a surface integral over the solvent accessible surface of the molecule on the basis of a local free energy surface density (FESD) ρ. This surface density function is represented in terms of a three-dimensional scalar field which is comprised of a sum of atomic increment functions to describe lipophilicity in the molecular environment.[14] The empirical model parameters are obtained by a least squares procedure with experimental $\log P$ values as reference data. It is found that the procedure works not only for the prediction of unknown partition coefficients but also for the localization and quantification of the contribution of arbitrary fragments to this quantity. In addition, the

formalism can also be used to obtain an estimate of the hydrophobicity index, a hypothetical log P value which depends on the actual molecular conformation.

The FESD approach as described in the preceding text can be used effectively to predict the unknown partition coefficients of molecules with a given structure. This fact has been demonstrated recently.[14]

The FESD data can be used directly to map local hydrophobicity onto the molecular surface and thereby generate patterns which then form the basis for a linguistic characterization of hydrophobic and hydrophilic parts of the molecule. This approach was recently adopted by Lichtenthaler and co-workers,[15, 16] who studied the structure–activity (sweetness) relationship of a variety of carbohydrates and other molecules. These workers followed the classical approach in which the sweet taste of organic compounds presumes the existence of a common AH-B-X glycophore (a proton donor, a proton acceptor B, and a hydrophobic group X arranged in a triangle) in all sweet substances. The sweet response is elicited via interaction with a complementary tripartite AH-B-X site in the taste bud receptor. They presumed that this "sweetness triangle" concept holds only when the hydrophobic X part is considered as an entire region with considerable flexibility rather than a specific corner of the "sweetness triangle." In sucrose and sucralose this region encompasses the outside area of the fructofuranose moiety, and in fructose the 1- and 6-CH_2 groups in either linked or separated form. This new concept[15, 16] has been developed on the basis of visual inspections of color-coded molecular electrostatic potential maps and FESD maps and has been tested on a variety of sweet compounds (see Fig. 2). Quite remarkably, the authors[16] found that the FESD profiles generated for the solid-state conformation of a variety of noncarbohydrate high-potency sweeteners, such as the sulfonamides, cyclamate, saccharine, and acesulfame, as well as structurally quite different dipeptides, e.g., aspartame, exhibit hydrophobicity distributions strikingly similar to those observed for the sugars (see Fig. 2).

D. Topographical Properties of Molecular Surfaces

Two molecules of complex structure may form a stable complex only when those parts that are important for the binding can come into close contact. From the point of view of the molecular surfaces, this means that both surfaces have to be complementary to some extent in the binding area. This surface complementarity can be identified in simple cases just by inspection of the computer-generated images, although this technique is not very useful for systematic searches. To achieve the latter, a formal classification is necessary. Several methods for the characterization of surfaces in topological terms have been proposed.[17–24] Mezey and co-workers[20] have established a method for topological analysis of contour surfaces and van der Waals surfaces represented by fused spheres. One approach of these authors is based on the calculation of a curvature parameter which is used for the classification of certain domains of the contour surface in terms of different curvature properties.

In work from our group, a numerical procedure for the calculation of the local and global canonical curvatures[17] at each point on a surface was developed which leads to domains of different quality. These domains can be characterized by *curvature profiles* which provide information about their topology. A comparison of the profiles of different molecules is helpful in the elucidation of docking procedures. The topological features of a surface can be quantified by two canonical curvatures at each surface point. Canonical curvatures are defined as the eigenvalues of the local Hessian matrix (i.e., the matrix of second derivatives). For example, if the surface is locally approximated by a paraboloid, the canonical curvatures are the curvatures along the principal axes of it. It should be noted that the curvatures of the fitted paraboloid are not identical to the "exact" canonical curvatures of the surface following the mathematical definition, which requires that we consider an infinitesimal neighborhood of a surface point. Only for the limiting case of infinite point density would the precise values be obtained. The approximate curvatures can nevertheless serve as a sufficiently accurate measurement of local surface topography as has been shown previously[17] with a few specific examples. Since we are interested in the characterization of large surface regions instead of the small areas considered so far, we introduce *global curvatures*, which are evaluated as follows. Instead of fitting a paraboloid to the direct neighbors of the reference point, now the paraboloid is matched to neighbor points at a predetermined distance on the surface (selection distance) from the reference center. These neighbor points are called selected neighbors. The curvatures of this paraboloid can be regarded as an average over the surface region bordered by the selected neighbors. The global curvatures may be thus interpreted as average curvatures of the corresponding surface region. They disregard the detailed shape resulting from atomic roughness. In other words, the surface can be smoothed, with the amount of smoothing depending on the choice of a selection distance to fix the area to be covered by the paraboloid. This enhancement of the concept of local canonical curvatures permits the classification and characterization of large surface regions, thus opening up the possibility of subdividing surfaces into domains specified by their area and curvatures (see Section III).

The comparison of two given molecular surfaces on the basis of curvature can be done interactively by the use of two-dimensional texture maps, with color coding of the two canonical curvatures calculated for different selection distances along the x and y coordinate of the texture map. However, the information from the curvature profile can be further reduced by introducing a surface topography index STI as demonstrated by

Heiden and Brickmann.[25] The surface topography index s may be defined on the basis of two global curvatures, c_1 and c_2, as follows:

$$s = \begin{cases} (c_1 - c_2)/c_1, & \text{if } c_1 > 0 \text{ and } c_2 > 0, \\ 1 + (1 - (c_1 + c_2)/c_1), & \text{if } c_1 > 0 \text{ and } c_2 \leq 0 \text{ and } |c_1| > |c_2|, \\ 2 + (c_1 + c_2)/c_2, & \text{if } c_1 > 0 \text{ and } c_2 \leq 0 \text{ and } |c_1| \leq |c_2|, \quad (3) \\ 3 + (1 - (c_2 - c_1)/c_2), & \text{if } c_1 \leq 0 \text{ and } c_2 < 0, \\ 0, & \text{if } c_1 = c_2 = 0. \end{cases}$$

The STI values vary within the interval $0 \leq s \leq 4$. When calculated from the relation of both global curvatures [each of which can be either concave $(+)$, flat (0), or convex $(-)$], where $c_1 \geq c_2$, the STI gives an expression for the regional shape of every surface point. The shape varies continuously among five basic shape descriptors, namely, bag $(+/+)$, cleft $(+/0)$, saddle $(+/-)$, ridge $(0/-)$, knob $(-/-)$, and as a special case, plateau $(0/0)$. However, information about the absolute curvature is lost during the process of STI calculation.

Based on the calculation of regional canonical curvatures, the surface topography index STI appears to give quite an accurate description of local shape and is able to relate surface regions to a set of five basic shape classes. As this definition of the STI—although continuous—certainly implies a discrete classification, the preceding method is well suited to completely automatic shape analysis algorithms. Shape analysis may be accomplished either by sharp contour cuts or—for a better characterization of local shape structures—by more sophisticated algorithms. The latter might be based, for example, on fuzzy logic strategies[30] (see next section).

A major advantage of the foregoing shape descriptor definition is the freedom of choice in the degree of globality, which is unfortunately associated with the major disadvantage of a rather large computational effort, which increases rapidly with increasing globality.

E. Surface Flexibility

Despite the great usefulness of the concept of surface topography descriptors, almost all such approaches suffer from a severe limitation: the flexibility of the molecular surfaces is not taken into account. There is no doubt, however, that a rigid surface model gives only a rough impression of the scenario involved, e.g., a ligand molecule approaching the surface of a protein. Particularly for the selectivity and specificity of enzymatic reactions, flexibility of the compounds is extremely important. Speaking in terms of the lock-and-key principle,[2] neither the lock nor the key is rigid;

however, they may accommodate each other in such a manner that optimal interaction is ensured.

In the work of Zachmann *et al.*,[26] new approaches to the quantification of surface flexibility have been suggested. The basis data for these approaches are supplied by molecular dynamics (MD) simulations. The methods have been applied to two proteins (PTI and ubiquitin). The calculation and visualization of the local flexibility of molecular surfaces is based on the notion of the *solvent accessible surface* (SAS), which was introduced by Connolly.[4,5] For every point on this surface a probability distribution $\rho(r)$ is calculated in the direction of the surface normal, i.e., the rigid surface is replaced by a "soft" surface. These probability distributions are well suited for the interactive treatment of molecular entities because the former can be visualized as color coded on the molecular surface although they cannot be directly used for quantitative shape comparisons. In Section IV we show that the ρ values can form the basis for a fuzzy definition of vaguely defined surfaces and their quantitative comparison.

III. FUZZY LOGIC STRATEGIES AND MOLECULAR RECOGNITION

In the last section we demonstrated how the molecular interactions that form the basis for all molecular recognition processes can be transformed into a scenario in which the eyeball technique can be applied by making use of the human sense of pattern recognition. It has been shown that in many cases microscopic information has to be averaged and thermodynamic concepts have to be extrapolated to a molecular scale to generate pictures that can be handled properly. All these efforts are reasonable as long as the scenario can be treated interactively with the eyeball technique. A typical example is provided by the sweetness triangle concept.[15,16] This strategy fails when there is a large variety of molecules under consideration. In this case the interactive treatment is no longer reasonable. One then has to deal with the question of how the principles, which lead to a limited recognition (from a human point of view), can be transferred to an algorithm that renders it possible to transfer the vaguely defined patterns to a computerized strategy. It is known that the dominant factor for the inhibition of the enzyme trypsin is the shape selectivity of a receptor site which forms a deep bag. A potential inhibitor either has to fit into that bag (like the benzamidine molecule) or it has to have a knob (like the natural trypsin inhibitor PTI, which has this feature). To define the class of molecules B which can be considered as possible inhibitors, one has to screen molecular shapes in a systematic manner. This can certainly be done by using deterministic algorithms,[17-24,27,28] but these algorithms are quite computer time consuming and the simple comparison of rigid surfaces may not be adequate to the problem.

Although we shall now focus our discussion on the recognition of molecular surfaces, the principles can be applied to a variety of different molecular properties (see Section II), as demonstrated by Heiden and Brickmann.[25]

To summarize the situation, there are obviously two questions in the field of automatic molecular shape recognition:

(1) How can the relevant characteristic properties of a molecular surface be classified and how can the surface be segmented into nonoverlapping patches such that shape complementarity can be formulated in a way similar to human recognition (e.g., a big knob fits into a big bag)?

(2) How can the vagueness inherently incorporated within the definition of molecular surfaces (e.g., that caused by the surface generation procedure or by molecular flexibility) be included in molecular matching strategies?

To deal with these questions one has to confront vaguely defined objects on the one hand and vaguely defined strategies to compare these objects on the other. It has been demonstrated[25] that at least parts of the answers can be given by using the technology of fuzzy logic. In this section we show that such an approach can, indeed, be adequately used to define topographical elements of the molecular surface, whereas Section IV deals with the question of how such elements can be quantitatively compared.

A. Fuzzy Logic and Linguistic Variables

The concept of the fuzzy set was introduced over 30 years ago by Zadeh.[29] After being neglected for many years, it was rediscovered in the mid-1980s for regulation in microelectronics, automatic process control, and operations research. At present, fuzzy set theory has many applications in a large variety of different fields. The reader is referred to the literature[3,30] for detailed examples.

One of the most important tools in applications of fuzzy set theory is the concept of *linguistic variables* (LV).[25] These are groups of fuzzy sets with (partially) overlapping membership functions over a common (crisp) basic variable x. To represent several classes within an LV, the membership functions should cover all the relevant definition space of the basic variable x. The overlaps of these functions define the fuzziness. A linguistic variable L, classified by n fuzzy sets A_i, can be defined as

$$L = \left\{ \left(x, \mu_{A_i}(x) \right), \ldots, \left(x, \mu_{A_n}(x) \right) \big| x \in X \right\}. \tag{4}$$

Usually, the information a decision should be based upon is given by crisp function values. For molecular surface segmentation, this may involve scalar qualities such as the surface topography index (STI), the electrostatic potential (ESP), or the free energy surface density (FSD), which are assigned to every node point on a triangulated surface. The decision itself again leads to a crisp value (in this case the binary decision between continuation or limitation of a surface domain). Decision making requires three steps:

(1) Fuzzification of crisp basic variables into linguistic variables (LVs)
(2) Fuzzy decision from different LVs using fuzzy operators (D_{LV})
(3) Defuzzification back to crisp values

B. Segmentation of Molecular Surfaces with Linguistic Variables

The segmentation of molecular surfaces can be carried out on the basis of different surface qualities.[25] Here we shall focus on a segmentation with respect to topographical criteria. The shape analysis is based on the surface topography index (STI) described previously. Following Heiden and Brickmann,[25] a six class linguistic variable

$$L = \left\{ (\text{bag}, \mu_B(s)); (\text{cleft}, \mu_C(s)); (\text{saddle}, \mu_S(s)); \right.$$

$$\left. (\text{ridge}, \mu_R(s)); (\text{knob}, \mu_N(s)); \left(\text{plateau}, \mu_p(\max(c_1, c_2))\right) \right\} \quad (5)$$

is introduced. The membership functions are schematically shown in Fig. 3.

Similarity/dissimilarity obviously plays an important role in all pattern recognition problems. The vagueness of this term itself suggests that there are numerous ways in which the dissimilarity D of two objects a and b may be defined, depending on the actual problem. Segmentation of molecular

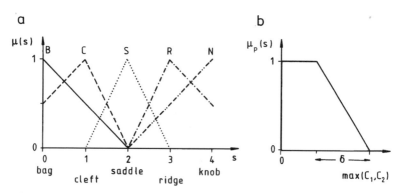

FIGURE 3   Membership functions for linguistic variables describing molecular shape.

surfaces by fuzzy dissimilarity criteria based on certain surface qualities first requires the translation of a scalar quality value into a linguistic variable. An adequate fuzzification of each quality is essential for the success of the whole procedure. The definition of membership functions should be guided by intuitive insight into the actual problem.

It has been demonstrated[25] that an automatic segmentation of molecular surfaces into distinct domains can be performed with a dissimilarity measure D_{LV} for linguistic variables which have the form

$$D_{LV}(A, B) = \frac{\sum_{i=1}^{n} w_i \left| \mu_{A_i}(x) - \mu_{B_i}(x) \right|}{\sum_{i=1}^{n} w_i \left(\mu_{A_i}(x) + \mu_{B_i}(x) \right)}, \tag{6}$$

where A and B are the linguistic variables

$$A = \left\{ \left(x, \mu_{A_i}(x) \right), \ldots, \left(x, \mu_{A_n}(x) \right) \right\},$$
$$B = \left\{ \left(x, \mu_{B_i}(x) \right), \ldots, \left(x, \mu_{B_n}(x) \right) \right\}, \tag{7}$$

and the w_i are weighting factors for the class i ($0 \leq w_i \leq 1$). From the dissimilarity definition given in Eq. (6) it follows that

$$0 \leq D_{LV}(A, B) \leq 1. \tag{8}$$

For $D_{LV} = 0$, the objects described by the linguistic variables A and B are of course identical.

In the actual calculations we started from a representation in which the molecular surfaces were given as a triangle mesh in three-dimensional space with location-dependent qualities assigned to each surface point (which is a node between adjacent triangles). The surfaces are divided into separate homologous domains. Neighboring domains differ with regard to a specified surface quality, whose value is characteristic for each domain (within a fuzzy limit). The algorithm is based on the growth of a surface domain, and starts from some characteristic reference point (for example, the point with the highest absolute STI not yet assigned to another domain). Linguistic variables are assigned in advance to each surface point and are updated continually to yield an average for the actual domain. Based on the neighborhood information given by the triangle mesh, the domain ends when the dissimilarity of a surface point to the domain average (or its direct neighbor within the domain) exceeds a given limit. The borders of other domains already defined also put an end to segment growth. By working its way sequentially through all triangle node points, the program eventually achieves complete segmentation of any triangulated surface. The result of the segmentation process is a set of surface patches of given surface area which can be uniquely related to the linguistic variables bag (B), cleft (C), saddle (S), ridge (R), knob (N), and plateau (P) [see Eq. (5)].

C. Application: Topographical Analysis of the Molecular Surfaces of the Proteins Trypsin and Trypsinogen

The use of the linguistic method can be demonstrated by the analysis of the digestive enzyme trypsin and its inactive precursor trypsinogen. Trypsin is known to cleave peptide bonds, specifically at the C-side of lysine (Lys) or arginine (Arg) side chains, and also to bind a Ca^+ ion with high affinity. The structures of the trypsin and its inactive precursor trypsinogen are very similar. However, some structural changes occur during the activation by cleavage of the N-terminal hexapeptide,[31] which obviously affects substrate affinity.

Segmentation by means of topography was carried out for trypsin (pure) and trypsinogen (three-dimensional structure without the activation hexapeptide). Trypsin was automatically divided into 36 domains, while for trypsinogen, 60 domains were found. However, disregarding insignificantly small domains (i.e., smaller than 25 Å^2 in size), there are 25 domains for trypsin and 30 for trypsinogen that show a good correlation (see Table I), i.e., in both proteins similar domains are found which have a clear one-to-one correspondence.

For each of the two molecular surfaces a large convex core domain containing more than 50% of the total surface area was found. Leaving these two domains, as well as the topographically insignificant ones, aside we have shown that homologies and differences in shape can be easily recognized. Comparing the pocket-shaped segments of both molecules, we have established that the change in shape between trypsinogen and trypsin can be seen easily[8, 25] (see Fig. 4). Another obvious structural change reveals itself in the domain pair numbered 6 (trypsin) and 10 (trypsinogen). This region lies near isoleucine 16 (Ile 16), which is the N-terminal residue in trypsin after the removal of six preceding residues from the pro-enzyme trypsinogen. It is not surprising that essential structural changes take place at this site. In fact, segment number 10 of trypsinogen, being built of the residues Ile 16-Gly 19, Ile 138-Ser 146, Pro 152, (Lys 156–158), Gly A188-Asp 194 and Val 213, Cys 220-Ala A221, can be correlated with the activation domain found by Huber and Bode,[31] which consists of four separate regions of the peptide chain: Ile 16-Gly 19, Gly 142-Pro 152, Gly A184-Gly 193 and Gly 216-Asn 223.

TABLE I Topographical Domains of Trypsin and Trypsinogen[a]

Trypsin		Trypsinogen		
Domain no.	Area ($\mathring{A}^2$)	Domain no.	Area ($\mathring{A}^2$)	Description
27	4506	30	4190	Core Domain
18	198	38	386	Isolated convex region (see Fig. 4b)
6	149	10	309	Cleft (trypsin), pocket (trypsinogen) (see Fig. 4b)
15	59	35	237	Cleft, two segments in trypsin
32	187			
14	202	14	201	Embedded plateau
31	158	33	113	Specificity pocket (see Fig. 4a)
3	151	6	126	Large pocket, two segments in
		9	61	trypsinogen
30	107	2	135	Declining cleft
12	126	5	126	Cleft branching out
4	124	32	93	Flat cleft
13	113	45	56	Cleft leading into a pocket,
		34	36	two segments in trypsinogen
2	100	16	88	Cleft
—	—	36	74	Without correlation; side chain protruding from core domain
7	71	13	61	Pocket
23	67	20	69	Flat cavity
19	64	22	66	Narrow cleft
5	60	31	62	Flat cavity
29	60	15	58	Ca^{2+} binding pocket (see Fig. 4)
20	59	1	51	Flat cavity
10	48	24	39	Narrow pocket

[a] Corresponding domains are oriented pairwise. The index numbers of the domains refer to the order of segmentation in the numerical procedure.

IV. MATCHING OF MOLECULAR SURFACES WITH FUZZY LOGIC STRATEGIES

We have shown that linguistic variables can be adequately used to discuss topographical differences of proteins in general and in the specific case of trypsin/trypsinogen. However, these linguistic descriptions could also be used for making initial conjectures in automatic docking procedures[32] (see Figs. 5 and 6). This will be discussed in the next subsection.

A. Rough Matching of Surface Patches

The positions of the linguistically characterized surface patches can be described by the coordinates of the centers of mass of these domains in

a

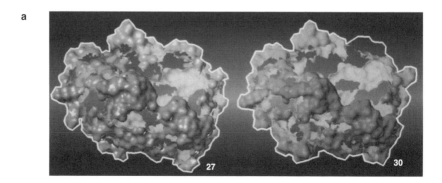

b

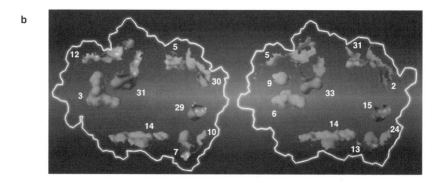

c

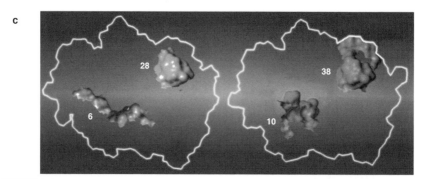

FIGURE 4 Automatic segmentation of the surface of trypsin (left) and trypsinogen (right) on the basis of the local shape (given by STI values). The numbers of the domains refer to Table I. Related domain pairs are named by domain numbers trypsin/trypsinogen, respectively. The outline of the complete surface is displayed as a white line. (a) Core domains of the two proteins. (b) Surface segments of significant sizes which can be pairwise related to each other in both proteins (see Table I). The specificity pocket is characterized by the domain pair (31/33); the Ca^{2+} binding pocket by (29/15). (c) One pair of corresponding segments (6/10) shows significant changes during trypsinogen activation.

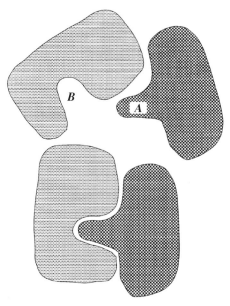

FIGURE 5 Docking of molecular surfaces according to shape complementarity: A big knob (*A*) fits into a big bag (*B*) (schematically).

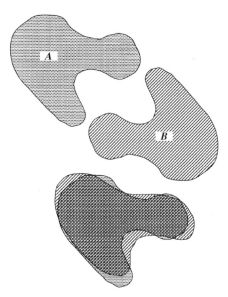

FIGURE 6 Shape fitting of similar molecules (*A*) and (*B*): One-to-one correspondence of surface elements of the same class (schematically).

space. The first trial fitting for a set of domains from a molecule A to a set for a molecule B with similar or complementary properties is performed by matching the points pairwise using the algebraic minimum distance least squares algorithm of Ferro and Hermans[33]—an algorithm commonly applied in standard molecular modeling procedures for atom/atom matching. Similarity and complementarity can thereby be completely expressed in linguistic terms. A simple example may serve to show this (see Fig. 7): A bag (at position x_1) in the neighborhood of a plateau (at x_2) and a knob (at x_3) is matched to a knob (at y_1) in the neighborhood of a plateau (at y_2) and a bag (at y_3). One cannot expect that this type of matching will lead to prediction of molecular complexes with atomic precision, but the effort involved in a systematic screening of possible arrangements of the two molecules (with three translational and three rotational degrees of freedom) can be drastically reduced. To optimize the positions of both molecules in space, one has to proceed with free energy minimization procedures. This is in any case very time consuming and is not discussed here. However, fuzzy logic strategies may successfully be applied in surface fine matching procedures as we now show.

B. Fine Matching of Surface Patches

Molecular surfaces defined on the basis of the Connolly algorithm[4, 5] (see Section II) can be effectively used in the interactive treatment of molecular scenarios. These surfaces are a representation of the repulsive

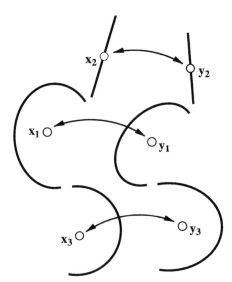

FIGURE 7   Rough matching of complementary surfaces by minimizing the squared deviations of the corresponding elements (schematically).

interaction of molecules, i.e., two molecules can be moved toward each other with a reasonable energetic effort as long as these surfaces do not substantially interfere. The situation of closest approach in which parts of the molecular surfaces are in close contact can be realized in an interactive treatment without major effort. However, there is no simple way to convert this strategy to an automatic procedure because of the following problems that arise:

(1) There is no a priori way of deciding which part of one surface should be compared with a particular part of another.
(2) The surfaces of molecules cannot be uniquely defined since a given molecule's surface is dependent on the properties of the probe molecule (whatever the surface generation procedure may be).
(3) There is no unique way to quantify the matching of two surfaces even if they are defined with arbitrary accuracy.
(4) A matching procedure should take into account the "softness" of molecular surfaces. If one knows that a certain part of a surface is quite flexible, one should not be concerned about local disagreement of the surfaces to be matched and should take this fact into consideration when designing the matching strategy.

Although there are a number of techniques that have been proposed to solve the surface matching problem,[27,28,34,35] they all suffer from the fact that an inherent uncertainty is replaced by ad hoc procedures. Even if the molecular surface concept is replaced by a one parameter family of isosurfaces,[4] this does not lead to a unique matching technique.

In this work a fuzzy matching procedure is suggested which takes the foregoing uncertainties into account at least in principle. The approach is based on a soft definition of a surface that is defined in terms of membership functions (see Fig. 8) These functions are $\mu_s(\mathbf{r})$ and $\mu_v(\mathbf{r})$ and they measure to what extent a given space point belongs to the surface and the bulk of a molecule, respectively. The matching of two molecules A and B can then be calculated in many different ways. In a first attempt[32] we used the intersection of two fuzzy sets

$$A \cap B = \{(x, \mu_{A \cap B}(x)) \,|\, x \in X\} \qquad (9)$$

with

$$\mu_{A \cap B}(x) = \min(\mu_A(x), \mu_B(x)) \qquad (10)$$

as a basis for the quantification of the surface matching of two molecules. The membership function in Eq. (10) is a limiting case of one suggested by Dubois and Prade.[36]

Maximal surface matching is obtained when the function

$$F(\mathbf{R}_A \, \mathbf{R}_B) = O_{AB}(\mathbf{R}_A, \mathbf{R}_B) - \lambda V_{AB}(\mathbf{R}_A, \mathbf{R}_B) \qquad (11)$$

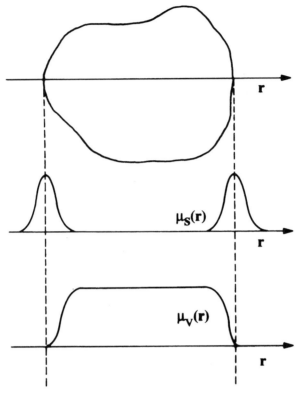

FIGURE 8 Membership functions $\mu_S(\mathbf{r})$ and $\mu_V(\mathbf{r})$ for the surface and the bulk of a molecule (schematically).

takes a global maximum. O_{AB} and V_{AB} are global measures for the surface overlap and the volume overlap, respectively, and λ is a positive control parameter. We thus have

$$O_{AB}(\mathbf{R}_A, \mathbf{R}_B) = \int \mu_{S_A \cap S_B} r \, dr^3, \qquad (12)$$

$$V_{AB}(\mathbf{R}_A, \mathbf{R}_B) = \int \mu_{V_A \cap V_B} r \, dr^3. \qquad (13)$$

Maximal surface complementarity between two molecules is reached when O_{AB} is maximal while V_{AB} takes a minimum value. The new technology has been tested in an initial application by matching the surfaces of the two flexible proteins tryspin and PTI.[32] In this application the membership functions μ_S and μ_V were calculated from molecular dynamic simulations similar to those reported earlier.[26] It turns out that the structure of the trypin-PTI complex is very close to that which was found in x-ray studies.

Studies on the refinement of the method especially in connection with domain decomposition are currently in progress.[37]

V. CONCLUSIONS

It has been demonstrated that the capacity of modern graphical workstations enables the chemist to view molecular scenarios from a molecule's point of view. Model experiments, such as the docking of a substrate to a receptor or the comparison of molecules of different structure but similar chemical activity (e.g., sweeteners), can be performed on a time scale that permits human interaction. It was shown that in particular the concept of molecular surfaces as screens for the representation of different properties is very helpful in the discussion of molecular recognition, i.e., specific intermolecular interactions. It was also established that the introduction of quantities describing local hydrophobicity, surface roughness, surface curvature, and surface flexibility lead to a better understanding of the "molecular language." The pattern recognition ability of human beings (the eyeball technique, which is of predominant importance in the field of interactive modeling) can be converted, at least in part, to a formal algorithmic concept by the use of linguistic variables and fuzzy logic strategies. This was confirmed by our comparison of the two proteins trypsin and trypsinogen. A method for the automatic segmentation of large molecular surfaces by means of topographical properties has also been presented. We conclude that the method can be used to extract those surface parts from the complete surface which are relevant to molecular recognition. Initial trials for the application of fuzzy strategies to the automatic docking of molecular surfaces have been reported as well. The use of fuzzy logic for molecular recognition problems is as yet restricted to a few applications, but the examples demonstrate the potential power of this method in this area. More systematic work, in particular the optimization of model parameters, could well lead to some promising new techniques.

ACKNOWLEDGMENTS

The author would like to thank Wolfang Heiden (Bonn) and Horst Vollhard and Carl-Dieter Zachmann (both at Darmstadt) for fruitful discussions and technical assistance. This work was supported by the Fonds der Chemischen Industrie, Frankfurt.

REFERENCES

1. M. Delaage, ed., *Molecular Recognition Mechanisms*, pp. 1–13. VCH, Weinheim/New York (1991).
2. E. Fischer, *Chem. Ber.* **27**, 2985 (1894).
3. H.-J. Zimmermann, *Fuzzy Set Theory and Its Applications*. Kluwer Academic, Dordrecht/Boston, 1991.
4. P. G. Mezey, *Reviews of Computational Chemistry*, Vol. I. *Molecular Surfaces* (K. B. Lipkowitz and D. B. Boyd, eds.), pp. 265–294. VCH, Weinheim/New York, 1990; P. G. Mezey, *Shape in Chemistry—An Introduction to Molecular Shape and Topology*. VCH, Weinheim/New York, 1993
5. M. Connolly, *Science* **211**, 709–713 (1983).
6. M. Waldherr-Teschner, T. Goetze, W. Heiden, M. Knoblauch, H. Vollhardt, and J. Brickmann, "MOLCAD—Computer Aided Visualization and Manipulation of Models in Molecular Science," in *Advances in Scientific Visualisation* (F. H. Post and A. J. S. Hin, eds.), pp. 58–67. Springer-Verlag, Berlin (1992); J. Brickmann, T. Goetze, W. Heiden, G. Moeckel, S. Reiling, H. Vollhardt, and C.-D. Zachmann, "Interactive Visualization of Molecular Scenarios with MOLCAD/SYBYL," in *Data Visualization in Molecular Science* (J. E. Bowie, ed.), pp. 84–97. Addison-Wesley, Reading, MA, 1995.
7. R. Langridge, T. E. Ferrin, J. D. Kunz, and M. L. Connolly, *Science* **211**, 661–666 (1981).
8. W. Heiden, M. Schlenkrich, and J. Brickmann, *J. Computer Aided Mol. Design* **4**, 255–269 (1990).
9. W. Heiden, T. Goetze, and J. Brickmann, *J. Comp. Chem.* **14**, 246 (1993).
10. J. C. Dill, *Computer Graphics* **15**, 153 (1981).
11. M. Waldherr-Teschner, C. Henn, H. Vollhard, S. Reiling, and J. Brickmann, *J. Mol. Graphics* **12**, 98 (1994).
12. A. H. Becket, *Fortsch. Arzneimittelforschung* **1**, 455 (1959).
13. G. Náray-Szabó, *J. Mol. Graphics* **7**, 76–81 (1989).
14. P. Pixner, W. Heiden, H. Merx, G. Moeckel, A. Möller, and J. Brickmann, *J. Mol. Inform. Comput. Sci.* **34**, 1309 (1994).
15. F. W. Lichtenthaler, S. Immel, and U. Kreis, in *Carbohydrates as Organic Raw Materials* (F. W. Lichtenthaler, ed.), pp. 1–32. VCH, Weinheim/New York, 1991; F. W. Lichtenthaler, S. Immel, and U. Kreis, **43**, 121 (1991); F. W. Lichtenthaler, S. Immel, D. Martin, and V. Mueller, in *Carbohydrates as Organic Raw Materials* (G. Descotes, ed.), Vol. 2. VCH, Weinheim/New York, 1993.
16. F. W. Lichtenthaler and S. Immel, "Sucrose, Sucralose and Fructose: Correlations between Hydrophobicity Potential Profiles and AH-B-X Assignments," in *Sweet Taste Chemoreception* (G. G. Birch, M. A. Kanters, and M. Mathlouti, eds.). Elsevier, Amsterdam, 1993.
17. C.-D. Zachmann, W. Heiden, M. Schlenkrich, and J. Brickmann, *J. Comput. Chem.* **13**, 76 (1992).
18. S. E. Leicester, J. L. Finney, and R. B. Bywater, *J. Mol. Graphics* **6**, 104 (1988).
19. P. Bladon, *J. Mol. Graphics* **7**, 130 (1989).
20. P. G. Mezey, *J. Comput. Chem.* **8**, 462 (1987); A. Arteca and P. G. Mezey, *J. Comput. Chem.* **9**, 554 (1988).
21. R. L. DesJarlais, R. P. Sheridan, G. L. Seibel, J. S. Dixon, I. D. Kuntz, and R. Venkataraghavan, *J. Med. Chem.* **31**, 722 (1988).
22. P. M. Dean, P. Callow, and P.-L. Chau, *J. Mol. Graphics* **6**, 28 (1988).
23. H. Nakamura, K. Komatsu, S. Nakagawa, and H. Umeyama, *J. Mol. Graphics* **3**, 2 (1985).
24. N. Colloc'h and J.-P. Mornon, *J. Mol. Graphics* **8**, 133 (1990).
25. W. Heiden and J. Brickmann *J. Mol. Graphics* **12**, 106 (1994).
26. C.-D. Zachmann, S. Kast, and J. Brickmann, *J. Mol. Graphics* **13**, 89 (1995).

27. P. L. Chau and P. M. Dean, *J. Mol. Graphics* **5**, 152 (1987).
28. F. Blaney, C. Edge, R. Phippen and C. Burt, *in Data Visualization in Molecular Science* (J. E. Bowie, ed.), pp. 99–129. Addison-Wesley, Reading, MA, 1995.
29. L. A. Zadeh, *Inform. Control* **8**, 338 (1965).
30. H. Schildt, *Artificial Intelligence Using C*. McGraw-Hill, New York, 1987.
31. R. Huber and W. Bode, *Acc. Chem. Res.* **11**, 114 (1978).
32. C.-D. Zachmann and J. Brickmann, unpublished.
33. D. R. Ferro and J. Hermans, *Acta Crystallogr. Sect. A* **33**, 345 (1977).
34. P. G. Mezey, *J. Comp. Chem.* **8**, 462 (1987).
35. G. A. Arteca, T. M. Gund, M. A. Hermmeier, V. B. Jaumal, P. G. Mezey, and J. S. Yadav, *J. Mol. Graphics* **6**, 45 (1988).
36. D. Dubois and H. Prade, *Int. J. Gen. Systems* **8**, 43 (1982).
37. T. Exner, J. Brickmann, unpublished.

7

The Use of Fuzzy Graphs in Chemical Structure Research

JUN XU

Oxford Molecular Group, Inc.
Baltimore, Maryland 21286

I. INTRODUCTION

One of the major tasks of chemistry is to determine the structures of various kinds of compounds. A relatively simple structure can usually be identified by inspection of its spectrum, such as its ^{13}C nuclear magnetic resonance spectrum (^{13}C NMR). In this simplest case, there is no ambiguity about the result and so additional evidence is not needed for the determination. Such an identification procedure can be classified as a crisp procedure. This means that the determination is made on the basis of traditional logic. For more complicated structures, some of their substructures can be identified by means of crisp rules. However, others cannot be identified by such rules, because their spectra are ambiguous, overlapping, and incomplete. To determine complicated structures, chemists have to make guesses based upon some kind of spectrum, such as the ^{13}C NMR spectrum. The need also arises to search for additional evidence from some other spectroscopy, such as infrared (IR) spectroscopy, to screen the guesses. This process involves putting the structural fragments together to generate candidate structures. Some of the guesses are based upon crisp rules, whereas others are based upon fuzzy rules. Fuzzy graph theory is a mathematical model that is widely used to describe the combination of crisp and fuzzy systems.

Modern spectral theory and technology have jointly brought chemical structure research into a new era. Mulidimensional NMR (nD-NMR) theory and instrumentation in particular have provided many new ways for

Fuzzy Logic in Chemistry
249

us to study various correlations among atoms in the same molecule or between molecules. New terminologies are continually being invented (see Table I) and new experiments are frequently being introduced.[1] Even a skilled spectral expect cannot interpret all routinely used spectra because each NMR spectrum has a special set of interpreting rules. Moreover, new experiments are still being developed and new rules are continually being introduced.[2] On the other hand, the interpretation of mass spectra and infrared (IR) results requires long-term skills. To solve this "information explosion" problem, a general theory is needed for computer-assisted structure research and development. The theory should be rigorous and flexible enough to tolerate ambiguity, incompleteness, redundancy, and impurity in the experimental data.[3]

Computer-assisted structure research can be divided into two parts:

(1) Computer-assisted primary structure elucidation (CASE), which focuses on relatively small molecules; and
(2) Computer-assisted three-dimensional structure determination (CASD), which is appropriate for biological macromolecules such as proteins.

In both CASE and CASD, there is a need to assign spectra to atoms or groups. An important difference is that CASE determines the primary structure through the assignments, and CASD makes the assignments based upon primary structure. CASD determines the three-dimensional structure by analyzing a spectrum which reflects the through-space correla-

TABLE I Abbreviations and Acronyms

Abbreviations	Meaning
2QF-COSY	2D double Quantum Filtered COrrelated SpectroscopY
CASD	Computer-Assisted Structure Determination
CASE	Computer-Assisted Structure Elucidation
COLOC	COrrelation spectroscopy via LOng-range Couplings
COSY	2D COrrelated SpectroscopY
DEPT	Distortionless Enhancement by Polarization Transfer
HMBC	Heteronuclear Multiple Bond Correlation spectroscopy
HMQC	Heteronuclear Multiple Quantum Correlation spectroscopy
HOHAHA	HOmonulear HArtmann–HAhn spectroscopy
INADEQUATE	Incredible Natural Abundance DoublE QUAntum Transfer Experiment
ISNet	Independent Spin coupling Network
NOESY	2D Nuclear Overhauser Effect SpectroscopY
ROESY	Rotating frame Overhauser Effect SpectroscopY
TOCSY	TOtal Correlation SpectroscopY

tions of atoms, such as the two-dimensional Nuclear Overhauser Effect SpectroscopY (NOESY). The problems CASE has also tend to be different from those of CASD. For example, CASE should generate primary candidate structures based upon a number of partially redundant, partially incomplete structural fragments from multispectra.[4] CASD's main problem on the other hand is to deal with situations where there is substantial overlap in the nD-NMR spectra.[5]

These two themes in chemistry have a great deal in common, however, and many of the algorithms developed for one can be applied with little modification to the other. Intrinsically, both CASE and CASD can be based upon fuzzy graph theory.[6]

II. FUZZY GRAPH THEORY

A crisp system is a fundamental system in the sense that it is used to make a decision. The basic rule in such a system may be represented as[7]

$$(P \subseteq B) \Rightarrow C, \tag{1}$$

where P is the test data set, B is the classification, and C is the conclusion. P can contain a set of discrete data, such as a listing of IR peaks, or a one-dimensional ^{13}C NMR peak list, or any combination of these. B is the set of descriptions of a class of subspectra, such as the expected peak positions, standard deviations, peak shapes, or multiplicities. C is either an atom or a substructure.

Equation (1) holds if the following assumptions are valid:

(1) The observed data are nonambiguous (no overlap);
(2) The observed data are complete (that is, one substructure can have more than one IR absorption peak; if all the peaks have been observed, the data are complete); and
(3) The observed data are not error-prone (due to impurity, strong noise, etc.)

Unfortunately, these assumptions do not always hold in practical structure elucidation, because observed spectral data are usually ambiguous, incomplete, and may contain artifacts or impurities. To take these considerations into account, Eq. (1) has to be modified to

$$(P \subseteq B) \Rightarrow (C, R_{P \to C}), \tag{2}$$

where $R_{P \to C}$ is the reliability ("fuzzy degree" or membership function) of the conclusion that P belongs to C. $R_{P \to C}$ can be any value between 0 and 1.

Systems that obey Eq. (1) are called crisp systems and systems that obey Eq. (2) are called fuzzy systems. The real world is actually a mixture of crisp and fuzzy systems.

To arrive at a conclusion, usually an attempt is made to use absolutely explicit evidence. If such evidence is not available, Eq. (2) may be used to rank the possible conclusions. The best ranked possible conclusion is not necessarily the correct conclusion. Therefore, to reach a reliable conclusion, additional evidence is needed. Fuzzinesss is a description or outcome of the real world, rather than a criterion used in making a decision. The fuzzy degree of a measurement is used to reduce a search space so that the correct answer may be reached more efficiently. For example, from the ^{1}H NMR spectrum of a peptide, a set of peaks (PS; parts per million) is observed as follows:

$$PS = \{8.30, 4.55, 4.13\}. \tag{3}$$

PS may be assigned to a glycine (Gly) or threonine (Thr), because PS is close to both the expected chemical shift data sets:

$$Gly = \{8.31, 3.74, 4.17\}, \tag{4}$$

$$Thr = \{8.30, 4.53, 1.15\}. \tag{5}$$

By comparing (3) against (4) and (5), it appears that $R_{PS \to Thr} > R_{PS \to Thr}$. However, we may not conclude that PS should be assigned to Thr. We have to search further "crisp" evidence to ascertain whether PS should be assigned to Gly or Thr. For example, if 4.55 and 4.13 are found to be correlated, so are 8.30 and 4.13. PS will then definitely be assigned to Gly because Thr should not have this correlation pattern; see Fig. 1.

Additional crisp evidence is to be found in the relationships among the same data set. Each peak in the data set PS can be considered as a node, whereas the relationship between two peaks is an edge. Therefore, the data set may be represented as a graph as shown in Fig. 1. This is a fuzzy graph and in such graphs the data points are not discrete, but constructed. To determine whether two fuzzy graphs belong to the same class, we have to apply both subgraph pattern recognition and numeric pattern recognition.[8]

A. Independent Spin Coupling Networks

A structure can be deduced from a set of substructures. The latter is derived from multispectra. The structure elucidation involves extracting the set of substructures from multispectra and then assembling them. Let us consider two-dimensional COrrelated SpectroscopY (COSY) as shown in Fig. 2. A cross-peak indicates that two protons in atoms have a spin coupling, which implies that they are separated by two or three bonds

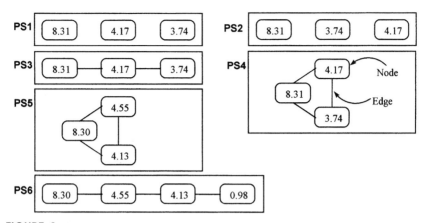

FIGURE 1 PS1 and PS2 are discrete data sets because they have no datum point correlation. PS1 = PS2. PS3 is different from PS4, although they have the exact same data points. PS5 can be assigned onto PS4, but not PS6.

(geminal or vicinal). Accordingly, given a structure, these proton spin coupling relations can be predicted. Each spin coupling pair corresponds to a COSY cross-peak. These spin coupling relations can be connected to form spin coupling topological networks [or alternatively, independent spin coupling networks (ISNets)]. A molecular structure may have more than one ISNet because its spin coupling topology may be disconnected. For the peptide fragment $\sim$ Lys $\rightarrow$ Val $\sim$, we show in Fig. 3 the relevant ISNets.

An ISNet may be defined as

$$\text{ISNet} = \{\text{CS}, \text{SC}\}, \tag{6}$$

$$\text{CS} = \{cs_1, cs_2, \ldots, cs_i, \ldots, cs_n\}, \tag{7}$$

$$\text{SC} \in \text{CS} \times \text{CS}, \tag{8}$$

where CS is a set of chemical shifts and SC is a set of spin couplings (edges). Here cs_i indicates the ith chemical shift in CS.

Before two-dimensional NMR experiments were invented, a set of discrete data points (chemical shifts) was usually assigned to a molecular structure with little or no knowledge of the correlations among the data points. There was no direct way to observe these correlations. Now multiple-dimensional NMR experiments have opened up opportunities for the analysis of the correlations between the atoms in a structure. The relationship of subspectra−substructure is no longer the relationship of the discrete data points and the substructure. Instead, it is the relationship of the graph (data points plus their correlations) and the substructure. Thus, multiple-dimensional NMR (nD-NMR) experiments offer a lot of advantages in solving spectral data overlap problems.

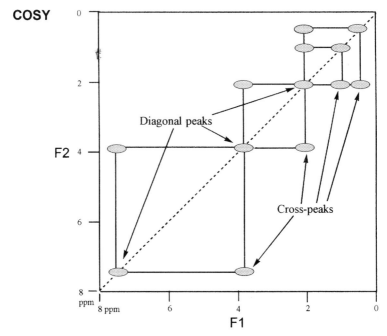

FIGURE 2 The COSY spectrum has two frequency domains, F1 and F2. Both of them represent one-dimensional ^{1}H NMR spectra. The cross-peaks have coordinates in the form (f_1, f_2). The coordinate indicates that the two protons f_1 and f_2 are coupled. This coupling can be a two- or three-bond coupling. The diagonal peaks are not informative for the two frequencies if their coordinates are equivalent. The diagonal peaks are usually crowded and cannot be easily interpreted.

The ISNet concept is different from the conventional description of a spin system. Just as an AX system implies that atom A couples with atom X, an ABX system implies that atom B couples with atom A and atom X. ISNets emphasize the global coupling network. However, a conventional spin system is usually just a fragment of an ISNet. More complicated ISNets have been discussed by Xu and Borer.[3] The cross-peaks in a three-dimensional spectrum (or three-dimensional hyperplane from a higher-dimensional spectrum) are represented by two edges sharing a common node, four-dimensional cross-peaks by three edges with two common nodes, etc. It should be noted that such a graph does not reflect the dimensionality of the nD-NMR signals. The formalism developed for two-dimensional NMR can be extended rather simply without needing to reinvent the two-dimensional formalism. The edges may also have a "color," e.g., a range of coupling constant values. The certainty of mapping observed ISNets onto the family of allowed ideal ISNets is thereby increased.[3]

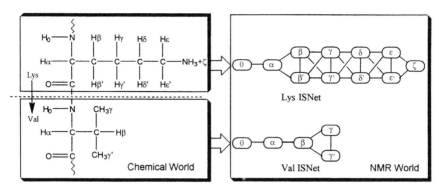

FIGURE 3 The Lys and Val residues are divided into two parts of a carbonyl group. The H^α of the Lys can couple with the H^0 of the Val. Theoretically, all the edges in Lys ISNet or Val ISNet can be observed as COSY cross-peaks. These edges represent through-bond couplings.

Figure 4 summarizes some two-dimensional NMR experiments and their interpretations from which homonuclear ISNets and heteronuclear ISNets can be extracted. Duddeck and Dietrich have discussed the details of these spectra.[9]

In practice, no experiment can give a complete peak set for a given compound. However, combining these spectra will give sufficiently complete ISNets for them to be mapped onto the entire structure. This procedure needs a number of graph-theoretical algorithms.[5]

As shown in Fig. 3, different substructures may have different ISNet patterns. In the ideal case, the assignment can be done by graphic pattern recognition. The same substructure may render a relatively different ISNet pattern for differing chemical environmental changes. The ISNet is a fuzzy graph. Fuzzy graph pattern recognition involves mapping an ISNet graph onto its cluster center.[5]

B. Cluster Centers

A structure can be considered as a supergraph SG that contains a set of spectrally observable substructures SS and an edge set E (a set of bonds which connect the substructures together) as follows:

$$SG = \{SS, E\}, \tag{9}$$

$$SS = \{S_1, S_2, \ldots, S_i, \ldots, S_n\}, \tag{10}$$

$$E \in SS \times SS. \tag{11}$$

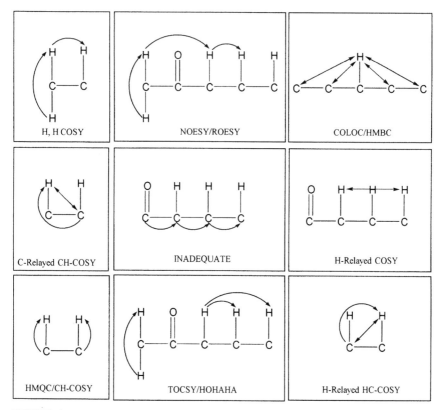

FIGURE 4   Spin coupling correlations observed from some two-dimensional NMR experiments. There are three ways to classify them: (1) short distance correlations (such as, COSY, H,C-COSY, 2D INADEQUATE) or long distance correlations (such as relayed COSY, TOCSY, HMBC); (2) homonuclear correlations (such as H,H-COSY, relayed H,H-COSY, TOCSY) or heteronuclear correlations (such as HMQC, HMBC); and (3) through-bond correlations (such as all types of COSY, TOCST, HMBC, 2D INADEQUATE) or through-space correlations (such as, NOESY, ROESY). 2D INADEQUATE is an ideal experiment to determine the carbon backbone of an unknown compound. However, its signal is too weak to be useful, as are C-relayed or H-relayed ^{1}H, ^{13}C-COSY. Long distance correlation is ambiguous, but provides good additional evidence for adding a new spin to a spin coupling system.

When NMR experiments R are applied to the structure SG, the structure is mapped onto a super-ISNet SI. Thus

$$R + SG \rightarrow SI, \qquad (12)$$

where

$$R \in \{COSY, TOCSY, INADEQUATE, \ldots\} \qquad (13)$$

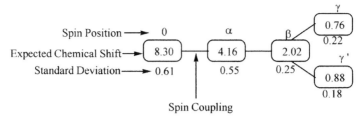

Spin Coupling

FIGURE 5 Valine ISNet cluster center. Data adapted from the paper of Gross and Kalbitzer.[11] The valine structure is shown in Fig. 3 on the left.

and

$$SI = \{ISNet_1, ISNet_2, \ldots, ISNet_i, \ldots, ISNet_n\}. \qquad (14)$$

Ideally, a given substructure corresponds to an ideal ISNet (such as supposing every geminal/vicinal proton coupling pair can be observed). See Fig. 3 for the ideal ISNets of Lys and Val. Four-bond coupling may be observed from a double bond or a "W" conformation of four σ bonds in a ring system.[10]

To assign an experimentally observed ISNet to its chemical structure, we need the ISNet cluster center of the structure. The ISNet cluster center of the valine residue is shown in Fig. 5. Accordingly, an ISNet cluster center (ICC) may be defined as

$$ICC = \{CS, SP, SD, SC\}, \qquad (15)$$

where SP is a set of the spin atom locations, such as α-H and β-H, SD is a set of standard deviations, which define the fuzzy degree of a given chemical shift, and SC and CS have been defined in Eqs. (6)–(8).

In practice, the observed ISNets are fuzzy because:

(1) The peaks may be missed or overlapped (as shown in Fig. 1, a peak represents a spin coupling; the coupling is an edge of an ISNet), and

(2) The observed chemical shift of a given spin (or atom) may have a larger deviation from the expected value due to a change of the chemical environment.

An experimentally observed ISNet is thus the fuzzy subgraph of the ISNet cluster center. Here, the word "fuzzy" in fuzzy graph pattern recognition has two meanings:

(1) The observed chemical shift value is spread around the expected value.

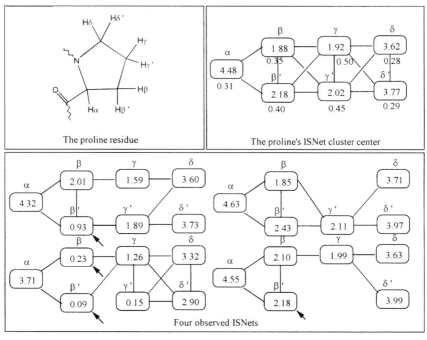

FIGURE 6   The proline residue and its ISNet cluster center (top). Four observed ISNets can be assigned onto the proline ISNet cluster center by means of fuzzy graph pattern recognition.[5] The four observed ISNets are the fuzzy subgraphs of the proline ISNet cluster center. The chemical shifts shown by arrows are far from the expected values. For example, the expected β-proton chemical shift value is 1.88 or 2.18 ppm; some of the observed values are 0.93, 0.23, and 0.09 ppm even though the correct assignments are still made through the spin coupling topologies. Normally, δ-δ' couplings cannot be observed for the cross-peaks because they are too close to diagonal peaks. This is why these peaks are missing.

(2) The observed ISNet is the topological subgraph of the ISNet cluster center because some peaks are missing.

For example, from protein basic pancreatic trypsin inhibitor (BPTI), four experimental ISNets are observed from a double quantum filtered COSY spectrum as shown at the bottom of Fig. 6. No one of them is completely matched onto its ISNet cluster center.

C. Fuzzy Graph Pattern Recognition for ISNet

To map experimental ISNets onto their ISNet cluster centers, the ISNets have to be subjected to two screening procedures, namely:

(1) A spin coupling topological match
(2) A fuzzy graph similarity calculation

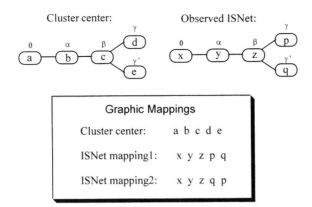

FIGURE 7 Graph match and mappings.

When an ISNet is matched to its cluster center, it may output a number of topological mappings as illustrated in Fig. 7. The best mapping will be used for the fuzzy graph similarity calculation.

Let MS be the mapping set, RM a match algorithm, and ISNet_i and ISNet_j two ISNet graphs. The graph matching procedure may then be represented as

$$\text{RM}(\text{ISNet}_i, \text{ISNet}_j) \rightarrow \text{MS}(\text{ISNet}_i \rightarrow \text{ISNet}_j); \quad (16)$$

MS is defined by

$$\text{MS} = \Sigma \, \text{MS}(y) \in \text{ISNet}_i(\text{CS}) \times \text{ISNet}_j(\text{CS}), \quad (17)$$

$$\text{MS}(y) = \left\{ (cs_{i1} \rightarrow cs_{j1}), (cs_{i2} \rightarrow cs_{j2}), \dots, (cs_{in} \rightarrow cs_{jn}) \right\}, \quad (18)$$

$$cs_i \in \text{ISNet}_i, \quad (19)$$

$$cs_j \in \text{ISNet}_j, \quad (20)$$

where n is the number of the spin nodes of ISNet_j.

If each observed chemical shift obeys a Gaussian distribution, the membership function of each mapping pair $\mu(cs_{ix} \rightarrow cs_{jx})$ may be calculated as

$$\mu(cs_{ix} \rightarrow cs_{jx}) = \exp\left\{ -\tfrac{1}{2}\left([cs_{ix} - cs_{jx}] \mid \delta_{jx} \right)^2 \right\}. \quad (21)$$

If $S_{MS(y)}$ is the similarity of the yth mapping, then $S_{MS(y)}$ can be calculated from

$$S_{MS(y)} = \sqrt[n]{\prod_{x=1}^{n} \mu(cs_{ix} \to cs_{jx})}, \tag{22}$$

$$= \sqrt[n]{\prod_{x=1}^{n} \mu(cs_{ix} \to cs_{jx})}, \tag{23}$$

$$= \lfloor \mu(cs_{ix} \to cs_{jx}) \rfloor_y. \tag{24}$$

It is interesting to analyze (22)–(24) individually. Equation (22) indicates that $S_{MS(y)}$ can be zero if the value of one of the member functions is zero. Equation (23) indicates that $S_{MS(y)}$ is a generalized Euclidean distance that will not be zero unless every μ value is zero. Equation (24) is a fuzzy similarity relationship that always takes the minimum μ value as the similarity, no matter what other μ values are.

It is critical to select one of functions (22)–(24) to calculate the graph similarity. The selection should be based upon the nature of the problem. If every node in the graph appears simultaneously, Eq. (22) should be chosen. If the similarity obeys fuzzy logical rules, then Eq. (24) should be chosen. If every node in the graph is not required to appear simultaneously, then Eq. (23) can be used. In our problems we selected Eq. (22) to calculate the graph similarities.

As previously indicated, the yth mapping has a similar value $S_{MS(y)}$. If m mappings have been found between $ISNet_i$ and $ISNet_j$, then the ISNet graphic similarity (that is, the correctness of $ISNet_i$ belongs to $ISNet_j$) is calculated as

$$S_{G(ISNet_i \to ISNet_j)} = \lceil SG(y) \rceil, \tag{25}$$

where $y = [1 \cdots m]$ and m is the number of the mappings. That is to say, the ISNet graphic similarity $S_{G(ISNet_i \to ISNet_j)}$ is the maximal mapping similarity $S_{MS(y)}$. For example, from a small peptide NAc-t21a 2QF-COSY subspectrum, we obtain an ISNet as shown in Fig. 8. The partial fuzzy pattern recognition results for the ISNet of Fig. 8 are listed in Table II. From review of Table II, we can conclude that the ISNet of Fig. 8 can be assigned to a valine residue, because $S_{(ISNet \to Val)} \gg S_{(ISNet \to Leu)}$, etc. However, if $S_{(ISNet \to Val)} \approx S_{(ISNet \to Leu)}$, this conclusion cannot be made unless additional evidence is forthcoming.

III. FUZZY GRAPH THEORY APPLICATIONS IN COMPUTER-ASSISTED BIOPOLYMER NMR ASSIGNMENT

A number of efficient computer programs for most aspects of the process of determining high-resolution structures of small biopolymers from nD-NMR have been reported.[12] Programs for the transformation

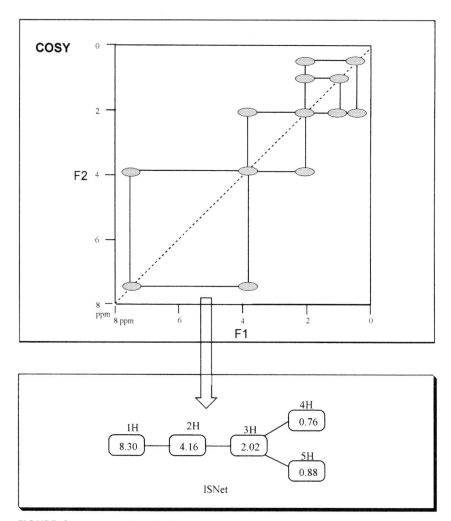

FIGURE 8 NAc-t21a 2QF-COSY subspectrum (top) and corresponding ISNet (bottom). Data adapted from Xu et al.[5] In the ISNet, each chemical shift is randomly named as 1H, 2H, etc. The fuzzy graph pattern recognition algorithm will assign atom names and positions to them by mapping operations and fuzzy logic.

from a time domain to a frequency domain, signal enhancement, peak picking, peak list accounting, the nuclear Overhauser effect and torsional constraint determination, distance geometry and restrained molecular dynamics calculations, and modeling have been fully developed and are commercially available. The computerized resonance assignment, which is a crucial step, however, is not well developed. Several computer-assisted proton assignment software packages, such as ANSIG,[13] EASY,[14] and

TABLE II Partial Fuzzy Pattern Recognition Results for the ISNet of Fig. 8

ICC[a]		1H		2H		3H		4H		5H		S_{MS}^{c}
Val	NH	0.77[b]	α H	0.95	β H	0.98	γ H □	0.91	γ H	0.96	0.77
	NH	0.77	α H	0.95	β H	0.98	γ H	0.67	γ H □	0.96	0.67
			Number of mappings = 16				$S_G = 0.77^{d}$				
Leu	NH	0.77	α H	0.86	β H □	0.50	γ H	0.18	β H	0.11	0.11
	NH	0.77	α H	0.86	β H □	0.50	β H	0.22	γ H	0.07	0.07
	NH	0.77	α H	0.86	β H	0.44	γ H	0.18	β H □	0.02	0.02
	NH	0.77	α H	0.86	β H	0.44	β H □	0.05	γ H	0.07	0.05
	NH	0.00	α H	0.00	γ H	0.17	δ H □	0.88	δ H	0.94	0.00
	⋮		⋮		⋮		⋮		⋮		⋮
			Number of mappings = 16				$S_G = 0.11$				
Glu	NH	0.74	α H	0.70	β H □	0.98	γ H □	0.00	γ H	0.00	0.00
	NH	0.74	α H	0.70	β H □	0.98	γ H	0.00	β H □	0.00	0.00
	NH	0.00	β H □ 0.00		γ H	0.62	γ H □	0.00	β H	0.00	0.00
	⋮		⋮		⋮		⋮		⋮		⋮
			Number of mappings = 24				$S_G = 0.00$				
Arg	NH	0.87	α H	0.70	β H	0.71	γ H □	0.21	γ H	0.12	0.12
	NH	0.87	α H	0.70	β H □	0.71	γ H □	0.21	β H	0.17	0.17
	NH	0.87	α H	0.70	β H □	0.71	γ H	0.25	γ H □	0.10	0.10
	α H	0.00	α H	0.00	γ H □	0.32	δ H □	0.00	δ H	0.00	0.00
	⋮		⋮		⋮		⋮		⋮		⋮
			Number of mappings = 116				$S_G = 0.17$				
	⋮		⋮		⋮		⋮		⋮		⋮

[a] ISNet cluster center.
[b] Calculated using (21).
[c] Calculated using (24).
[d] Calculated using (25).

CLAIRE,[15] have been reported for proteins. ANSIG is essentially an assignment support system, or "electronic drawing board."[15] The others are experience-based systems including a number of programs to help in the process of assigning two-dimensional proton NMR spectra of proteins. These experience-based systems emulate manual assignment procedures. They usually start at the amide-α proton region and then rank the spin patterns with experience-based scoring rules. If the amide-α cross peak is missing, these programs usually have trouble proceeding. For example, EASY cannot assign proline residues because they do not have amide protons. It is common that amide-α proton coupling peaks are absent when the sample is dissolved in D_2O, or amide protons are substituted, or the peaks are broadened. Also, some chemical shifts of protons may have such large deviations from the expected values that the built-in scoring rules may lead to incorrect assignments.

Most proton NMR resonance assignments of proteins are based upon Wüthrich's strategy.[16] In fuzzy logic and fuzzy graph theory, the strategy can be formalized into the following procedures:

(1) Generation of spin coupling patterns (ISNets) from COSY-type spectra
(2) Mapping ISNets onto possible residues
(3) Sequential assignments to connect protein residues

A. ISNet Generation

To generate an ISNet, we need to merge COSY cross-peaks belonging to the same ISNet. If we have two COSY cross-peaks $P_{c1}(\omega_i, \omega_j)$ and $P_{c2}(\omega_{i'}, \omega_k)$, where the ωs are the chemical shift values, then given a tolerance t (usually ~ 0.01 ppm), if $|\omega_i - \omega_{i'}| < t$, we postulate that P_{c1} and P_{c2} share the same frequency. Therefore, we may merge P_{c1} and P_{c2} at ω_i. Rigorously, P_{c1} and P_{c2} may still belong to two different ISNets, because P_{c1} and P_{c2} may overlap at ω_i. So, we may not merge P_{c1} and P_{c2} at ω_i unless we can prove this merge by additional evidence. Total COrrelation SpectroscopY (TOCSY) is one of the NMR experiments that can provide the additional evidence.

Let us consider a heavy overlap situation involving three COSY cross-peaks $P_{c1}(\omega_i, \omega_j)$, $P_{c2}(\omega_{i'}, \omega_k)$, and $P_{c3}(\omega_{i''}, \omega_l)$, and two TOCSY cross-peaks $P_{T1}(\omega_{j'}, \omega_{k'})$, and $P_{T2}(\omega_{j''}, \omega_{l''})$. These are all in the amide-α proton coupling region (the primes and double primes denote close similarity, e.g., ω_i, $\omega_{i'}$, and $\omega_{i''}$, have nearly the same frequency value, as shown in Fig. 9).

In Fig. 9, with a given tolerance T_m, P_{c1}, P_{c2}, and P_{c3} overlap at ω_i. Three possible interpretations need to be examined (see Fig. 9 on the right):

(1) If $(d_1 = |\omega_i - \omega_{i'}| < T_m)$, $(d_2 = |\omega_j - \omega_{j'}| < T_c)$, and $(d_3 = |\omega_k - \omega_{k'}| < T_c)$, then P_{c1} and P_{c2} belong to the same ISNet.
(2) If $(d_4 = |\omega_i - \omega_{i''}| < T_m)$, $(d_5 = |\omega_j - \omega_{j''}| < T_c)$ and $(d_6 = |\omega_1 - \omega_{1''}| < T_c)$, then P_{c1} and P_{c3} belong to the same ISNet.
(3) P_{c1}, P_{c2}, and P_{c3} may belong to three different ISNets.

If both P_{c2} and P_{c3} are suggested to be merged with P_{c1}, then two criteria are applied to decide which merge is accepted, namely:

(1) If one of the merges produces the ISNet which cannot be matched onto any ISNet cluster center, the merge will be rejected;
(2) Otherwise, the one which has a better merge degree (MD) will be accepted. The MDs are calculated as follows:

$$MD_1 = \left\{ 1 - \sqrt[3]{(\delta_1/T_m) \times (\delta_2/T_c) \times (\delta_3/T_c)} \right\}, \qquad (26)$$

$$MD_2 = \left\{ 1 - \sqrt[3]{(\delta_4/T_m) \times (\delta_5/T_c) \times (\delta_6/T_c)} \right\}. \qquad (27)$$

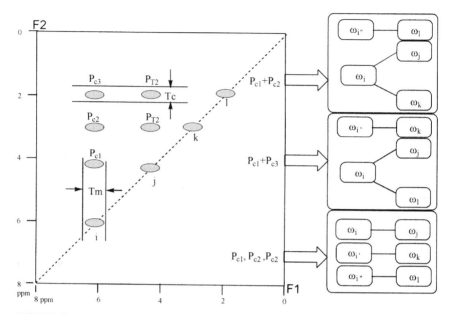

FIGURE 9   The interpretation of a COSY-type spectrum and a TOCSY-type spectrum. Only half side peaks are listed. Two spectra are overlaid. T_m is the merger tolerance for comparing two peaks in the same spectrum (here a COSY spectrum); T_c is the comparison tolerance for comparing two peaks from different spectra (here COSY and TOCSY spectra).

B. Integration of ISNets

Because of missing peaks, an ISNet may be broken into a number of ISNet fragments. These fragments may be reconnected to a complete ISNet by means of long distance spin coupling correlations from TOCSY-type spectra. Missing peaks may also be found from different types of correlated nD-NMR spectra. An example is shown in Fig. 10. In this example, the MeBmt residue has one ISNet cluster center. However, we get three ISNet fragments from a COSY spectrum. By using long distance correlation, such as (2.669, 1.745) in a five-bond coupling observed from TOCSY, these fragments have been integrated.

Some ISNet fragments cannot be connected owing to the nature of the structure itself. Tertiary carbons (no attached protons) or NMR-silent nuclei, for instance, can cut off spin coupling topology. By combining different types of two-dimensional NMR spectra, these intrinsically disconnected ISNets can be integrated and mapped to the corresponding structure. Figure 11 shows another example of using two-dimensional heteronuclear NMR spectra to integrate ISNet fragments.

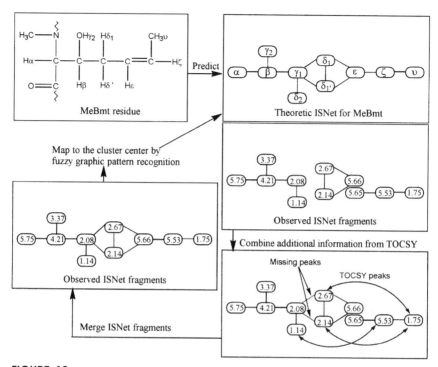

FIGURE 10 Integration of ISNet fragments using long distance coupling correlations (dual arrow lines) from TOCSY-type spectra.[17]

C. Sequence-Specific Assignments

A protein or peptide consists of a number of amino acid residues (substructures). The same type of residue may repeatedly appear in different places in the protein. For example, BPTI (basic pancreatic trypsin inhibitor) has six arginines (Arg), four prolines (Pro), two aspartic acids (Asp), four phenylalanines (Phe), six cysteines (Cys), etc.[5] The sequence-specific assignment involves determining which ISNet belongs to which specific residue, e.g., assigning the 54th ISNet to Arg1, the 55th ISNet to Pro2, the 33rd ISNet to Asp3, etc. Since some ISNets can be observed only partially (missing peaks), the ISNet may have more than one candidate assigning residue.

Direct and indirect methods are used to make sequence specific assignments. The direct method is to use a number of larger than two-dimensional hetero NMR spectra to generate the heteronuclei-based interresidue backbone atom correlations.[18] One of the schemes of the direct method is shown in Fig. 12. These three-dimensional NMR experiments need not only [13]C- and ^{15}N-enriched samples, but also very expen-

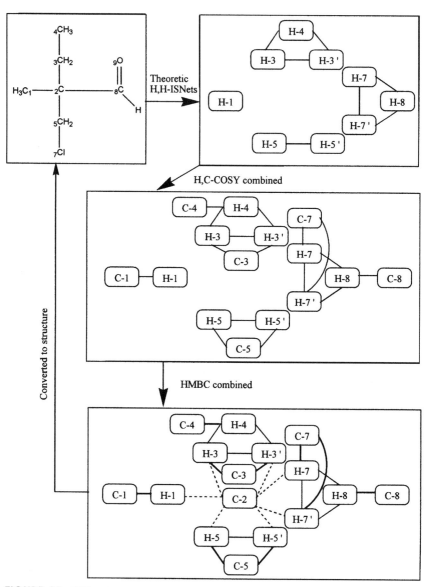

FIGURE 11 ISNet fragments are integrated by combining heternuclear spin coupling correlations and long distance coupling correlations. Thick lines represent H-C one-bond coupling correlations. Thin lines represent two-bond or three-bond coupling correlations. Dashed lines represent three-bond long distance H-C coupling correlations. The ISNet containing C and H atoms is called a hetero ISNet. The ISNets shown in this figure are predicted; the experimentally observed ISNets are the fuzzy subsets of the ISNet cluster centers.

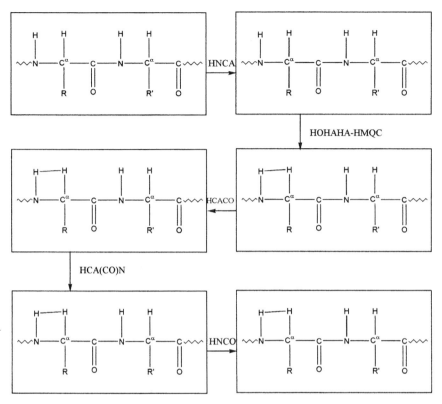

FIGURE 12 Direct bonded coupling correlation scheme for sequence-specific assignment using three-dimensional heteronuclear NMR spectra. HNCA: a three-dimensional NMR experiment that generates spin coupling correlations among amide protons, isotoped nitrogen, and α-carbons; HOHAHA: homonuclear Hartmann–Hahn spectroscopy, apparently TOCSY-type spectroscopy; HMQC: heteronuclear multiple quantum correlation that creates the correlation of a carbon and its attached proton; HOHAHA-HMQC: three-dimensional NMR combining the correlations of HOHAHA and HMQC spectra; HCACO: three-dimensional NMR creates correlations among the α-proton, α-carbon, and the carbon atom of a carbonyl group; HNCO: three-dimensional NMR creates the correlations among amide protons, isotoped nitrogen, and the carbon atom of the carbonyl group at the next residue.

sive computing resources. This approach is usually used in world-class NMR laboratories for relatively larger proteins, such as the proteins with ~ 100 residues.[19]

The indirect method is based upon two-dimensional nuclear Over-hauser effect spectroscopy (NOESY) experiments. For each residue, the H^0 (amide proton), H^α, and H^β are called backbone protons. Based upon the statistics of short proton–proton distances in protein crystal structures, Wüthrich and co-workers[16] summarized the relationships between the

TABLE III Statistics of Short ^{1}H-^{1}H Distances in Protein Crystal Structures (after Wüthrich[16])

Type	Distance (Å)	NOESY peak intensity	Probability (%) $(j - i) = 1$
$d_{NN(i,j)}$	≤ 2.4	Strong	98
	≤ 3.0	Medium	88
	≤ 3.6	Weak	72
$d_{\alpha N(i,j)}$	≤ 2.4	Strong	94
	≤ 3.0	Medium	88
	≤ 3.6	Weak	76
$d_{\beta N(i,j)}$	≤ 2.4	Strong	79
	≤ 3.0	Medium	76
	≤ 3.6	Weak	66

distances of backbone protons and the probabilities of these protons belonging to two neighboring residues (see Tables III and IV).

In Table III, d_{NN} denotes the distance from one H^0 (amide proton) to the other residue's H^0, with $d_{\alpha N}$ the distance from one H^α to the other residue's H^0, and $d_{\beta N}$ the distance from one H^β to the other residue's H^0. When two NOESY evidences of d_{NN}, $d_{\alpha N}$, and $d_{\beta N}$ are found at the same time, the probability increases as shown in Table IV.

Due to the geometrical restriction, $d_{\alpha N(i,j)}$ and $d_{N\alpha(i,j)}$ are mutually exclusive. That is, if a proton $H^\alpha(i)$ is observed to be close to the next residue's amide proton $H^N(i + 1)$, we should not observe $H^N(i)$ to be close to $H^\alpha(i + 1)$. As shown in Fig. 13, the correlations represented by single direction arrows and the ones represented by dashed single direction arrows cannot appear simultaneously.

To make sequential assignments by using NOESY peaks, the backbone protons have to be recognized. This can be done by fuzzy graph pattern recognition. An ISNet may be recognized as the candidate of Val or Lys, etc. This relation is called the ISNet graph-to-residue graph (GTR). The GTR graph can be converted to a residue-to-graph graph (RTG).

TABLE IV Combined Interresidue Probability Estimation Based on NOESY Cross-Peaks (after Wüthrich[16])

Type	Distance (Å)	NOESY peak intensity	Type	Distance (Å)	NOESY peak intensity	Probability (%) $(j - i) = 1$
$d_{\alpha N(i,j)}$	≤ 3.6	$\geq$ Weak	$d_{NN(i,j)}$	≤ 3.0	$\geq$ Medium	99
$d_{\alpha N(i,j)}$	≤ 3.6	$\geq$ Weak	$d_{\beta N(i,j)}$	≤ 3.4	$\geq$ Weak	95
$d_{NN(i,j)}$	≤ 3.0	$\geq$ Medium	$d_{\beta N(i,j)}$	≤ 3.0	$\geq$ Weak	90

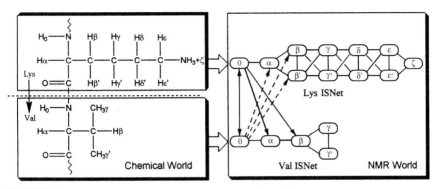

FIGURE 13 Sequential nuclear Overhauser effects (NOEs). If the Val residue is next to the Lys residue, the dashed arrows will not be observed as shown in the figure on the right, and vice versa.

Both GTR and RTG are supergraphs. For example, cyclosporin A is a small peptide consisting of 11 residues. From its 2GF-COSY and TOCSY spectra, 11 ISNets have been found.[17] After fuzzy graph pattern recognition, its GTR and RTG are generated as shown in Fig. 14.

The NOESY-based sequential assignment procedure involves searching a path in the RTG graph on the ISNet candidate side. At the beginning, the backbone protons (usually, there are amide-α, and -β protons) of ISNets are recognized and labeled. If a chemical shift can be

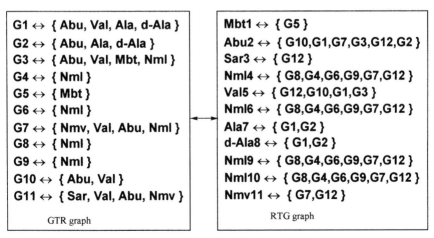

FIGURE 14 The GTR graph and RTG graph of cyclosporin A. Mbt, Abu, Sar, Nml, and Nmv are the names of nonstandard amino acids.[20] Data adopted from the paper of Xu, Weber, and Borer.[17]

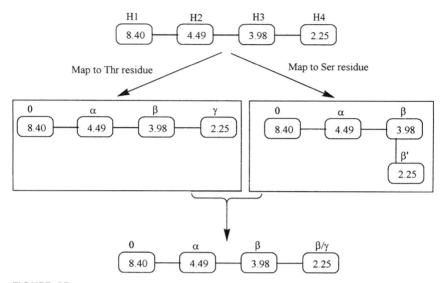

FIGURE 15   Backbone proton labelling. If the ISNet is mapped to threoline (Thr), spin H4 is not a backbone proton, but if it is mapped to serine (Ser), spin H4 is a backbone proton. We will assume that H4 is a possible backbone proton and search for it against NOESY peaks to prove its existence.

labeled either as a backbone proton or a non-backbone proton, it should be labeled as a backbone proton. By means of this labeling, we ensure that all possible backbone NOESY correlations are taken into account. Also, we avoid any search of non-backbone NOESY peaks to reduce the computing complexity. This backbone labeling procedure is illustrated by Fig. 15 with an example. The sequence-specific assignment based upon a NOESY spectrum can be made by the constrained tree search algorithm.[17]

IV. STRUCTURE ELUCIDATION RESEARCH BASED UPON MULTIPLE SPECTRA

Much of the research and development on computer-assisted structure elucidation (CASE) has been reported in recent decades. The key problem with CASE is that the system should integrate different pieces of information from multiple spectra (such as one- and two-dimensional ^{1}H, ^{13}C NMR spectra, IR spectra, and mass spectroscopy spectra), filter redundant information, generate a reasonable number of structural candidates, verify the candidates, and eventually suggest the best structure(s) for the chemist.

There is thus a need to investigate the interpretation rules for different types of spectra and the corresponding data structures of various knowledge bases.

A. Rules in Multiple Spectral Knowledge Bases

To formalize a rigorous structure elucidation procedure, the data structures for multiple spectra should be well defined. The data structure represents the relationship between structure and spectrum (or spectra). Based upon the data structure, we can build CASE applications. The data characteristics of spectra routinely used for structure elucidation are summarized in Table V.

1. Characteristics of ^{1}H NMR and ^{13}C NMR Substructure–Subspectrum Rules

The ^{1}H chemical shift range is $-1 \sim 15$ ppm, and the ^{13}C chemical shift range is $-40 \sim 230$ ppm.[21] The one-dimensional ^{1}H NMR spectrum has a greater overlap than the ^{13}C NMR spectrum because the ^{1}H NMR spectrum has a much narrower chemical shift range. A one-dimensional ^{1}H or ^{13}C NMR cluster center (substructure–subspectrum rule) consists of the following seven components:

(1) A substructure graph (SSG)
(2) An assigned atom (AA) (that is, a ^{1}H atom or a ^{13}C atom)
(3) A number of assigned atoms (NA) (^{1}H NMR only)
(4) An expected chemical shift value (ECS) that obeys some type of statistical distribution
(5) An expected tolerance range (ETR) [lower limit $\sim$ upper limit (in parts per million)]
(6) A standard deviation (δ)
(7) A multiplicity (MP)

An example to illustrate these rules is shown in Fig. 16.

TABLE V   The Structure–Spectra Relationships in Different Spectra

Spectroscopy	Atom(s)[a]	Peak(s)[a]	Spectral unit	Spectral peak description
1D NMR	1	1	ppm (or Hz)	Position, volume, integral area, splitting, phase
nD-NMR ($n \geq 2$)	≥ 1	1	ppm (or Hz)	Coordinates, volume, integral area, splitting, phase, correlations
IR	> 1	≥ 1	cm^{-1} (or cm)	Position, intensity, half-width (shape), multiplicity
MS	> 1	1	m/Z	Position, intensity
UV	> 1	≥ 1	nm (or Å)	$\lambda_{max(nm)}$, ε_{max} (%), solvent

[a] The relationship between atoms and observed peaks in the spectrum.

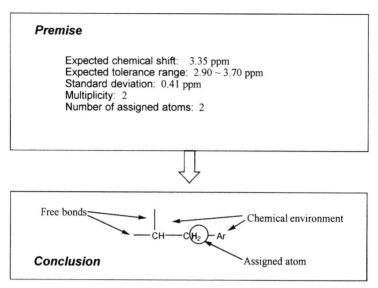

FIGURE 16 A premise–conclusion pair (cluster center) of ^{1}H NMR spectral knowledge.

Based upon this knowledge representation for a given ^{1}H peak p, if p is in the expected tolerance range, the conclusion substructure graph ssg can be deduced. The membership of p belonging to $ssg\,\mu(p \to ssg)$ can be calculated using Eq. (21). Due to severe peak overlaps, different deductions may have the same $\mu(p \to ssg)$ value. It is also possible that an incorrect deduction can render a better $\mu(p \to ssg)$ value. Therefore, the conclusion cannot be based only upon a single membership function value.

The substructure graph (SSG) part consists of three components:

(1) Free bonds (FB): the bonds may connect to other substructure fragments (a free bond type can be single, double, aromatic, etc.)
(2) A fixed chemical environment (FCE)
(3) Variable chemical environments (VCE) (such as Markush groups)

All these components are important constraints in structure generation.

2. Characteristics of IR Substructure–Subspectra Rules

Usually an infrared peak represents the information on a bond or a group of bonds, that is a substructure, rather than a single atom. A substructure may show more than one infrared peak. Therefore, an infrared cluster center is a multiple–multiple relation.

An infrared substructure–subspectra rule may contain more than one spectral band. The band has the following components:

(1) An expected wave number (EWN) (that is, an expected infrared peak position)
(2) An expected tolerance range (ETR) [lower limit ~ upper limit (per centimeter)]
(3) A half-width (expected half-width of the peak)
(4) A standard deviation (δ)
(5) An intensity (expected peak intensity)
(6) A type: vibration type; and
(7) An assigned bond (where the vibrations come from).

The conclusion part of the rule may also have a Markush group. An example is shown in Fig. 17.

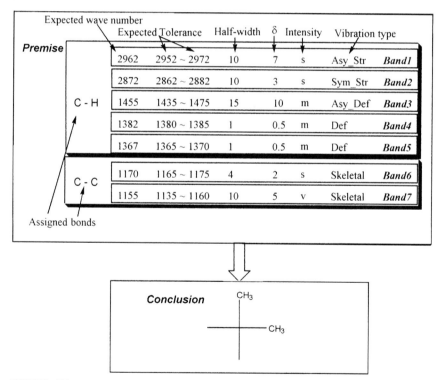

FIGURE 17 A premise–conclusion pair (cluster center) of IR spectral knowledge. Data partially adopted from IR Mentor (BIO-RAD Laboratories, Sadtler Division, 1994). To identify the conclusion substructure, seven IR peaks should be observed in the listed seven bands simultaneously. Intensity is indicated as s, strong, m, medium, and v, variable.

Let IRP(i) be the ith set of infrared peaks and $i = [1..n]$. Based upon the infrared rule, defined for n bands, a concluded substructure G is deduced. If $\mu_{\text{IRP}(i) \to G}$ be the membership function of IRP(i) belonging to substructure G, $\mu_{\text{IRP}(i) \to G}$ is calculated as

$$\mu_{\text{IRP}(i)} = \sqrt[2]{\prod \exp - \tfrac{1}{2} \left[(P_j - C_j) \,|\, \delta_j + |I_j - CI_j| + |W_j CW_j| \right]^2}, \quad (28)$$

where $j = [1..n]$, n is the number of spectral peaks in the band, P_j is the peak observed in the jth band, C_j is the cluster center of the jth band (expected value), δ_j is the standard deviation of the jth band, I_j is the intensity of P_j, CI_j is the expected intensity of P_j, W_j is the half-width of P_j, and CW_j is the expected half-width of P_j. It should be noted that the $\mu_{\text{IRP}(i) \to G}$ value includes the differences of peak positions, intensities, and peak shapes. If one of the infrared bands is not found, the membership function $\mu_{\text{IRP} \to G}$ will become zero. In practice, for instance, in a mixture, peaks may be hidden in strong bands. Equation (28) may be modified as follows to allow for this kind of missing peak(s):

$$\mu_{\text{IRP}(i)} = \sqrt[2]{\sum \exp - \tfrac{1}{2} \left[(P_j - C_j) \,|\, \delta_j + |I_j - CI_j| + |W_j - CW_j| \right]^2}. \quad (29)$$

Equations (28) and (29) are generalized. They can work well even if the infrared peak has no intensity or peak shape measurement, i.e., when $|I_j - CI_j| = 0$ or $|W_j - CW_j| = 0$. This feature is required because it may be hard to measure infrared peak intensity or peak shape.

3. Characteristics of Mass Spectroscopy Substructure–Subspectrum Rules

There are two types of mass spectroscopy rules that concern (1) common fragment ions and (2) common fragments lost. Both of these rules can be defined in the same type of data structure. The components of the data structure are:

(1) Mass-to-charge ratio m/z (for ions or molecular ion minus); and
(2) Inference substructure(s).

Here, the m/z is a premise and the inference substructure is the conclusion. An m/z premise may lead to more than one conclusion substructure. These conclusion substructures may all be correct or partially correct. The structures will be different than those obtained from nuclear magnetic resonance (NMR) and infrared (IR) spectra. In NMR, one peak is correlated to one atom, whereas in IR multiple peaks are correlated to one substructure. In mass spectroscopy, one peak can be correlated to more than one substructure. Figure 18 shows an example.

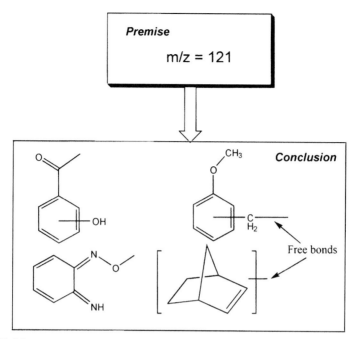

FIGURE 18   A premise–conclusion pair (cluster center) of mass spectrometry knowledge. Data adopted from Silverstein, Bassler, and Morrill.[22] The terpene substructure has only one free bond, but it may appear anywhere in the substructure. The OH and CH_2 groups may be connected to any position on the benzene ring. These Markush structures can be more complicated.

In mass spectroscopy, two more undetermined factors exist in peak–substructure correlation rules:

(1) A mass spectroscopy peak can be correlated to a number of substructures; there is no priority in this set of subgraphs. This is why a complex chemical structure cannot be determined by mass spectroscopy peaks only. These multiple conclusions have to be screened by the information from other spectra, such as IR or NMR.

(2) The free bond position can be undetermined. For example, in the terpene substructure of Fig. 18, the free bond position is fixed at a specific position. The position cannot be fixed until other evidence is provided.

Again, in IR and NMR, there are always peak overlaps. However, in mass spectroscopy, substructure overlap occurs very often.

B. Structure Deduction from Multiple Spectra

As previously discussed, different types of spectra have different data structure and interpretation rules. To deduce the structure from these multiple spectra, we need to choose one of the spectra as the base spectrum and then start our analysis. Typically, the base spectrum should have an explicit correlation between substructure and subspectra. For instance, the ^{13}C NMR spectrum provides explicit structure–spectrum relations. We now select the one-dimensional ^{13}C NMR spectrum as the base spectrum to show one of the structure elucidation strategies. Other spectra, such as infrared or two-dimensional NMR spectra, are used as constraints to reduce the search space. This strategy is outlined in Fig. 19.

In this scheme, the most computationally time-consuming procedure is the connection of the atoms or substructures to generate structure candidates. We illustrate this scheme via elucidation of the structure of gibberellic acid (GBA). The molecular formula of gibberellic acid was determined as $C_{19}H_{22}O_6$ from mass spectroscopy. By using a ^{13}C distortionless

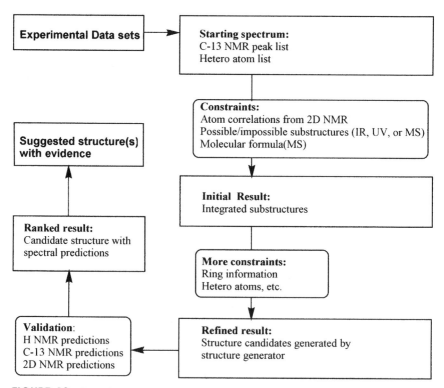

FIGURE 19   One of the general schemes for structure elucidation from multiple spectra.

enhancement by polarization transfer (DEPT) spectrum, the initial assignments may be made as shown in Table VI. The bold atoms in the column titled Related Substructures are correlated to the NMR assignments. The assignment is based upon a substructure prediction from the ^{13}C NMR chemical shift. The predicted substructure has a central atom (bold letter); the atoms and bonds surrounding the central atom are the chemical environment.

By adding ^{13}C, ^{1}H-COSY spectral peaks, Table VI becomes Table VII. Also, more entries are filled. The H, C atomic connections are generated, although they are far from complete. Combined with ^{1}H, ^{1}H-COSY peaks, Table VII becomes Table VIII. More C, C atomic connections are generated by COSY. In the column Bond Adjacency, the bond information is recorded. For example, at row 4, column 6, "5/2, 9/1" means that the

TABLE VI Deduction of the Gibberellic Acid Structure: Generated from Mass Spectroscopy and DEPT

No.	Atom	Protons	^{13}C shift	^{1}H shift	Connection	Related substructure	2D NMR peaks	IR peaks
1	C	0	179.09			$-CO-O-$, $-CO-OH$		
2	C	0	173.39			$-CO-O-$, $-CO-OH$		
3	C	0	157.18			$CH_2=C\big\langle$		
4	C	1	133.29			$\big\rangle C=CH-$		
5	C	1	131.42			$\big\rangle C=CH$		
6	C	2	106.51			$\big\rangle C=CH_2$		
7	C	0	90.17			$-C(=O)-O-C(-\big\langle$		
8	C	0	76.86			$\big\rangle C(OH)-$		
9	C	1	68.67			$\big\rangle CH-OH$		
10	C	0	53.30			$\big\rangle C\big\langle$, etc.		
11	C	1	52.42			$-CH-R$, etc.		
12	C	1	50.84			$-CH-R$, etc.		
13	C	1	50.66			$-CH-R$, etc.		
14	C	0	49.78			$\big\rangle C\big\langle$, etc.		
15	C	2	44.40			$-CH_2-$, etc.		
16	C	2	42.90			$-CH_2-$, etc.		
17	C	2	38.62			$-CH_2-$, etc.		
18	C	2	16.75			$-CH_2-$, etc.		
19	C	3	14.43			$-CH_3$		
20	O							
21	O							
22	O							
23	O							
24	O							
25	O							

TABLE VII   The Deduction of the Gibberellic Acid Structure: ^{13}C, ^{1}H-COSY Peaks Combined

No.	Atom	Hs	^{13}C shift	^{1}H shift	Connection	Related substructure	2D-NMR peaks	IR peaks
1	C	0	179.09			$-CO-O-$, $-CO-OH$		
2	C	0	173.39			$-CO-O-$, $-CO-OH$		
3	C	0	157.18			$CH_2=C\big\langle$		
4	C	1	133.29	5.876		$\big\rangle C=CH-$		
5	C	1	131.42	6.334		$\big\rangle C=CH$		
6	C	2	106.51	5.190, 4.885		$\big\rangle C=CH_2$		
7	C	0	90.17			$-C(=O)-O-C(-\big\rangle\big\langle$		
8	C	0	76.86			$\big\rangle C(OH)-$		
9	C	1	68.67	3.970		$\big\rangle CH-OH$		
10	C	0	53.30			$\big\rangle C\big\langle$, etc.		
11	C	1	52.42	3.131		$-CH-R$, etc.		
12	C	1	50.84	2.597		$-CH-R$, etc.		
13	C	1	50.66	1.911		$-CH-R$, etc.		
14	C	0	49.78			$\big\rangle C\big\langle$, etc.		
15	C	2	44.40	1.911, 1.696		$-CH_2-$, etc.		
16	C	2	42.90	2.315, 2.187		$-CH_2-$, etc.		
17	C	2	38.62	1.987, 1.759		$-CH_2-$, etc.		
18	C	2	16.75	1.835, 1.606		$-CH_2-$, etc.		
19	C	3	14.43	1.149		$-CH_3$		
20	O							
21	O							
22	O							
23	O							
24	O							
25	O							

fourth carbon is connected to the fifth carbon with a double bond and is connected to the ninth carbon with a single bond.

By adding infrared spectral peaks, Table VIII becomes Table IX. Thus, substructures $-CO-O-$, $-COOH$, $>C(OH)-$ have more confirmation. More two-dimensional NMR spectra, such as the heteronuclear multiple bond correlation (HMBC) spectrum, may add more backbone connections to Table IX. By depiction technology, the connection table in Table IX is converted to a chemical structure picture for the

TABLE VIII Deduction of the Gibberellic Acid Structure:
^1H, ^1H-COSY Peaks Combined

No.	Atom	Hs	^{13}C shift	^1H shift	Connection	Related substructure	2D NMR peaks	IR peaks
1	C	0	179.09			$-CO-O-$, $-CO-OH$		
2	C	0	173.39			$-CO-O-$, $-CO-OH$		
3	C	0	157.18			$CH_2=C\big\langle$		
4	C	1	133.29	5.876	5/2, 9/1	$\big\rangle C=CH-$	5.876−6.335	
							5.876−3.970	
5	C	1	131.42	6.334	4/2	$\big\rangle C=CH$	5.876−6.335	
6	C	2	106.51	5.190,	16/2	$\big\rangle C=CH_2$	5.190−2.187	
				4.885			5.190−2.315	
							4.885−2.315	
7	C	0	90.17			$-C(=O)-O-C(-\big\langle$		
8	C	0	76.86			$\big\rangle C(OH)-$		
9	C	1	68.67	3.970	4/1	$\big\rangle CH-OH$	5.876−3.970	
10	C	0	53.30			$\big\rangle C\big\langle$, etc.		
11	C	1	52.42	3.131	12/1	$-CH-R$, etc.	3.131−2.597	
12	C	1	50.84	2.597	11/1	$-CH-R$, etc.	3.131−2.597	
13	C	1	50.66	1.911		$-CH-R$, etc.		
14	C	0	49.78			$\big\rangle C\big\langle$, etc.		
15	C	2	44.40	1.911,		$-CH_2-$, etc.		
				1.696				
16	C	2	42.90	2.315,	6/2	$-CH_2-$, etc.	5.190−2.187	
				2.187			5.190−2.315	
							4.885−2.315	
17	C	2	38.62	1.987,		$-CH_2-$, etc.		
				1.759				
18	C	2	16.75	1.835,		$-CH_2-$, etc.		
				1.606				
19	C	3	14.43	1.149		$-CH_3$		
20	O							
21	O							
22	O							
23	O							
24	O							
25	O							

chemists' review. A structure generation algorithm will connect the free bonds implied in Table IX and produce suggested structure candidates. Figure 20 shows the partial structure generation results from Table IX. A spectral prediction program is used to evaluate the candidate structures S1 and S2. The efficiency of the structure generator has been a problem.

TABLE IX   Deduction of the Gibberellic Acid Structure: IR Peaks Combined

No.	Atom	Hs	^{13}C shift	^{1}H shift	Connection	Related substructure	2D NMR peaks	IR peaks
1	C	0	179.09		20/2,21/1	—CO—O—, —CO—OH		1755, 2913
2	C	0	173.39		22/2,23/1	—CO—O—, —CO—OH		1755, 2913
3	C	0	157.18			CH_2=C<		
4	C	1	133.29	5.876	5/2,9/1	>C=CH—	5.876–6.335 5.876–3.970	
5	C	1	131.42	6.334	4/2	>C=CH	5.876–6.335	
6	C	2	106.51	5.190, 4.885		>C=CH_2		
7	C	0	90.17			—C(=O)—O—C(—)<		
8	C	0	76.86		24/1	>C(OH)—		3620, 1152
9	C	1	68.67	3.970	4/1,25/1	>CH—OH	5.876–3.970	3620, 1110
10	C	0	53.30			>C<, etc.		
11	C	1	52.42	3.131	12/1	—CH—R, etc.	3.131–2.597	
12	C	1	50.84	2.597	11/1	—CH—R, etc.	3.131–2.597	
13	C	1	50.66	1.911		—CH—R, etc.		
14	C	0	49.78			>C<, etc.		
15	C	2	44.40	1.911, 1.696		—CH_2—, etc.		
16	C	2	42.90	2.315, 2.187		—CH_2—, etc.		
17	C	2	38.62	1.987, 1.759		—CH_2—, etc.		
18	C	2	16.75	1.835, 1.606		—CH_2—, etc.		
19	C	3	14.43	1.149		—CH_3		
20	O	0			1/2			
21	O	1			1/1			
22	O	0			2/2			
23	O	1			2/1			
24	O	1			8/1			
25	O	1			9/1			

Bangov and Simova[23] reported a structure generator algorithm that uses fuzzy logic and two-dimensional NMR information to enhance the performance.

V. SUMMARY

Modern spectral experiments, i.e., nD-NMR, IR, mass spectroscopy, and UV, produce many different types of substructural information. This information is incomplete, redundant, ambiguous, impure, and in different

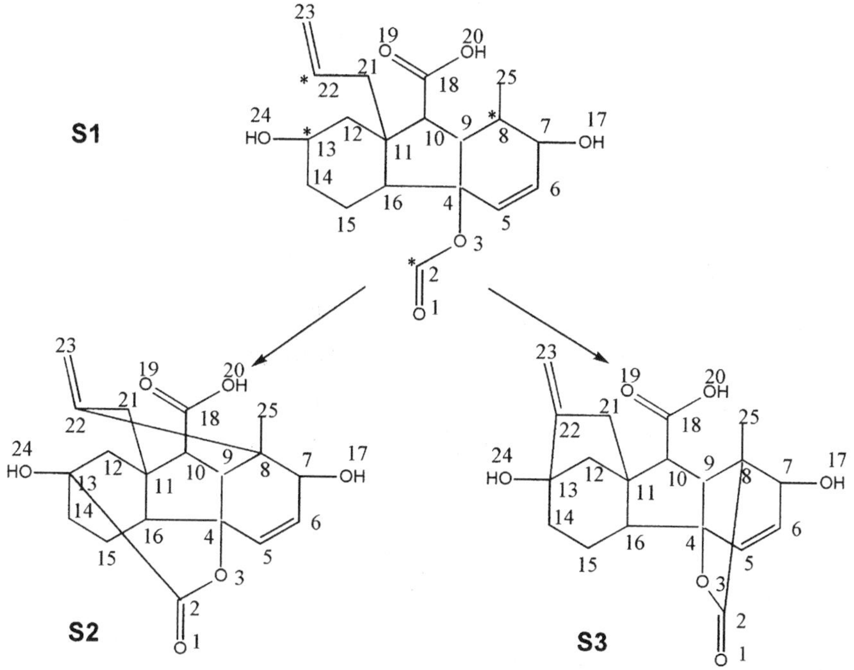

FIGURE 20 Structures generated from an incomplete structure for the gibberellic acid structure elucidation. S1: incomplete structure generated from Table IX; S2: incorrect candidate structure; S3; the gibberellic acid structure. The asterisks denote free bonds.

forms. CASE or CASD research involves building a logical program system to filter incorrect and redundant information, integrate incomplete information, generate structure candidate(s), and verify the candidate(s). This system should be based upon a rigorous mathematical foundation. Otherwise, the chemist will not trust the software system. One of the mathematical foundations may well be fuzzy graph theory.

ACKNOWLEDGMENTS

Dr. B. A. Woods is acknowledged for valuable advice in preparing the manuscript. The NMR spectra of gibberellic acid were provided by Dr. I. Pelczer and the infrared spectrum comes from BIO-RAD Laboratories, Sadtler Division

REFERENCES

1. S. W. Homans, *A Dictionary of Concepts in NMR*. Oxford Univ. Press, Oxford, 1989.
2. J. Xu, B. C. Sanctuary, and B. N. Gray, *J. Chem. Inf. Comput. Sci.* **33**, 475 (1993).
3. J. Xu and P. N. Borer, *J. Chem. Inf. Comput. Sci.* **34**, 349 (1994).

4. N. A. B. Gray, *Computer-Assisted Structure Elucidation*. Wiley, New York, 1986.
5. J. Xu, S. K. Straus, B. C. Sanctuary, and L. Trimble, *J. Chem. Inf. Comput. Sci.* **33**, 668 (1993).
6. A. Kaufmann, *An Introduction to the Theory of Fuzzy Subsets*, Vol. 1. Academic Press, San Diego, 1975.
7. N. J. Nilsson, *Principles of Artificial Intelligence*. Tioga, Portola Valley, CA, 1980.
8. J. Xu and M. Zhang, *Tetrahedron Computer Methodology* **2**, 75–83 (1989).
9. H. Duddeck and W. Dietrich, *Structure Elucidation by Modern NMR: A Workbook*. Springer-Verlag, New York, 1989.
10. J. K. M. Sanders and B. K. Hunter, *Modern NMR Spectroscopy, A Guide for Chemists*. Oxford Univ. Press, London, 1988.
11. K. H. Gross and H. R. Kalbitzer, *J. Magn. Reson.* **76**, 87 (1989).
12. R. R. Ernst, *in Computational Aspects of the Study of Biological Macromolecules by Nuclear Magnetic Resonance Spectroscopy* (J. C. Hoch, F. M. Poulsen, and C. Redfield, eds.), Vol. 1. Plenum, New York, 1991.
13. P. J. Kraulis, *J. Magn. Reson.* **88**, 601 (1989).
14. C. Eccles, P. Güntert, M. Billeter, and K. Wütherich, *J. Biomol. NMR* **1**, 111 (1991).
15. G. J. Kleywegt, R. Boelens, M. Cox, M. Llinas, and R. Kaptein, *J. Biomol. NMR* **1**, 23 (1991).
16. K. Wüthrich, *NMR of Proteins and Nucleic Acids*. Wiley, New York, 1986.
17. J. Xu, P. L. Weber, and P. N. Borer, *J. Biomol. NMR* **5**, 183–192 (1995).
18. T. L. James and V. J. Basus, *Ann. Rev. Phys. Chem.* **42**, 501 (1991).
19. M. Ikura, L. W. Kay, and A. Bax, *Biochemistry* **29**, 4659–4667 (1990).
20. H. Kessler, H. R. Loosli, and H. Oschkinat, *Helv. Chim. Acta* **68**, 661 (1985).
21. E. Brietmaier, *Structure Elucidation by NMR in Organic Chemistry*. Wiley, New York, 1993.
22. R. M. Silverstein, G. C. Bassler, and T. C. Morrill, *Spectrometric Identification of Organic Compounds*, 5th ed. Wiley, New York, 1991.
23. I. Bangov and S. Simova, *J. Chem. Inf. Comput. Sci.* **34**, 546 (1994).

8

Fuzzy Logic in Computer-Aided Structure Elucidation

IVAN P. BANGOV*

Institute of Organic Chemistry
Bulgarian Academy of Sciences
Sofia 1113, Bulgaria

I. WHY IS FUZZY LOGIC NECESSARY?

Computer-aided structure elucidation (CASE) has been developed over a period of 30 years. Several major projects that were initiated and carried out during this time consist, in the main, of interdisciplinary efforts toward the automatic solution of problems arising from the everyday practice of chemists. The DARC system of Dubois et al.,[1-6] the Heuristic DENDRAL project developed at Stanford University,[7-16] and the CHEMICS system of Sasaki et al.,[17-23] as well as the developments of Munk and co-workers[24-28] and Bremser,[29] may be considered among the pioneers in this field. The DENDRAL system is still referred to as one of the most promising successes of artificial intelligence (AI).

Despite the many extensive investigations carried out over the past three decades, the ultimate goal of this research has not yet been achieved. This goal can be formulated as *the determination by computer of the structure of an unknown compound by feeding into the computer appropriate spectral information.* The problems arising from the automated inference of the structure of an unknown molecular moiety from varied spectral information seemed to be extremely difficult. Hereafter we discuss these problems and offer a strategy of our own for their solution.

The concept of molecular structure, consisting of atoms and their mutual connections, is fundamental in chemical theory. Knowledge about

---

*Present address: Sadtler Division BIO-RAD, 3316 Spring Garden Street, Philadelphia, PA 19104-2596.

Fuzzy Logic in Chemistry
283

the structure of a given compound provides chemists with significant information concerning its properties. As expressed by Munk,[30] the relevant chemical relationship may be formally written as

$$\text{property} = f(\text{structure}). \tag{1}$$

If the property is a spectrum, Eq. (1) can be rewritten as

$$\text{spectrum} = f(\text{structure}). \tag{2}$$

Hence, determination of the chemical structure of an unknown compound can be represented by the inverse of relationship (2):

$$\text{structure} = f^{-1}(\text{spectrum}). \tag{2'}$$

Let the spectrum be described mathematically by a specific vector S with components s_i which take the values 0 and 1, with the dimension of this vector being the range of the spectrum, given in corresponding units. In the case of a ^{13}C NMR spectrum, for instance, an integer vector of dimension 241 may be used to indicate the range 0–240 ppm, each element representing 1 ppm. The values of the vector elements represent the presence or absence of signals as follows:

$$S_i = \begin{cases} 1, & \text{a signal is present,} \\ 0, & \text{no signal is present.} \end{cases}$$

Such an approximate vector representation is depicted in Fig. 1. The vector components have only two meanings, which correspond to the two *truth values* in classical propositional logic: *true* (1) or *false* (0).

On the other hand, chemical structure is algebraically treated in chemical graph theory[31, 32] by the so-called *molecular graph G*. The

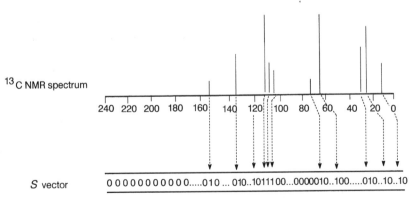

FIGURE 1 Vector representation of ^{13}C NMR spectrum.

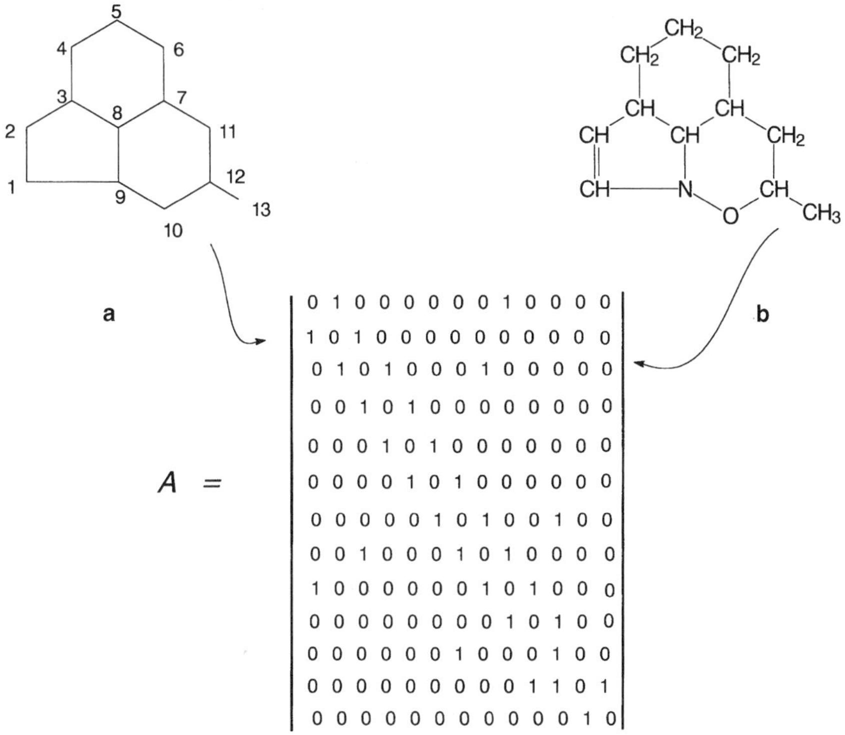

$$A = \begin{vmatrix} 0 & 1 & 0 & 0 & 0 & 0 & 0 & 0 & 1 & 0 & 0 & 0 & 0 \\ 1 & 0 & 1 & 0 & 0 & 0 & 0 & 0 & 0 & 0 & 0 & 0 & 0 \\ 0 & 1 & 0 & 1 & 0 & 0 & 0 & 1 & 0 & 0 & 0 & 0 & 0 \\ 0 & 0 & 1 & 0 & 1 & 0 & 0 & 0 & 0 & 0 & 0 & 0 & 0 \\ 0 & 0 & 0 & 1 & 0 & 1 & 0 & 0 & 0 & 0 & 0 & 0 & 0 \\ 0 & 0 & 0 & 0 & 1 & 0 & 1 & 0 & 0 & 0 & 0 & 0 & 0 \\ 0 & 0 & 0 & 0 & 0 & 1 & 0 & 1 & 0 & 0 & 1 & 0 & 0 \\ 0 & 0 & 1 & 0 & 0 & 0 & 1 & 0 & 1 & 0 & 0 & 0 & 0 \\ 1 & 0 & 0 & 0 & 0 & 0 & 0 & 1 & 0 & 1 & 0 & 0 & 0 \\ 0 & 0 & 0 & 0 & 0 & 0 & 0 & 0 & 1 & 0 & 1 & 0 & 0 \\ 0 & 0 & 0 & 0 & 0 & 0 & 1 & 0 & 0 & 0 & 1 & 0 & 0 \\ 0 & 0 & 0 & 0 & 0 & 0 & 0 & 0 & 0 & 1 & 1 & 0 & 1 \\ 0 & 0 & 0 & 0 & 0 & 0 & 0 & 0 & 0 & 0 & 1 & 0 \end{vmatrix}$$

FIGURE 2 Mathematical representation of chemical structure as: (a) a graph; (b) a molecular graph; (c) an adjacency matrix.

mathematical notion of a graph G is usually defined in terms of two sets,

$$G = (V, E), \tag{3}$$

where V is the set of *vertices* and E is the set of *edges*. Classical set theory was first formulated by Cantor. In colloquial usage the term "set" is taken to mean a collection of things or objects which in some sense belong together or are akin.[33] The things or objects s_i which form the set S are called *elements of the set* and one writes $S_i \in S$. In contrast, an item that is not an element of a given set is denoted as $S_i \notin S$. Accordingly, we can write the sets of vertices and edges as $v_i \in V$ and $e_{ij} \in E$. As one can see from Fig. 2, there is one-to-one correspondence between the mathematical definition of a graph and the classical notion of chemical structure, consisting of a set of atoms (A) and a set of bonds (B). Hence, the

TABLE I   One-to-One Correspondence between Some
Graph-Theoretical and Chemical Notions and the Relevant Type
of Information[a]

Graph-theoretical notion	Notation	Chemical notion	Notation	Type of information
Set of vertices	V	Gross (empirical) formula	A	Crisp
Vertex	$v \in V$	Atom	$a \in A$	Crisp
Vertex color	—	Atom type in a defined hybridization state or of defined atom attribute	—	Fuzzy
Bonding site	BS	Free valence	f_v	Crisp
Vertex degree	$d = \Sigma BS^a$	Atom valency	$n = \Sigma f_v^a$	Crisp
Edge (arc)	$e_{ij} = (v_i, v_j) \in E$	Chemical bond	$b_{ij} = (a_i, a_j) \in B$	Fuzzy
Graph	$G = (V, E)$	Molecular structure	$S = (A, B)$	Fuzzy
Colored graphs	G	Structures with heteroatoms or atoms of different atom attributes	S	Fuzzy
Subgraph	G^s	Chemical group	—	Crisp
		Fragment	F	Fuzzy
Extension	G'	Partial structure obtained after adding a new vertex to a subgraph	—	Fuzzy

[a] Reprinted with permission from *J. Chem. Inf. Comput. Sci.* (ref. 87), 1992, American Chemical Society and *Discrete Appl. Math* (ref. 80), 1996 Elsevier Science-NL.

molecular graph can be rewritten as

$$G^M = (A, B). \qquad (4)$$

The correspondence between mathematical graph theory and classical chemical structure theory is manifested in Table I. A widely used mathematical representation for graphs is the adjacency matrix (A). The rank of this matrix equals the number of the vertices (atoms), and its entries a_{ij} are equal to either 0 or 1:

$$a_{ij} = \begin{cases} 1, & \text{a bond is present between atoms } i \text{ and } j, \\ 0, & \text{no bond is present between atoms } i \text{ and } j. \end{cases}$$

Representations of a graph, molecular graph, and their adjacency matrix are shown in Figs. 2(a), (b), and (c), respectively. It is apparent that the elements a_{ij} may be treated, again, as *Boolean truth values* of classical logic. Hence, the relationship $(2')$ may be written in matrix form as

$$
\begin{bmatrix}
a_{11} & a_{12} & \cdots & a_{1n} \\
a_{21} & a_{22} & \cdots & a_{2n} \\
\vdots & \vdots & \vdots & \vdots \\
a_{n1} & a_{n2} & \cdots & a_{nn}
\end{bmatrix}
= f^{-1}
\begin{bmatrix}
s_1 \\
s_2 \\
\vdots \\
s_n
\end{bmatrix}.
\tag{5}
$$

Thus formalized, the structure elucidation problem consists of determining the A matrix elements (0 and 1) from the S vector elements (also 0 and 1). From information about the presence ($s_i = 1$) or absence ($s_i = 0$) of signals in different places within the spectrum range, we shall draw conclusions about the presence ($a_{ij} = 1$) or absence ($a_{ij} = 0$) of bonds between the atoms i and j within the chemical structure. Note that in relationship (5) we have values that are sharply defined (1 or 0) for both the spectrum parameters and the chemical bonds. Such truth values and the corresponding logic have been designated as *crisp*.[34, 35]

Although the formalized definition (5) of our structure elucidation problem appears straightforward and may look extremely useful for automation of the problem, it does not match real-world cases for at least two reasons:

First, relationships (1) and (2) may be considered correct problems from a purely mathematical point of view; i.e., the number of knowns is equal to or greater than the number of unknowns. However, from the matrix representation (5) of the inverse problem $(2')$ (as for most inverse problems), one can see that this is not mathematically correct: the number of unknowns (a_{ij}) is much greater than the number of knowns (s_i). The information content of any spectrum, except perhaps for the two-dimensional (2D) NMR spectra of some molecules, is not sufficient for a complete determination of the molecular structure of a unknown compound. This derives from the very essence of the information provided by different spectral methods. Thus, some chemical groups can be derived only from the most sophisticated infrared (IR) spectra. The same may be stated for proton nuclear magnetic resonance (^{1}H NMR) spectra. The mass spectrum can give information both about the molecular ion (in the most favorable cases), which is related to the empirical molecular (gross) formula, and about the masses of some fragments. However, most of these masses are not specific to a given structure or class of structures. Thus, the most frequently observed peak at m/e 18 due to the CH_2 group does not provide any useful information because these groups are present in each organic compound. Even the information richer carbon-13 nuclear magnetic resonance (^{13}C NMR) spectra cannot provide for a complete determination of molecular structure. In fact, in the most favored case of no signal overlap, the broad-band proton decoupled spectrum gives the number of carbon atoms in the studied molecule, which coincides with the number of signals along with their chemical shifts. This in turn means that we obtain information about the subset of carbon atoms in the set of all atoms of a given structure (the vertex set A of the molecular graph), but no or rather vague information (derived from the chemical shifts) about their mutual connnectivities, i.e., about the bond set B. The latter appears more important for the solution of the structure elucidation problem.

Two-dimensional nuclear magnetic resonance (2D NMR) techniques are the richest sources of structural information. They allow Eq. (5) to be formally expressed as

$$
\begin{bmatrix}
a_{11} & a_{12} & a_{13} & \cdots & a_{1n} \\
a_{21} & a_{22} & a_{23} & \cdots & a_{2n} \\
a_{31} & a_{32} & \cdots & \cdots & a_{3n} \\
\vdots & \vdots & \vdots & \vdots & \vdots \\
a_{n1} & a_{n2} & a_{n3} & \cdots & a_{nn}
\end{bmatrix}
= f^{-1}
\begin{bmatrix}
s_{11} & s_{12} & s_{13} & \cdots & s_{1n} \\
s_{21} & s_{22} & s_{23} & \cdots & s_{2n} \\
s_{31} & s_{32} & \cdots & \cdots & s_{3n} \\
\vdots & \vdots & \vdots & \vdots & \vdots \\
s_{n1} & s_{n2} & s_{n3} & \cdots & s_{nn}
\end{bmatrix}. \quad (6)
$$

Here, the spectrum has the form of a matrix. One or more 2D spectra may be used for a complete determination of the structure of the unknown compound. Such a combination may be the use of C,H HETCOR (*Carbon,* H*ydrogen* HET*eronuclear* COR*relation* spectroscopy), providing the C—H adjacent bonding, and H,H COSY (H*ydrogen,* H*ydrogen* CO*rrelation* S*pectroscop*Y), giving the vicinal H—H coupling. However, there is also uncertainty inherent in these spectra. As in the case of H,H COSY spectrum, it is difficult to discriminate between the vicinal and longer range H—H couplings. Even the most powerful *inadequate* technique presents uncertainty, consisting of a lack of signals, overlapping problems, and so on.

Second, even if the number of knowns in Eq. (5) were greater than or equal to the number of unknowns, a new type of uncertainty would be observed. As stated previously, we set the values of the S vector to $(0, 1)$, i.e., the parameters derived from the spectrum are crisp. From the values of these parameters assigned to the vertices, the structure of the unknown molecule (given by the elements a_{ij} of the adjacency matrix) is to be deduced. From a graph-theoretical point of view these vertices may be considered as *colored* by different spectral parameters, e.g., chemical shifts from NMR spectra, or by frequencies in IR spectroscopy, etc. This can be done by assignment of these parameters to the corresponding atoms. The assignment procedure implies that the experimental values are compared with values assumed or evaluated from the structure. Insofar as these parameters are functions of the atom connectivity environment, they are widely used to elucidate structural connectivity. Several severe problems again emerge from this approach: (1) We can evaluate the spectral parameters only after knowing the structure, i.e., if we have the solution of the direct problem Eq. (2), which is not the case. (2) The region of the ^{1}H chemical shifts is about 14 ppm, the region of the ^{13}C chemical shifts is 240 ppm, and the region of the IR frequencies is about 3000 cm^{-1}. All these regions encompass an immense variety of connectivities in billions of billions of plausible structures; hence the overlapping and the presence of alternative structures are inevitable. (3) The influence of a given nucleus on the signals of its neighbors is reduced, in most cases, to its near environment of no more than several bonds. On the one hand, the notion *several* is not strictly defined. It changes from structure to structure and strongly depends on the kind of adjacent atoms and the presence of heteroatoms, multiple bonds, conjugation, and hyperconjugation. Longer range effects are so small that they disappear into the informational noise. This partitions the studied structure into small pieces, which often overlap. (4) As a consequence of all these effects, the parameters are no longer fixed within a given place in the spectrum. Usually they form parameter ranges with rather obscure borders. All this makes the structural information derived from the different spectra vague and uncertain. Hereafter we call such information *fuzzy*. It is obviously quite difficult to assign such crisp values as true (0) or false (1) to assertions about connectivity (the a_{ij} element of the adjacency matrix) derived on the basis of the signal s_k.

Chemical practice in the process of structural inference is to assume a plausible structure or structural fragment and then to compare it against the spectral information available. Evidently, this approach requires some initial information about the studied compound. This information can be derived from a preliminary inspection of the spectral information available and/or data pertaining to its preparation, synthesis, similar natural products, etc. Very significant parts in the inference process are also the *expectation* of a given result, the *experience* of the chemist, and, not least, his *intuition*. These features of human reasoning cannot be sharply defined. They reveal, again, a rather vague, fuzzy nature and make structure elucidation into something of an art as well as a science. There is a profound discrepancy between the fuzzy nature of human reasoning and the explicit results which are required—one or several crisp structures.

Digital devices such as computers work with sharply defined logical elements—bits which take the values of 1 (true) and 0 (false); i.e., their work is based on classical prepositional logic or its mathematical equivalent, Boolean algebra. The fuzzy nature of human reasoning, on the other hand, makes things much more flexible, but may lead to irreproducible results and hence be a source of error. For instance, two chemists may draw different conclusions about a structure derived from the same spectral information from the same unknown compound measured on the same instrument. In contrast, the *classical logic* of computers always leads to the same result, whether correct or not (hardware and software errors are not considered here), whatever the number of trials or the types of computer used. Evidently, structure elucidations carried out by humans and by computers are qualitatively different. To obtain a reliable result, these two methods must converge to yield the correct structure.

A key problem in CASE is estimation of the spectral parameters. A straightforward approach might be based on the methods of quantum chemistry. However, their use entails severe and even insuperable problems. The first and most severe is that these methods are extremely intensive in computing time. As will be discussed in subsequent text, all the strategies for the solution of the inverse problem require the tracing of a whole *space of solutions*, i.e., the tracing of a great number of structures and for each an evaluation of its spectral parameters. This makes the problem unmanageable in real time. However, let us suppose a solution of this problem is obtainable, even though the results are not particularly useful. Calculated values are in most cases shifted from the experimental values. In quantum chemistry, an accurate result will be expected if we have an infinite basis of orbitals—something that cannot be achieved by modern quantum chemistry methods. However, imagine that even this is achieved; we still have to confront the classical behavior of computers versus the fuzzy behavior of the real world. Whereas the computed values for a given structure are in all cases (different trials, different computers, different scientists using the computers) the same up to the computer word precision, the experimental values for the corresponding compound, even measured on the same instrument and under the same conditions, provide values that exhibit *statistical* behavior—they appear dispersed. Accordingly, spectral parameters computed by rigorous quantum methods can hardly be used to elucidate the structure of an unknown in the near future. Up to now they have been used only in theoretical chemistry, for the support of some hypotheses to explain the observed trends in the chemical behavior of studied molecules.

Some attempts to employ much less rigorous empirical schemes in the estimation of the electronic distribution for the considered structure and its correlation with spectral parameters such as ^{13}C chemical shifts were suggested by Bangov.[36–41] The charge densities of each candidate structure may be calculated by a fast empirical scheme based on either full[42] or partial[43] equalization of orbital or atomic[44] electronegativity and further correlated with the chemical shifts from the ^{13}C spectrum of the query structure. The structure providing the best correlation is considered to be the correct solution for the unknown compound. However, this approach, as subsequently discussed, also yields rather vague results.

Another approach to the evaluation of spectral parameters is the application of a variety of empirical additivity schemes.[45, 46] Such schemes have been developed mainly for ^{1}H and ^{13}C NMR spectra. Parameters for the α, β, γ, δ, steric, etc., contributions have been derived from the statistical treatment of sets of compounds and their spectra.[47, 48] The results from their application to the immense variety of real cases are also uncertain. They present, again, the crisp character of all computed values against the fuzzy character of the measured, real-world values. Moreover, empirical additivity parameter schemes were developed for complete structures and hence are not flexible enough to characterize chemical groups and fragments.

In most structure elucidation systems, pieces of the query structure are derived by comparing experimental spectrum parameters with the spectral parameters compiled in data bases consisting of either complete structures[49] or structural fragments.[23, 28] These approaches have their limitations, which also arise from the uncertainty of the spectral information. Thus, in the cases of data bases of compiled fragments and chemical groups, the number of possible fragments is infinite and no data base encompassing all possible cases can be constructed. The same problem exists with data bases of complete chemical structures. The derivation of structural fragments from complete structures is additionally complicated by the necessity of employing substructure matching procedures. The separate substructures have different environments in different structures, and so have different values for the spectral parameters. It is impossible to measure and evaluate these parameters for separate fragments.

The other parameters that may be derived from spectral information show the same behavior. Thus, the signals in a mass spectrum may be assigned to complete fragments rather than to single atoms. Generally speaking, the theory of breaking of molecular bonds in the mass spectrum experiment is not well formalized; hence relationship (2) appears rather unclear. The different techniques providing direct $C-H$ coupling in NMR, such as off-resonance, APT, and DEPT, are very useful for the determination of adjacent $C-H$ bonding. However, this is not really conducive to molecular skeleton elucidation. Moreover, employment of the longer range $H-H$ and $C-H$ coupling constants from ^1H and ^{13}C NMR spectra, respectively, generates a great deal of uncertainty. As previously mentioned, it is very difficult to discriminate between vicinal and longer range couplings. Again, overlapping appears to be a serious problem.

The intersection of different types of spectral information appears to be an extremely promising approach. However, it does not necessarily lead to a full determination of the query compound structure, because frequently the different methods provide overlapping information about certain structural units and no information about others.

To summarize, it may be said that the available spectral information is always insufficient for complete determination of the query structure. Hence, the inverse problem given by Eqs. (2') and (5) cannot be solved in purely mathematical form, i.e., no explicit expression for the function f can be derived. Accordingly, f is usually represented by one or several procedures as forming the structure elucidation system. These procedures model the logic that leads to the determination of one (in the most favored case) or several chemical structures from the best matching to the spectral information available. The structural information derived from the available spectra is rather fuzzy, which additionally makes the CASE problem more complicated. Figuratively speaking, the creation of a CASE system resembles squaring the circle. We might attempt that by inscribing polygons, which have a sharply defined (crisp) structure, but the quadrature is always outside these structures in the infinite limit.

Two major problems face those who initiate the development of a computer-based system of procedures for automated structure elucidation: (1) to find some way to solve the inverse problem [Eqs. (2)–(6)] with (2) to model the fuzzy character of the spectral parameter/structure relationship. The former is usually solved by devising different strategies. The latter can be achieved by employing special representations of the spectral information and chemical structure matching their fuzzy nature; this includes modeling the fuzzy character of human reasoning as well by using the methods of fuzzy logic.[34, 35]

II. COMPUTER-AIDED STRUCTURE ELUCIDATION STRATEGIES

The automatic treatment of the process of structure elucidation is based on a *heuristic search scheme* that was defined in 1968.[15, 50] This scheme treats the collection of all possible solutions (structures in our case) as a *space*. The correct solution, the query structure, or some final structures in a given problem (an empirical formula in our case) may be found by applying a series of heuristic methods (inference rules) to guide the search in the solution space. This is illustrated in Fig. 3. Generally, there are two kinds of solution spaces used in different CASE approaches: (1) sets of structures derived from structure/spectral data bases and (2) sets of structures generated from any preliminary information (empirical formula, chemical groups, and structural fragments). Notwithstanding the great variety of methods and approaches used as inference rules in the search process, all share the inverse problem given in (2′) and (5), which may be reduced to multiple direct problems (2), which are mathematically exact. Each direct problem is associated with a candidate structure from the solution space and consists of evaluation of its spectrum and then correlation of it with the experimental spectrum of the unknown compound. The structure(s) providing the best fit is (are) considered the correct solution(s). This process may be divided into two parts: formation of the solution space and obtaining the most likely structures from all the

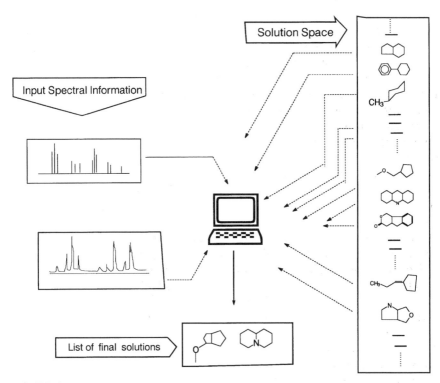

FIGURE 3 Depiction of the heuristic search procedure which employs spectral information to reduce the solution space of structures to one or several solutions.

solutions. It is apparent that the approaches used in the second part must allow for all the shortcomings originating from the insufficiency and the fuzzy character of the information discussed in the preceding text. However, the insufficiency of information has implications for the formation of the solution space too. Thus, two such consequences may result if the solution space is derived from a data base. In the case of a very small data base, the probability of the query structure being found in it becomes small, and, vice versa, if the base is very large, it appears impossible that all the candidate solutions can be traced in real time. In those cases where a limited amount of spectral information is available, the number of structures matching this information sharply increases. The more precise and detailed the information, the smaller the number of structures. This may be viewed as a peculiar manifestation of the well-known uncertainty principle.

The same situation obtains when the solution space is formed by candidate structures generated from any preliminary known information, e.g., an empirical formula. With any increase in the number of atoms in this formula or an increase in their variety (the presence of different heteroatoms, multiple bonds, rings, etc.), the number of generated structures increases exponentially. Thus, the generation process can lead to a so-called combinatorial explosion for most real-world problems; i.e., the number of structures is so enormous that it cannot be processed in real time. Moreover, even if we succeed in limiting this number to thousands of structures, this by no means solves the CASE problem, because it is extremely difficult for the user to trace and consider all of them. Consequently, the *"generate and test"* approach[51] is not conducive to obtaining satisfactory results. This approach consists of the following steps: (1) derivation of structural information from spectral data, (2) generation of the complete set of candidate structures, and (3) ranking of the candidate structures by comparing their simulated spectra with the experimental spectra.

It is obvious that any available spectral information can constrain the generation of an enormous number of structures, but its fuzzy character will greatly complicate the problem. Note that generators are created mostly for a simple enumeration of chemical structures.[52-55] Therefore, they produce a crisp set of crisp structures. Each such structure is correlated to spectral information of a fuzzy nature.

The use of fragments sharply constrains the combinatorial explosion. Accordingly, the structure generator CONGEN[14] in the first version of the DENDRAL project has been designed to treat both single atoms and fragments. However, the problem that emerges from this approach is that the fragments that may be derived from any spectral information are not well defined; they are either alternative or overlapping. Here again we observe the crisp character of the CASE approach versus the fuzzy behavior of the real-world cases. This conflict called forth the creation of the GENOA structure generator,[56] which had been constructed in such a way as to enable treatment of combinatorially overlapping and alternative input substructures. Overlapping structures have a common structural pattern. Alternatives are those structures where a common pattern appears in a different environment, e.g., the two double bond input fragments $CH-C=O$ and $CH=C-O$. The overlapping regions are perceived by using maximal common substructure search procedures, and a special combinatorial algorithm has been devised to produce all possible alternative structures. Obviously, the main difficulty comes from the derivation of structural fragments from the spectral information of the query structure. In the early versions of DENDRAL, this action was left to the user. Some procedures for the treatment of both mass[57] and ^{13}C NMR[58] information have been further developed.

The overlapping problem does not arise in the CHEMICS system.[22] At the heart of this system is a small data base formed of basic units called components, such as the chemical groups $-CH$, CH_2, NH and their spectral parameters: IR frequencies and ^{1}H and ^{13}C chemical shifts. The ^{1}H and ^{13}C NMR data base, for example, consists of 630 appropriate data.[59] A smart procedure has been developed that extracts alternative sets of substructures from the data base, each complying with the input empirical formula and the input spectral information. The generation process produces the solution space of generated structures. However, as discussed previously, only small substructures can be extracted from such spectral data. This leads to generation of an immense number of structures —a problem that still remains unresolved. The generation process in CHEMICS may be constrained by the input of macrocomponents, which are practically molecular fragments. A substructure search procedure based on the set reduction algorithum was developed to trace all the generated structures and check for the presence or absence of the input macrocomponent.

Munk *et al.* have developed a series of approaches.[24-28] Their CASE system is also based on substructures. A data base of about 5100 atom-centered fragments (ACFs) has been constructed.[60] Several types of structure generation programs, such as ASSEMBLE[26] and COCOA,[27] have been developed. Whereas the ASSEMBLE program combines the ACFs in a way similar to that of other generators, the generator COCOA starts with generation from a hypergraph instead of from an empty graph (a graph with no bonds). A hypergraph is a graph that has all possible connections present between the different vertices (atoms). Hence, the generation is a combinatorial process of breaking separate edges (bonds) to reduce the hypergraph to the connectivity of the real molecule. It may be argued that this circumvents the need for the treatment of alternative and overlapping structures.

The utilization of 2D NMR information is also incorporated into these systems, but the uncertainty within this information is not treated. The first work in this direction was reported by Lindley *et al.*[61] The structural fragments were user-deduced from the 2D INADEQUATE spectrum and included as input into GENOA. The 2D INADEQUATE automated analysis has also been optionally included in the CHEMICS system.[59] The recent development of SESAMI by Christie and Munk[62] is similarly based on the use of 2D NMR spectral information. The task of this system is divided into two parts: spectrum interpretation (carried out by the procedure INTERPRET) and structure generation using the COCOA generator). INTERPRET is a two-track procedure: the first track, PRUNE, produces a short list of concentrically one-layered atom-centered fragments (ACFs); the other track, INFER, produces substructural inferences that serve as constraints on the structure generation process. Finally, the COCOA generator generates all the alternative structures. All of these approaches work provided the input 2D spectral information is crisp.

An extremely interesting approach which may be related to the fuzzy character of the available spectral information (although not in terms of *fuzzy logic* theory) has been developed by Dubois and Carabedian.[63, 64] It is based on the mutual α-environmental dependence of ^{13}C chemical shifts. This dependence may be represented by a two-dimensional map with chemical shift values for each dimension. A third dimension at each position where two chemical shift values cross provides the probability that the corresponding two carbon atoms are adjacent. This probability is formed by examining each pair of chemical shifts in a large data base by counting the cases where they appear adjacent. Thus the corresponding carbon–carbon connectivities are derived. The connectivity yielding the highest probability is selected as the most likely. The structure is constructed by forming one bond after the other. Of course, some connections having similar probabilities are also alternatively formed. This is, perhaps, the best matching of the fuzzy character of both the spectral information and human reasoning. The problem remains that a comparatively large data base must be present and a statistical treatment is required for each pair of chemical shifts during the structure elucidation process.

All the methods outlined here are directed mainly toward the solution of the inverse problem. However, some studies on the use of fuzzy logic have been carried out. Sasaki and co-workers applied both the symbolic logic[65] and the membership function[66] approaches in the automated treatment of IR spectral information. It is evident that with the increase of complexity in the CASE targets, the uncertainty will also increase. The structure of a compound with a low molecular mass is well defined. In our terminology it may be considered crisp. In contrast, the structures of polymers, biopolymers, mixtures, coals, etc., may be considered as being fuzzy. They will certainly require employment of either stochastic or fuzzy methods. Foulon[67] developed a stochastic method for structure generation in the analysis of coal products, and Xu *et al.*[68] employed the elements of fuzzy mathematics and fuzzy graphs in assignment of 2D NMR signals to proteins.

Here it is worth mentioning some other trends in the development of alternative approaches to the solution of structure recognition problems. Extensive studies have been carried out on the application of pattern recognition methods to CASE problems.[69] One of the most intensively studied methods at present is that of *neural networks*. This method of artificial intelligence (AI) attempts to model the functioning of the human brain in the process of decision making. It is well known that the human cortex consists of neurons and synapses, which may be mathematically modeled by artificial neurons, where the input signals are weighted and connections are made between the neurons. The set of neurons forms the neural net. The solution of a neural network problem consists practically of working out a model of transformation of the input signals to output information. Such a model is obtained by teaching the machine with a set of examples. The most interesting feature of this approach is that the machine treats fuzzy patterns instead of crisp symbols and rules based on Boolean logic. We shall not discuss the details here. The reader is referred to an excellent review[70] for a more comprehensive presentation of neural network fundamentals and applications to chemical problems. However, it should be emphasized that, in spite of extensive investigation on the use of this approach to the treatment of IR[71] and ^{13}C NMR[72] spectral information, such methods have not reached the stage of practical application in complete CASE systems, including those discussed in the preceding text. The possibility of coupling neural networks with the methods of fuzzy logic also has been discussed.[73]

An extremely interesting approach that is now under rapid development involves the use of *genetic algorithms*[74] for the solution of CASE problems. These algorithms have been developed to mimic the processes of fission and recombination in genetics. They can be used as powerful problem solvers in the case of large solution spaces. The underlying principle here, as well as in genetics, is "the survival of the fittest."[75] A membership or other evaluation function may be used for ranking the successful solutions. The elementary acts of fission and recombination are carried out and the process then goes on to the next ranking. Instead of generating the complete set of solutions, this method strongly constrains the generation process to several generations.

III. FUZZY SETS, FUZZY LOGIC, AND FUZZY GRAPHS

The concept of the fuzzy set was introduced by Zadeh[34, 35] to deal with entities that cannot be precisely defined. A more detailed description of these ideas is presented in specialized literature[73, 76, 77] and in chapter 2 of this book.

Let the OH group be an element of the set of chemical groups [NH, NH_2, CH, CH_2, CH_3, OH, Ph, CO]. The membership of this element in the set is sharply defined and may be either true or false (see Table I). Hence, it is subject to classical propositional logic, based on the *two-value principle*. According to this principle, every proposition is either true or false (and so takes the Boolean values of either 1 or 0) and these values are called *truth values*. Unfortunately, the cases where one can draw such clear values are not very frequent. In the real world, the affiliation of an element to a set of things is in most cases rather uncertain, ambiguous, and vague. Thus, if one has to specify the relationship of a OH group to a set of spectral parameters, one often runs into some difficulties. The OH groups of different compounds resonate at different frequencies in the infrared region and they have chemical shifts in different regions of the

NMR spectra. The idea of fuzzy sets has been developed to deal with just such sets of elements whose boundaries are not precisely determined. It is clear that the classical prepositional logic of two values, namely, true and false, is replaced by the fuzzy logic of an infinite set of real values in the range $[0, 1]$. For example, instead of asserting that the ^{13}C NMR signal that has a chemical shift of 64 ppm corresponds to a $C—O$ carbon atom is true, we may say that it is 0.72 (72%) true, i.e., $m(C—O) = 0.72$, whereas the concurrent assertion that this chemical shift originates from a $C—N$ carbon atom is (0.22) 22% true, i.e., $m(C—N) = 0.22$, etc. It is obvious that this way of representing the sets better matches the fuzzy nature of the information and its logical treatment. Additionally, fuzzy logic defines a minimum set of operations to deal with the uncertainty. Fuzzy logic uses the same set of operators that classical Boolean logic uses: fuzzy AND (f-AND), fuzzy OR (f-OR), probability AND (p-AND), and probability OR (p-OR), and the unitary operation fuzzy-NOT (f-NOT). These operations are defined in Table II.[78] For example, in the foregoing case of the ^{13}C NMR chemical shift of 64 ppm, the *fuzzy union* will be

$$m(C—O) \ (\text{f-OR}) \ m(C—N) = \max(0.72, 0.22) = 0.72.$$

This means that there is a 0.72 *possibility* that the signal of 64 ppm corresponds to either a $C—O$ or a $C—N$ group. However, this disregards the accumulation of the two factors. In contrast, after the application of the p-OR operation, the result will be

$$m(C—O) \ \text{p-OR} \ m(C—N) = 0.72 + 0.22 - (0.72 \times 0.22) = 0.78.$$

The possibility that the group corresponding to the signal of "64 ppm is not $C—O$" is

$$\text{f-NOT} \ 0.72 = 1 - 0.72 = 0.28.$$

Consider the case where the ^1H NMR signal of 4.5 ppm is assigned to a $CH=C$ proton with likelihood $m_H(CH=C) = 0.68$ and the ^{13}C NMR signal of 85.5 ppm is assigned to one of the $CH=C$ carbons with $m_C(CH=C) = 0.15$. Then, the likelihood that the two signals indicate the

TABLE II Fuzzy Set Operations

Fuzzy set operation	Application and result	Name
f-AND	$m(a) \ \text{f-AND} \ m(b) = \min(m(a), m(b))$	Fuzzy intersection
f-OR	$m(a) \ \text{f-OR} \ m(b) = \max(m(a), m(b))$	Fuzzy union
f-NOT	$\text{f-NOT} \ m(a) = 1\text{-}m(a)$	Fuzzy complement
p-AND	$m(a) \ \text{p-AND} \ m(b) = m(a) \times m(b)$	Probability intersection
p-OR	$m(a) \ \text{p-OR} \ m(b) = m(a) + m(b) - (m(a) \times m(b))$	Probability union

presence of a $CH{=}C$ fragment is given by the fuzzy intersection

$$m_H(CH{=}C) \text{ f-AND } m_C(CH{=}C) = \min(0.68, 0.15) = 0.15.$$

The p-AND operation results in

$$m_H(CH{=}C) \text{ p-AND } m_C(CH{=}C) = 0.68 \times 0.15 = 0.10.$$

These operations may of course be used in compound statements. It is reported that such capabilities have been added as a shell to TURBO-Prolog programming language,[78] but the usual if $\cdots$ then $\cdots$ else capabilities of other high level programming languages can also be used to this end.

The definition of a graph given in Eq. (3) shows that it is formed of two sets: a set of vertices V and a set of edges E. This definition can be formulated in the context of fuzzy set theory[79] with both sets V and E represented by fuzzy sets (respectively V^f and E^f to constitute the *fuzzy topological graph* $G^f = (V^f, E^f)$. This means that each vertex $v_i \in V$ and each edge $e_{ij} \in E$ may be associated with the membership functions $m_v(v_i)$ and $m_e(e_{ij})$, which map these two sets on the range of real values $[0, 1]$. The membership function $m_e(e_{ij})$ may be defined[51, 73] as

$$m_e(e_{ij}) \leq \min\big(m_v(v_i), m_v(v_j)\big).$$

The entries of the adjacency matrix corresponding to the fuzzy set E^f of edges will then contain the values of the associated membership functions. Here E is called the *support set* of E^f. By associating a membership function with each of the members in the set of vertices and in the set of edges, the classical graph is transformed into a *fuzzy graph*. A fuzzy graph with its adjacency matrix is represented in Fig. 4.

In the same way that a molecular graph was defined by Eq. (4), we can define the notion of *fuzzy molecular graph*. The notion of a fuzzy molecular graph may be sought in the mathematical representation of the alternativeness of the derived structural information. For example, consider a carbon nucleus that resonates at 166 ppm in the ^{13}C NMR spectrum and has membership functions with the values $m_v(C{=}O) = 0.38$, $m_v(C{=}C) = 0.22$, and $m_v(Ph) = 0.40$ for the carbonyl, vinyl, and phenyl carbon atoms, respectively. Evidently, these three assignments may be considered alternative. The same applies to the E^f set of edges (bonds). One can see from Fig. 4 that the $m_e(e_{ij})$ function gives the likelihood of the bond being substantiated by the spectral information available. Thus, consider a cross-peak of the two-dimensional H,H COSY spectrum. It may originate from couplings of different $C{-}H$ atoms; most probable are the vicinal couplings, but longer range couplings are also possible. The different possibilities produce different alternative connectivities within the molecular graph of the studied compound. Their likelihood may be esti-

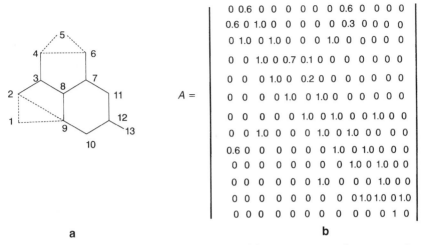

a   **b**

FIGURE 4   Fuzzy graphs: (a) a fuzzy graph drawing; (b) an adjacency matrix representation.

mated by developing the membership function $m_e(e_{ij})$ related to the particular type of information.

One can see that the truth values in fuzzy logic strongly resemble the stochastic values from the theory of probabilities. However, methods based on the use of statistics are not considered fuzzy by the orthodox fuzzy theory protagonists. Instead of using *probability* values, fuzzy theory works with *possibility* values. It is argued that both values are substantially different and that the latter have to be evaluated by methods other than statistical. Our understanding, however, is that at a very fundamental level, both values have essentially the same nature.

IV. A NOVEL STRATEGY FOR COMPUTER-AIDED STRUCTURE ELUCIDATION

Our efforts have been directed toward the development of a new strategy for the construction of a CASE system that incorporates both treatments of the uncertainty inherent in the inverse problem. The uncertainties arise from the combinatorial explosion and the fuzzy character of the spectral information. It is clear that the process of structure elucidation must start from preliminary information · considered more or less certain. This information is necessary to constrain the number of candidate structures within the solution space; otherwise, this number will be infinite. In more general terms, each piece of information may serve as a constraint on the number of possible candidate structures. It is very important how this information is incorporated within the CASE system. The strategy based on the generate and test approach, namely, first generate the structures forming the solution space and then compare them with the available spectral information, proved to be highly inefficient because of the immense number of structures that are generated. Accordingly, the heuristic search program had to be modified in such a way that the heuristic search is carried out during the formation of the chemical structures. Thus, the solution space is constrained to the most likely structures from the spectral information point of view. To cope with the two problems discussed in the previous sections, the following structure generation and elucidation scheme has been developed.

First, it is accepted that the basic elements forming the structure are atoms (vertices of graphs). Each atom has its own attributes[80]; i.e., the vertices are colored (here the term "labeled" may also be used) to represent these attributes. The attributes characterize different aspects and features of the atom within chemical structures. Practically, the input spectral information is transformed into these attributes, with the latter guiding the generation process toward the correct solution(s). The atom attributes may be classified as follows:

(1) *Atom kind* (C, O, N, Br, etc.) and *atom valence* (the number of bonding sites).

(2) *Assigned signals* (^{13}C chemical shift and C — H direct multiplicity) and/or a list of coupled signals from the 2D NMR spectra.

(3) *Hybridization state* (sp^3, sp^2, sp) and *α-environment* (types of adjacent atoms).

(4) List of the ⊕-type bonding sites (saturation sites) associated with each atom ←-type bonding site (saturating valences).

A. Determination of the Atom Kind and Atom Valence Attributes

This type of attribute can be derived from the empirical (gross) formula. It provides the number and the kind of atoms. Every CASE approach is basically constrained by this information. The gross formula can be automatically estimated from the molecular ion of the query compound mass spectrum either by using a generation algorithm of all possible formulas corresponding to the molecular ion mass[81] or by intersecting information from both the ^{13}C NMR broad-band (BB) proton decoupled spectrum and the mass spectrum.[82-84] The signals from the ^{13}C NMR BB spectrum determine the number of carbon atoms. The signal overlapping can be treated by using the signal intensities, but only within separate sets of resonating nuclei of uniform multiplicity: quaternary, CH,

CH$_2$, or CH$_3$ carbon atoms.[83] It is implied that the resonating nuclei within each of these sets have the same relaxation times; hence their intensities may be considered proportional to the number of their nuclei.

Once the empirical formula is estimated, the atom kind and the atom valence attributes may be considered as crisp values (see Table I). They are further required for determination of the chemical features of the atoms forming the query structure. The atom valence attribute is dynamically determined during the generation process by the number of bonding sites (BSs), i.e., the locations on any atom where there are free valences. Such attributes are also considered crisp. The mathematical representation of structure within this approach employs *directed graphs*. This type of representation is illustrated in Fig. 5. Every bond in a directed graph has a direction, which is indicated by an arrow. This makes the BSs of a given atom different. One can see from Fig. 5 that there are two types of BSs: the ← type and the ⊕ type. In our previous papers,[80, 85−88] these bonding sites were called saturating valences (SVs) and saturation sites (SSs), respectively. Hereafter both names are used. All atoms except the first have $(n − 1)$ ⊕-type BSs and one ← -type BS. The first atom has all its n BSs of ⊕ type, where n is the atom valence. The atom with a closed cycle has one more ⊕-type BS transformed into an ← -type BS. In the case of

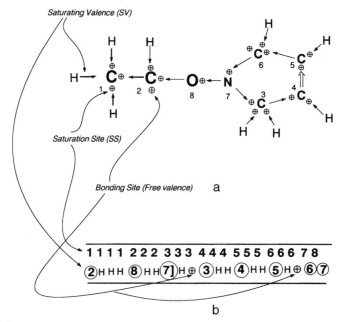

FIGURE 5   Directed graphs: (a) a directed graph with the corresponding ← -type (SV) and ⊕-type (SS) bonding sites; (b) two-row matrix representation.

free valences, one or more $\oplus$-type BSs are left unsaturated. Our two-row array mathematical representation of a directed graph is presented in Fig. 5b. More detailed discussion on this representation may be found in Refs.[88-93] Obviously, the differentiation between $\oplus$- and $\leftarrow$-type BSs (SVs and SSs) is a mathematical simplification and has no relation to real chemical behavior.

B. Determination of the Assigned Signal Parameters Attributes

These attributes are very important for further treatment of the structure/spectral information relationship. Here, however, the fuzzy nature of this relationship is manifested. The intuitive way of thinking about the signal assignment inherent in the everyday practice of spectroscopists, namely, assume a structure and then assign the signals to it, proves to be inefficient. The development of such an automatic assignment procedure is accompanied by major structure/spectrum matching and combinatorial difficulties.[94-96] These can be avoided by applying our *arbitrary assignment* approach.[89, 90] This approach is based on the following assumptions: all the free (not participating in fragments) carbon atoms of the empirical gross formula may be considered equivalent. Thus, each carbon atom from the gross formula sequence of atoms is arbitrarily assigned one ^{13}C NMR signal from the sequence of signals. It thereby assumes two additional attributes: *chemical shift* and *multiplicity*. The multiplicity information is also considered crisp. The latter either leaves the carbon atom as it is, in the case of multiplicity 1, or transforms it into a CH (multiplicity 2), CH_2 (multiplicity 3), or CH_3 (multiplicity 4) chemical group. It is worth stressing here that the arbitrariness of the assignment does not hold for the cases of *fixed* fragment connectivities (subsequently discussed), since their atoms are no longer indistinguishable. They can be singled out by their connectivity within the fragment. A special assignment procedure was developed to this end for the particular case of multiple bonds.[90]

The arbitrary signal assignment proves to be very useful in the case of 2D NMR spectral information. A list of correlations with the other signals derived from the 2D maps is assigned to each signal. A one-to-one correspondence between the carbon atoms and the assigned signals is conducive to correlations between the atoms themselves. Accordingly, the generation of chemical structures may be regarded as the generation of different combinations of possible signal correlations, with the structure elucidation process being transformed into selection of the correct correlation.

Arbitrary signal assignment leads to assignment of the third type of attributes: hybridization and α-environment.

TABLE III Signal and Atom α-Environment and Hybridization Attributes[a]

| Signal attributes | Atom attributes | | Description |
	Hybridization	α-Environment	
SP3_C	SP3	C_C	sp^3 Carbon atom attached to carbon atom
SP3_O	SP3	C_O	sp^3 Carbon atom attached to oxygen atom
SP3_N	SP3	C_N	sp^3 Carbon atom attached to nitrogen atom
SP3_X	SP3	C_X	sp^3 Carbon atom attached to another heteroatom
SP2_CC	SP2	C_C	sp^2 Carbon atom attached to carbon atom
SP2_CC_O	SP2	C_O	sp^2 Carbon atom attached to oxygen atom
SP2_CC_N	SP2	C_N	sp^2 Carbon atom attached to nitrogen atom
SP2_CC_X	SP2	C_X	sp^2 Carbon atom attached to another heteroatom
SP2_ary	ary	C_C	sp^2 Carbon atom attached to another heteroatom
SP2_ary_O	ary	C_O	Aryl carbon atom attached to oxygen atom
SP2_ary_N	ary	C_N	Aryl carbon atom attached to nitrogen atom
SP2_ary_X	ary	C_X	Aryl carbon atom attached to another heteroatom
SP2_CO	CO	any	Carbonyl carbon atom attached to any atom
SP2_COO	COO	any	Carboxyl carbon atom attached to any atom
SP_CC	SP	C_C	sp Carbon atom attached to carbon atom
SP_CC_O	SP	C_O	sp Carbon atom attached to oxygen atom
SP_CC_N	SP	C_N	sp Carbon atom attached to nitrogen atom
SP_CN	CN	any	Nitryl carbon atom attached to any atom

[a]Reprinted with permission from AIP Conference Proceedings 330, E.C.C.C.1, Computational Chemistry, FECS Conference. 1994 American Institute of Physics.

C. Determination of the Hybridization/α-Environment Attributes

Our intention has been to give an automatic assessment of the nature of each signal chemical shift/multiplicity pair of values based on fuzzy theory. By *signal nature* we mean the hybridization state and the α-environment of the resonating carbon nucleus producing the signal. The possible hybridization state/α-environment signal attributes are displayed in Table III. Their determination is achieved by making use of the fuzzy sets $F_{k.}$, defined by the membership function

$$m_k(\delta_i) = P + \frac{H}{1 + \exp(-a(\delta_i - b))}. \tag{7}$$

For each ^{13}C NMR signal i having multiplicity M and chemical shift δ_i, the value of the membership function $m_k(\delta_i)$ is an estimate of the degree of support for the hypothesis that this atom is indeed a member of the fuzzy set F_k, i.e., possesses the corresponding hybridization state and α-environment. A graphical representation of the membership functions given by Eq. (7) for the particular case of multiplicity 2 and hybridization sp^2 is presented in Fig. 6. The membership functions for the different attributes have been built up by comparing the distribution curves (histograms) of chemical shifts observed for each case (hybridization state and α-environments) in a collection of spectra. All the details of their construction may be found in our previous papers.[90, 91] The $m_k(\delta_i)$ values are automatically evaluated for each input pair of signal chemical shift/multiplicity values, and a table of all possible membership values is displayed

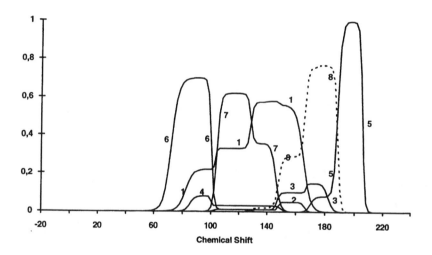

FIGURE 6 Membership functions of different α-environments for multiplicity 2 and sp^2 hybridization state. Reprinted from *Analitica Chim. Acta*, **298**, I. P. Bangov, I. Laude, D. Cabrol-Bass, *Combinatorial problems in the treatment of fuzzy ^{13}C NMR spectra in the process of computer aided structure elucidation: Estimation of the carbon atom hybridization and α-environment states*, pp. 33–52, (1994) with kind permission of Elsevier Science-NL, Sara Burgerhartstraat 25, 1055 KV Amsterdam, The Netherlands.

(see Fig. 7). The user is invited to select one or more (up to four) of these alternative *possibility values*.

The equivalence between the carbon atoms and the attached signals leads to the automatic assignment of the hybridization state/α-environment attributes. The latter provide new characterizations of the given carbon atom in addition to the chemical shift and C—H multiplicity attributes. One can see from Table III that each signal hybridization state/α-environment attribute is split into two atom attributes: the hybridization state and the α-environment. The hybridization attribute transforms each carbon atom into an sp^3 (coded as C), $sp^2(=$ C), or sp (#C) hybridized carbon atom.[90] The atom attributes are slightly different from the signal attributes. Thus, in contrast to common chemical sense, the carbonyl, carboxyl, and nitryl groups are considered as dummy CO, COO, and CN hybridization states that are different from the sp^2 hybridization state. This comes about because of the different treatment of the chemical groups and fragments (such as C=C and C≡C). Thus, while a C=C fragment is formed by two C atoms of hybridization attribute SP2, the CO

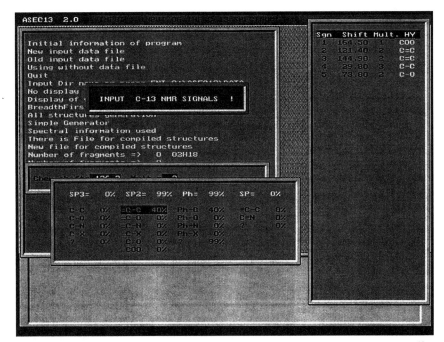

FIGURE 7   User-driven hybridization state/α-environment attribute selection for each ^{13}C NMR signal. Reprinted from *Analitica Chim. Acta*, **298**, I. P. Bangov, I. Laude, D. Cabrol-Bass, *Combinatorial problems in the tratment of fuzzy* [13]*C NMR spectra in the process of computer aided structure elucidation: Estimation of the carbon atom hybridization and α-environment states*, pp. 33–52, (1994) with kind permission of Elsevier Science-NL, Sara Burgerhartstraat 25, 1055 KV Amsterdam, The Netherlands.

hybridization attribute instructs the program to form a carbonyl group. The hybridization attribute of each atom makes the system construct a specific two-row representation before starting the generation process. The α-environment attribute is effective during the real process of structure generation within the SS-selection procedure (discussed in subsequent text). According to this scheme, each carbon atom can be assigned one to four alternative hybridization attributes and each hybridization state can be assigned one to four α-environment attributes. The presence of several alternatives is conducive to the generation of alternative structures. Thus if (say) a signal of 160 ppm is entered, the program will treat the corresponding carbon atom once as a carbonyl atom, building the corresponding representation, and the second time as an ethene $=C-O$ sp^2 carbon atom bonded to oxygen, forming a new representation. In the first case, the carbon atom automatically forms a carbonyl group with one of the oxygen atoms present. In the second case, the olefin carbon atom forms an olefin fragment with another $=C$ atom.[90] The C_O α-environment attribute, however, implies that one BS of this atom must be bonded to one or more oxygen atoms. This constraint reduces sharply the number of generated structures. Note that, on the other hand, the combinatorial problem may be complicated with more hybridization attribute alternatives being selected, but, on the other hand, the problem is alleviated in the process of selection of specific α-environment attributes. It is clear that the *sharper* the hybridization and α-environment attribute information, the fewer alternatives are required, which means fewer combinations and hence fewer candidate structures. Conversely, the fuzzier this information, the more hybridization and α-environment attribute alternatives should be selected, making the combinatorial problem more severe. In any case, however, this approach is more flexible and less redundant than the brute-force generation of all possible structures with a further discrimination of the most unlikely ones from the spectral point of view.

The one-to-one correspondence between the ^{13}C NMR signals and the carbon atoms and the assignment of possibility values through the membership function to each signal transforms the crisp set of vertices V from definition (3) of the graph into a fuzzy set V^f, forming a fuzzy molecular graph.

D. Determination of the List of ⊕-Type BSs (SSs) Associated with Each Atom ←-Type (SV) Attribute

The list of ⊕-type BSs is an attribute of great importance for directing the generation process toward the most likely structures. As stated, each atom (except the first) has an ←-type (SV) free valence. A list of ⊕-type BSs (SS) may be associated with each such SV either during the input of fragments or during the generation process. As long as a chemical bond is always formed by juxtaposing an ← - with a ⊕-type BS, this list represents the possible connectivities of the given atom. Accordingly, the list of possible connectivities may contain two types of input connectivities: fixed and alternative. Fixed connectivities have been derived on the basis of definitive information about the connectivity within an input substructure (fragment). This representation may be deemed the crisp part of the input structural information. Alternative connectivities provide more than one possible connectivity; they reflect the alternative fuzzy part of the molecular graph. The two types of connectivities are user-entered within the fragment input. A third type of connectivity, which we call a selected connectivity, is the most uncertain and fuzzy. It is also represented by an associated $\oplus$-type BS (SS) list and is formed during the structure generation process. The three types of connectivity are presented in Fig. 8. A mathematical representation of the information in Fig. 8 is given by the two-row matrix representation shown in Fig. 9. All atom BSs are in the first row, and the $\leftarrow$-type and the $\oplus$-type free valence bonding sites are in the second row. Juxtaposing the two rows gives the connectivity within the structure.

The fixed type of connectivities form the crisp part within two types of input substructure: chemical groups and/or fragments. Because the development of a CASE system requires a stronger formalization of all notions, we have been compelled to differentiate them. Each substructure may be regarded as formed from two sets: one of atoms and the other of bonds which are equivalent (in mathematical language *isomorphic*) to a part of the structure. As long as some of the substructure atoms possess unsaturated free valences, the substructure may be mathematically defined as

$$G^s = (A^s, B^s, F^s). \qquad (8)$$

Here $a^s \in B^s$ is the set of substructure atoms, $b^s \in B^s$ is the set of substructure bonds, and $f^s \in F^s$ is the set of substructure free valences. Bearing in mind this definition, the chemical groups and the fragments may be differentiated[90, 92]:

DEFINITION. A chemical group is a substructure having all its free valences equivalent, whereas a fragment is a substructure having two or more nonequivalent free valences.

As a result, the substructures $-CH_3$, $-CH_2-$, $-CH-$, $-NH-$, $C=O$, Ph, etc., may be considered chemical groups. In contrast, substructures such as $O-C=O$, $HO-Ph-$, and an unlimited variety of larger substructures are considered as fragments. This differentiation is necessitated by the different ways of treating these two types; the number of nonequivalent free valences plays a significant role in the process of structure generation. Obviously, this is a working definition which is not related to standard chemical conceptions.

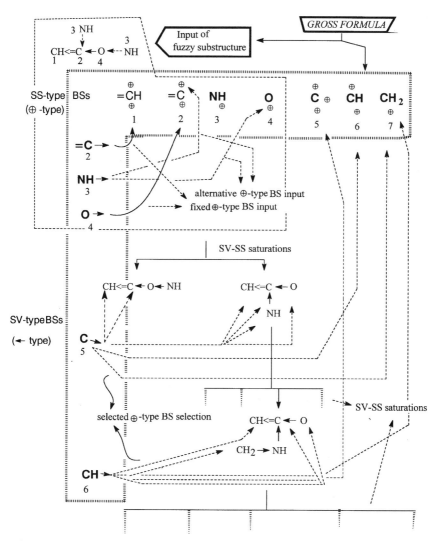

FIGURE 8 Treatment of fixed, alternative, and selected ⊕-type BS lists and multilevel guided generation.

Usually information on the presence or absence of chemical groups is more precise. Thus, concerning the presence or absence of (say) OH (the same applies for NH, CN, CO, etc.) groups, we may have the answers presented in Table IV. The answers N, $n_{OH} = 0$, and $n_{OH} = n$ provide obviously crisp information. The answer Y may be considered fuzzier; i.e., there have been observed signals for OH groups either in the IR or in the ^{1}H NMR spectra, but their number is not explicitly known. This means

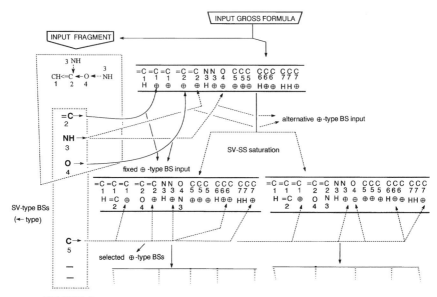

FIGURE 9   Mathematical representation of the guided multilevel generation.

that at least one OH group must be present in the generated candidate structures. The answer ? is the fuzziest—a complete lack of information about the presence or absence of OH groups. All possibilities for candidate structures either without OH groups (ether structures) or with a varying number of OH groups (limited by the number of oxygen atoms in the empirical formula) are generated. The treatment of the various kinds of answers is different in our structure generation scheme.

An additional and extremely efficient constraint on the generation process may be the input of sizable fragments. However, their extraction from the spectral information is a separate and complicated CASE problem. Thus, consider an empirical formula of (say) 20 heavy atoms. At least 15 of them must be atoms of one or more fragments in order for the CASE problem to be efficiently constrained. The derivation of a substruc-

TABLE IV   Input of Chemical Groups

Input	Meaning
$n_{OH} = n$	n OH groups are present (n is the precise number of chemical groups)
N or $n_{OH} = 0$	No OH groups are present
Y	There are OH groups, but their number is not known
?	There may be OH groups, but there may not be (fully ambiguous)

ture of more than 10 atoms is also a separate structure elucidation problem. Hence, it is very important that the structure generation and elucidation process be carried out without the use of any sharply defined structural fragments. If we consider the fragment structure as a bulk, then it has a crisp nature and will not match the fuzzy character of the structure/spectra information. By contrast, viewing a substructure as a set of atom–atom connectivities represented by the associated ⊕-type SSs list provides us with an extremely flexible representation of both structural fragments and whole structures. Accordingly, the generation of chemical structures is thus carried out in a uniform way by using fragments, chemical groups, and single atoms.

E. Guided Structure Generation

Our structure generation scheme[80, 85–93] is based on the following principles. Because all the nonbonded ←-type (SV) BSs form a set of levels, the generation process is a multilevel procedure. Two steps are carried out at each level: the first is the formation of the ⊕-type BS list (SS-selection step); the second is a saturation of each SS from the list with the current level SV (SV-SS saturation step). Obviously, if the ⊕-type BS list of the current SV level is formed from input fixed or alternative ⊕-type BSs, the SS-selection step is simply skipped. Otherwise, the ⊕-type BS list is formed by a SS-selection step, which consists of tracing all the second row ⊕-type BSs, comparing their atom attributes with the possible SV-SS bondings, and selecting the most probable bondings. Thus the selected connectivities in the list of ⊕-type SSs are formed. A SV-SS saturation step follows each SS-selection step. After such a saturation, an extension of the existing partial structure is obtained. The process of structure generation is depicted in Figs. 8 and 9.

The generation process may be conducted in two ways.[97] The first is *depth-first*, where each branch of the generation tree is formed from its beginning (from its root) to its end, followed by a backtracking procedure to the previous node, with a new branch being formed. This approach is a successive execution of the steps of depth-first forming of branches and backtracking. It is depicted in Fig. 10. The previous versions of our generators[85–93] were also developed in this way.

The second generation process is the *breadth-first* approach. All the nodes of a given level are fully traced and then the next level is treated. In this approach, no backtracking procedure is carried out. The generation goes breadthwise and forward only. The breadth-first approach has been implemented in the current realization of our structure generation method[41, 80] and is depicted in Fig. 11.

It should be emphasized that the SS-selection procedure is rather important for guiding the generation process toward the most likely connectivities. The selection is carried out following a series of rules based

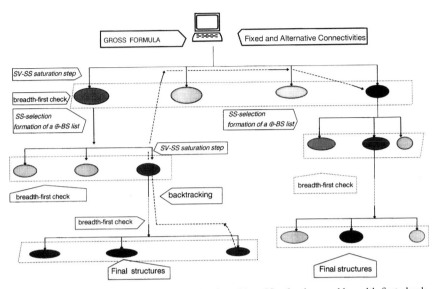

FIGURE 10 Depth-first generation with backtracking, SS-selection, and breadth-first check.

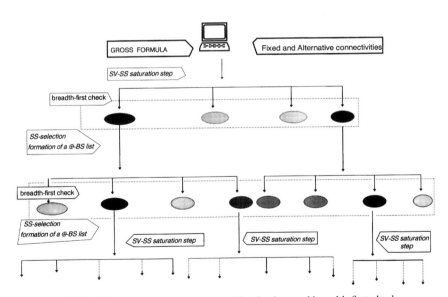

FIGURE 11 Breadth-first generation, SS-selection, and breadth-first check.

on the atom attributes. Some of these rules guide the generation process toward the formation of chemically consistent structures, e.g., an ⊕-type BS (SS) selection from the same atom that the current SV originates from is prohibited. This rule prevents an atom from being bonded to itself. Similar rules are developed to prohibit the generation of different types of unsaturated structures: those not conjugated (N answer to the input query), conjugated only (Y answer), or the generation of both conjugated and not conjugated (? answer). No ⊕-type BS of SP2 hybridization attributes for the C=C fragment SV is selected for the first case, whereas ⊕-type BSs of SP2 hybridization attributes are selected only for the second case, and both ⊕-type BSs SP3 and SP2 hybridizations are selected for the third case. Similar selection rules are applied to the formation of different chemical groups, such as OH and NH, and ring systems of different sizes.

Another group of selection rules refers to the spectrum/structure relationship. The system of these rules forms an a priori selection. There is also an a posteriori selection, which will be further discussed.

Thus, the α-environment connectivity attribute directs the SS-selection procedure toward the selection of SSs that are appropriate to this attribute only. Consider that the SP3_N and SP3_O attributes assigned to the current SV atom have different $m_v(\text{SP3\_N})$ and $m_v(\text{SP3\_O})$ values. The generation process is carried out by selecting ⊕-type BSs from the nitrogen and oxygen atoms only. As a result, a great number of branches of the generation tree are a priori pruned and the combinatorial problem is greatly alleviated. The same comments refer to the selection of the associated ⊕-type list of heteroatom SVs. Saturation sites from carbon atoms having the corresponding α-environment attributes are selected only for each heteroatom SV. The bonds formed by generating the different extensions at a given level may be weighted by their m_v values. Thus, the fuzzy set E^f of the fuzzy graph is formed. Apparently, the efficiency of this constraint depends on the reliability of the α-environment attributes of an atom.

Furthermore, all the generated extensions at a given level undergo a retrospective selection that is carried out by a breadth-first check procedure[40, 89, 91] that traces all generated extensions at a given level by comparing each with the spectral information assigned to its carbon atoms (^{13}C chemical shift/multiplicity attribute). Every extension generated at the current level is associated with a set of possibility values, which give its assessment.

In a previous paper,[91] a new possibility value associated with each newly generated extension was introduced. The chemical shifts δ^{calc} of the carbon atoms may be evaluated by using any of the additivity parameter schemes[98] according to the expression

$$\delta^{\text{calc}} = \delta_0 + \Delta\delta_\alpha + \Delta\delta_\beta + \Delta\delta_\gamma. \tag{9}$$

Here δ_0 is a fixed constant for each type of compound (the alkanes, alkenes, etc.) and $\Delta\delta_\alpha$, $\Delta\delta_\beta$, $\Delta\delta_\gamma$ are additive parameters[47, 48] for the substituents at the α, β and γ positions with respect to the evaluated carbon atom. For each extension generated after a bond (ij) is formed, Hamilton's agreement (R) factor[99]

$$\text{HAF} = R_{ij} = \frac{\sqrt{\Sigma_k \left(\delta_k^{\text{calc}} - \delta_k^{\text{exp}} \right)^2}}{\sqrt{\Sigma_k \left(\delta_k^{\text{exp}} \right)^2}} \tag{10}$$

is calculated, where δ_k^{calc} and δ_k^{exp} are the calculated and experimental chemical shift values. Whereas the latter, as discussed previously, are attributes of the carbon atoms and are constant during the whole generation process, the former change with each new extension. All the R_{ij} generated at a given level are summed (summation over S) and, for each extension, the following real number is generated:

$$m(ij) = \frac{1/R_{ij}}{\Sigma_s 1/R_s}. \tag{11}$$

Each new bond (i, j) created at the current level is thus associated with a fuzzy number $m(ij)$. Equation (11) defines a membership function associated with every new bond formed during the generation process. Precautions have been taken in the program so that the term $1/R$ will not introduce an error in those cases where the R_{ij}s approach zero (which hardly ever occurs).

The structure elucidation process is carried out by eliminating the extensions that have lower $m(ij)$ values and retaining for the next level of structure generation those extensions with higher $m(ij)$ values. Thus, a real-number vector that has the dimension of the two-row matrix can be formed. Furthermore, whereas the two-row matrix in Fig. 5 represents the support of $\mathbf{M}^f$, this vector will represent the fuzzy part $\mathbf{M}^f$. In fact, the elements $m(ij)$ reflect, in a way, the chemical shift/structure relationship. Note that according to Eq. (9), these possibility values estimate the influence of environments (β-, γ-) deeper than the α-environment attributes. Unfortunately, additivity parameter schemes were developed for complete structures, although here they have been employed for partial structures. This makes the result rather uncertain and fuzzy. Hence, instead of using additivity parameter schemes to automatically discard the less possible extensions, all the generated extensions are listed with their associated possibility values and the final decision is left to the user.

The δ_K^{calc} values also may be evaluated by using the simple correlation between ^{13}C chemical shifts and the charges of the carbon atoms[36–40]:

$$\delta_{13C} = Aq_c + B. \tag{12}$$

The coefficients A and B are obtained by applying a least-squares procedure, and the standard approximation error (SAE)

$$SAE = \sqrt{\frac{\left(\delta_{13C}^{obs} - \delta_{13C}^{calc}\right)^2}{N-2}} \tag{13}$$

is calculated for each extension. The membership function $m(ij)$ is then calculated from Eq. (11).

Series of inference rules can be developed for estimating and ranking the separate extensions generated at each level with respect to the available spectral information. For example, such a system of selection rules has been developed in the case of availability of 2D NMR information. The various 2D NMR techniques may be used as powerful structure inference tools.

Our breadth-first check procedure allows for a more flexible examination of the connectivity of each newly generated extension and its comparison with the corresponding couplings from the 2D NMR maps. Thus, in the combination of C,H HETCOR and C,C COSY spectra, the former provides direct connectivity between the carbon atoms i, j and the hydrogen atoms m, n, respectively (see Fig. 12), whereas the latter gives the couplings between the hydrogen atoms themselves. Some simple rules may be derived from the theory of this technique: if carbon atom i is adjacent to hydrogen atom m, in the C,H HETCOR spectrum, and atom j is adjacent to atom n, and atoms m and k are coupled in the H,H COSY spectrum, then one may consider the carbon atoms i and j as being adjacent. It is obvious that by using this simple rule, the whole carbon spectrum can be formed. Hence, this rule would have been programmed as a SS-selection rule. Unfortunately, the information provided by the H,H COSY method is in many cases ambiguous. It is not possible to distinguish between the vicinal and the longer range couplings, which, although they are of lower probability, still appear. In some rare cases, a vicinal coupling

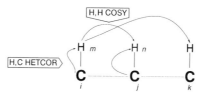

FIGURE 12 C — C connectivity elucidation by using H,C HETCOR and H,H COSY 2D NMR spectra.

may even be absent, but longer range couplings are present. Moreover, the quaternary carbon atom and heteroatom connectivities cannot be treated in this way because they require that the inference rules based on this information produce fuzzy rather than crisp values.

Accordingly, the following procedure has been developed.[89] All extensions generated at a given level are scanned through a breadth-first check procedure. A set of (N, Y, ?) messages has been developed and each extension of a given level is associated with a message by checking all the connectivities formed by the current level SV with the other atom SSs. If the current SV atom forms a connectivity corresponding to the 2D signal pattern, a Y (yes) message is associated with this extension. In the case of a carbon heteroatom or a quaternary carbon–carbon extension, a ? (ambiguity) message is provided, and in the case of the lack of a corresponding cross-peak for a carbon–carbon linkage, an N (no) message is associated with the generated extension. Additionally, a *possibility index* (PI) was devised to indicate the possibility distribution between the alternative solutions at the current level. It is based on rather formal rules, which are listed in Table V.

The cardinality of all possibilities is formed by summing the PIs of all extensions at the current level, thereby forming a membership function $m(ij)$ associated with each generated extension. Thus, $m(ij)$ gives the possibility distribution of the generated extensions (given as a percentage).

TABLE V   Possibility Index (PI) Values Related to 2D NMR Spectral Information for Different Cases of SV-SS Connectivities

PI	For the case when
0.5	The current SV-SS extension is formed by a carbon–heteroatom bond (either SV or the SS atoms are heteroatoms)
0.5	Either the current SV or the SS atoms forming the extension bond are quaternary carbon atoms, i.e., have a $^{13}C - ^{1}H$ direct multiplicity equal to 1
1.0	The 2D NMR spectra indicate either direct or longer range connectivity between the corresponding SV-SS carbon atoms and such a connectivity is observed in the generated extension
0.1	The 2D spectra indicate the presence of connectivity between the corresponding SV and SS carbon atoms, but it is not found in the generated extension, while a longer range C—C connectivity (through two or three atoms) is observed.
0.03	The 2D analysis rejects the presence of connectivity between the current SV and the SS carbon atoms, but such connectivity is observed in the extension. A small possibility is left for the rare cases where a signal does not appear.

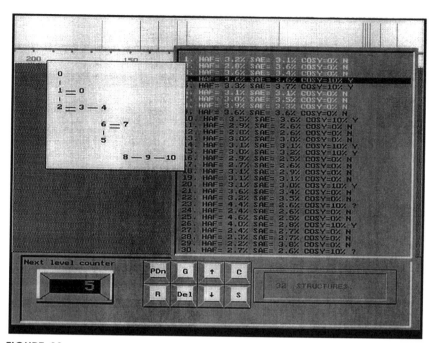

FIGURE 13   Depiction of the on-screen selections of the most likely extensions using fuzzy membership values. Reprinted with permission from *J. Chem. Inf. Comput. Sci.* (ref. 87), 1992 American Chemical Society.

Although rather formal, this $m(ij)$ provides extremely valuable information about the 2D spectrum/substructure relationship within the set of generated extensions at a given level.

Accordingly, the breadth-first check procedure is carried out by using a series of membership function values that weight each extension and allow selection of the most likely and discard the less likely extensions. Up to now the selection has been carried out by the user, as depicted in Fig. 13, but with the further development of more reliable membership functions, the fuzzy set operations discussed in the previous section may be applied to render the CASE process fully automatic.

ACKNOWLEDGMENTS

This work was supported by the National Fund for Scientific Investigations from the Ministry of Education, Science and Technologies, Sofia, Bulgaria (Contracts X96 and X412). Many of the ideas on the use of membership functions were jointly developed with Professor Daniel Cabrol-Bass and Isabell Laude from LARTIC, Université de Nice Sophia-Antipolis, France, during the author's stay as a NATO and EEC guest Senior Researcher. Their contributions and NATO and EEC support are gratefully acknowledged.

REFERENCES

1. E. Dubois, D. Laurent, and H. Viellard, *C.R. Acad. Sci. Paris Ser. C* **263**, 764 (1966); **264**, 348, 1019 (1967).
2. J. E. Dubois, *Entropie* **27**, 1 (1969).
3. J. E. Dubois and H. Viellard, *Bull. Soc. Chim. Fr.* 900, 905, 913 (1968); 839 (1971).
4. J. E. Dubois, J. P. Anselmeni, M. Chastrette, and F. Hennequin, *Bull. Soc. Chim. Fr.* 2439 (1969).
5. J. E. Dubois, *J. Chem. Doc.* **13**, 8 (1973).
6. J. E. Dubois, in *Computer Representation and Manipulation of Chemical Information* (W. T. Wipke, S. R. Heler, R. J. Feldmenn, and E. Hyde, eds.), Chap. 10, p. 239. Wiley, New York, 1974.
7. J. Lederberg, *DENDRAL-64*, Part I, NASA Report CR-57029, STAR No. 65-13158 (1964).
8. J. Lederberg, G. L. Sutherland, B. G. Buchanan, E. A. Feigenbaum, V. A. Robertson, A. M. Duffield, and C. Djerassi, *J. Am. Chem. Soc.* **91**, 2973 (1969).
9. J. Lederberg, Topology of Cyclic Graphs, *DENDRAL-65*, Part II, NASA Report CR-68898, STAR No. N66-14074 (1965).
10. A. M. Duffield, A. V. Robertson, C. Djerassi, B. G. Buchanan, G. L. Sutherland, E. A. Feigenbaum, and J. Lederberg, *J. Am. Chem. Soc.* **91**, 2977 (1969).
11. G. Schorll, A. M. Duffield, C. Djerassi, B. J. Buchanan, G. L. Sutherland, E. A. Feigenbaum, and J. Lederberg, *J. Am. Chem. Soc.* **91**, 7444 (1969).
12. A. Buchs, A. M. Duffield, G. Schroll, C. Djerassi, A. B. Delfino, B. G. Buchanan, G. L. Sutherland, E. A. Feigenbaum, and J. Lederberg, *Helv. Chim. Acta* **55**, 1394 (1970).
13. H. Eggert and C. Djerassi, *J. Am. Chem. Soc.* **95**, 3710 (1973).
14. L. Masinter, N. S. Sridharan, J. Lederberg, and D. H. Smith, *J. Am. Chem. Soc.* **96**, 7702 (1974).
15. D. H. Smith, L. M. Masinter, and N. S. Sridharan, in *Computer Representation and Manipulation of Chemical Information* (W. T. Wipke, S. R. Heler, R. J. Feldmenn, and E. Hyde, eds.), Chap. 12, p. 287. New York, 1974.
16. R. E. Carhart, D. H. Smith, H. Brown, and C. Djerassi, *J. Am. Chem. Soc.* **97**, 5755 (1975).
17. S. Sasaki, H. Abe, Y. Hirota, Y. Ishida, Y. Kudo, S. Ochiai, K. Saito, and T. Yamasaki, *Sci. Rep. Tohoku Univ. I* (*Japan*) **60** (4), 153 (1978).
18. T. Yamasaki, H. Abe, Y. Kudo, and S. Sasaki, in *Computer-Assisted Structure Elucidation* (D. Smith, ed.), ACS Symposium Series, Vol. 54. American Chemical Society, Washington, DC (1977).
19. T. Oshima, Y. Ishida, K. Saito, and S. Sasaki, *Anal. Chim. Acta* **122**, 95 (1980).
20. S. Sasaki, I. Fujiwara, H. Abe, and T. Yamasaki, *Anal. Chim. Acta* **122**, 87 (1980).
21. S. Sasaki and H. Abe, in *Computer Applications in Chemistry* (S. R. Heller and R. Potenzione, Jr., eds.), p. 185. Elsevier, Amsterdam, 1983.
22. S. Sasaki and Y. Kudo, *J. Chem. Inf. Comput. Sci.* **25**, 252 (1985).
23. K. Funatsu, A. Del Carpio, and S. Sasaki, *Fresenius Z. Anal. Chem.* **324**, 750 (1986).
24. D. B. Nelson, M. E. Munk, K. B. Gash, and D. L. Herald, *J. Org. Chem.* **34**, 3800 (1969).
25. C. A. Shelley and M. E. Munk, *J. Chem. Inf. Comput. Sci.* **19**, 247 (1979).
26. C. Shelley, T. Hays, M. Munk, and R. Roman, *Anal. Chim. Acta* **103**, 121 (1978).
27. B. D. Christie and M. E. Munk, *J. Chem. Inf. Comput. Sci.* **20**, 27 (1988).
28. A. H. Lipkus and M. A. Munk, *J. Chem. Inf. Comput. Sci.* **28**, 9 (1988).
29. W. Bremser and W. Fashinger, *Magn. Reson. Chem.* **23**, 1056 (1985).
30. M. E. Munk, *J. Chem. Inf. Comput. Sci.* **32**, 263 (1992).
31. D. H. Rouvray, in *Chemical Applications of Topology and Graph Theory* (R. B. King, ed.). Elsevier, Amsterdam, 1983.

32. N. Trinajstić, *Chemical Graph Theory*. CRC Press, Boca Raton, FL, 1992.
33. W. Gellert, H. Küstner, M. Hellwich, and H. Käsner, eds. *Mathematics at a Glance, A Compendium*, Chap. II (14), p. 320.
34. L. A. Zadeh, *Inform. Control* **8**, 338 (1965).
35. L. A. Zadeh, *J. Fuzzy Sets Syst.* **1**, 3 (1978).
36. I. P. Bangov, *Commun. Dept. Chem.* **17**, 459 (1984) (in Bulgarian).
37. I. P. Bangov, *Microchim. Acta (Wien)* **2**, 281 (1986).
38. I. P. Bangov, *Commun. Dept. Chem.* **21**, 194 (1988).
39. I. P. Bangov, *Anal. Chim. Acta* **109**, 29 (1988).
40. I. P. Bangov, *XXVI CSI Selected Papers* **8** (1990).
41. I. P. Bangov, I. Laude, and D. Cabrol-Bass, in *E.C.C.C.1 Computational Chemistry, AIP Conference Proceedings* (B. Bernardi and J. L. Rivail, eds.), Vol. 330, Chap. V, p. 740. API Press, Hong Kong.
42. R. T. Sanderson, *Science* **114**, 670 (1951).
43. J. Gasteiger and M. Marsili, *Tetrahedron* **36**, 3219 (1980).
44. M. A. Kirpichenok and N. Zefirov, *Zh. Strukt. Khim.* **22**, 673 (1987).
45. L. Lah, M. Tusar, and J. Zupan, *Tetrahedron Comp. Techn.* **2**, 5 (1989).
46. H. N. Cheng and J. Ellingsen, *J. Chem. Inf. Comput. Sci.* **23**, 197 (1983).
47. E. Pretsch, T. Clerc, J. Seibl, and W. Simon, *Tables of Spectral Data for Structure Determination of Organic Compounds*. Springer-Verlag, Berlin, 1989.
48. D. W. Brown, *J. Chem. Educ.* **65**, 209 (1985).
49. W. Robien, *Mikrochim. Acta (Wien)* **2**, 271 (1986).
50. E. A. Fegenbaum, *in Information Processing 68*. North-Holland, Amsterdam, 1968.
51. D. Cabrol-Bass, I. Laude, T. Laidboeur, and I. P. Bangov, *in E.C.C.C.1 Computational Chemistry AIP Conference Proceedings* (B. Bernardi and J. L. Rivail, eds.), Vol. 330, Chap. IV, p. 610. API Press, Hong Kong.
52. R. Grund, A. Kerber, and R. Laue, *Commun. Math. Chem.* **27**, 87 (1992).
53. M. L. Contreras, R. Valdivia and R. Rozas, *J. Chem. Inf. Comput. Sci.* **32**, 223 (1992).
54. M. L. Contreras, R. Valdivia, and R. Rozas, *J. Chem. Inf. Comput. Sci.* **32**, 483 (1992).
55. V. Kvasnicka and J. Pospichal, *Chemom. Intell. Lab. Syst.* **18**, 171 (1993).
56. R. A. Carhart, D. H. Smith, N. A. B. Gray, J. G. Nourse, and C. Djerassi, *J. Org. Chem.* **46**, 1708 (1981).
57. J. Lederberg, G. I. Sutherland, B. G. Buchanan, E. G. Feigenbaum, A. V. Robertson and A. M. Duffield, *J. Am. Chem. Soc.* **91**, 2973 (1969).
58. N. A. B. Gray, *Prog. Nucl. Magn. Reson. Spectrosc.* **15**, 201 (1982).
59. K. Funatsu, Y. Susuta, and S. Sasaki, *J. Chem. Comput. Sci.* **29**, 6 (1989).
60. M. Munk, R. Lind, and M. Clay, *Anal. Chim. Acta* **184**, 1 (1986).
61. M. Lindley, J. Schoolery, D. Smith, and C. Djerassi, *Org. Magn. Reson.* **21**, 405 (1983).
62. B. D. Christie and M. Munk, *J. Am. Chem. Soc.* **113**, 3750 (1991).
63. M. Carabedian and J.-E. Dubois, *J. Chem. Inf. Comput. Sci.* **31**, 557 (1991).
64. J.-E. Dubois and M. Carabedian, *J. Chem. Inf. Comput. Sci.* **31**, 564 (1991).
65. K. Funatsu, Y. Susita, and S. Sasaki, *Anal. Chim. Acta.* **220**, 155 (1989).
66. Y. Ishida and S. Sasaki, *Computer Enhanced Spectrosc.* **1**, 173 (1983).
67. J.-L. Foulon, *J. Chem. Inf. Comput. Sci.* **34**, 1204 (1994).
68. J. Xu, S. K. Straus, C. Sanctuary, and L. Trimble, *J. Chem. Inf. Comput. Sci.* **33**, 668 (1993).
69. P. C. Jurs, *in Computer Representation and Manipulation of Chemical Information* (W. T. Wipke, S. R. Heler, R. J. Feldmann, and E. Hyde, eds.), Chap. 11, p. 265. Wiley, New York, 1974.
70. J. Zupan and J. Gasteiger, *Anal. Chim. Acta* **248**, 1 (1991).
71. D. Ricard, C. Cachet, and D. Cabrol-Bass, *J. Chem. Inf. Comput. Sci.* **33**, 202 (1993).
72. J. W. Ball and P. C. Jurs, *Anal. Chem.* **65**, 505 (1993).

73. M. Otto, *Anal. Chem.* **62**, 797 (1990).
74. D. E. Goldberg, *Genetic Algorithms in Search, Optimization and Machine Learning.* Addison-Wesley, Reading, MA, 1989.
75. R. Wehrens, C. Lucasius, L. Buydens, and G. Kateman, *J. Chem. Inf. Comput. Sci.* **33**, 245 (1993).
76. M. Otto, *Chemom. Intell. Lab. Syst.* **4**, 101 (1988).
77. M. Otto, *Anal. Chim. Acta* **235**, 169 (1990).
78. B. L. Richards, *BYTE* **April**, 285 (1988).
79. A. Rosenfeld, *in Fuzzy Sets and their Applications to Cognitive and Decision Processes* (L. A. Zadeh, K.-S. Fu, K. Tanaka, and M. Shimura, eds.), p. 77. Wiley, New York, 1974.
80. I. P. Bangov, *Discrete Appl. Math.* **67**, 27 (1996)..
81. A. Fürst, J. T. Clerc, and E. Pretsch, *Chemom. Intell. Lab. Syst.* **5**, 329 (1989).
82. C. A. Shelley and M. E. Munk, *Anal. Chem.* **50**, 1522 (1978).
83. I. Fujiwara, T. Okuyama, T. Yamasaki, H. Abe, and S. Sasaki, *Anal. Chim. Acta* **133**, 527 (1981).
84. K. Funatsu, H. Katsumi, and S. Sasaki, *Computer Enhanced Spectrosc.* **3**, 87 (1986).
85. I. P. Bangov, *J. Chem. Inf. Comput. Sci.* **30**, 277 (1990).
86. I. P. Bangov, *Commun. Math. Chem.* **27**, 3 (1992).
87. I. P. Bangov, *J. Chem. Inf. Comput. Sci.* **32**, 167 (1992).
88. I. P. Bangov, *J. Chem. Inf. Comput. Sci.* **34**, 318 (1994).
89. I. P. Bangov, S. Simova, I. Laude, and D. Cabrol-Bass, *J. Chem. Inf. Comput. Sci.* **34**, 546 (1994).
90. I. P. Bangov, I. Laude, and D. Cabrol-Bass, *Anal. Chim. Acta* **298**, 33 (1994).
91. T. Laidboeur, I. Laude, D. Cabrol-Bass, and I. P. Bangov, *J. Chem. Inf. Comput. Sci.* **34**, 171 (1994).
92. I. P. Bangov and K. D. Kanev, *J. Math. Chem.* **2**, 31 (1988).
93. I. P. Bangov, *Commun. Math. Chem.* **14**, 235 (1983).
94. I. P. Bangov, *Org. Magn. Reson.* **16**, 296 (1981).
95. I. P. Bangov and R. Redeglia, *Org. Magn. Reson.* **21**, 443 (1983).
96. I. P. Bangov, T. Steiger, and R. Radelia, *Z. Phys. Chem.* **268**, 891 (1987).
97. N. L. Nilsson, *Problem-Solving Methods in Artificial Intelligence*, Chap. 3, p. 53. McGraw-Hill, New York, 1971.
98. D. M. Grant and E. G. Paul, *J. Am. Chem. Soc.* **86**, 2984 (1964).
99. W. C. Hamilton, *Acta Cryst.* **18**, 502 (1965).

9

Fuzzy Hierarchical Classification Methods in Analytical Chemistry

DAN-DUMITRU DUMITRESCU

Faculty of Mathematics and Computer Sciences
Babes-Bolyai University
RO-3400 Cluj-Napoca, Romania

I. INTRODUCTION

Classification involves basically a process in which an object (or an observation) is assigned to a class representing some natural category. The central problem in classification is thus to establish the natural grouping for any given data set. Generally speaking, by a *natural grouping* we understand some homogeneous but well-separated structures within the data. Whenever data can be partitioned in this way, the classes formed can be viewed as *clusters* or *clouds* of data points. The classification of objects belonging to a given category (or collection) affords the essential link between observational or experimental data and scientific hypotheses. The attainment of a classification actually represents the first step toward a theory concerning a collection of objects. As a result of classification, for instance, it may be possible to structure experimental observations into a hierarchy of meaningful categories. The concepts associated with each of these categories can be useful in helping us to explain our data and to predict which concepts apply to which objects.

Classification may thus be viewed as a process by which a structure can be detected from the chaos in the data that are presented. In chemistry and other sciences the ideal form of the classification process is the transformation of experimental or observational data into scientific laws. Throughout the history of chemistry many empirical laws have been discovered by analyzing and classifying experimental data. In general, classification needs to be performed on newer kinds of objects or on novel

Fuzzy Logic in Chemistry

observational data to achieve a new understanding of the categories. The classification edifice is founded on observational data and gives rise to a set of classes. Each object must belong to one or more of such classes. The belonging relationship may be absolute or fuzzy. In the latter case, we obtain a set of membership values for each object that is classified.

If the classification is done properly, elements belonging to the same class can be considered as similar objects and characterized by the same concept. Classification is also a means of refining scientific language in that it introduces new concepts or notions associated with a new class of objects. In this sense, object classification may be seen as the fundamental tool in observational sciences. Classification is very widely used in analytical chemistry to assist us in understanding the structure of large sets of objects and their mutual relationships. Classification can suggest fruitful hypotheses concerning chemical phenomena and processes and can be used to verify conjectures about a class of objects. Each object (sample, point) is specified by the values of certain characteristics (features, variables). Generally, a characteristic results from a measurement process, and the values of characteristics are expressed by real numbers. An object is therefore often represented as a point (or a vector) in a multidimensional data space.

The algorithms employed for clustering data points need not be used only to cluster similar points; they can also be used to reduce the dimensionality of the data space. Dimensionality reduction techniques enable us to eliminate non-relevant characteristics. In this way we may achieve important data compression by retaining only useful information. The relative success of any classification is determined by the so-called reduction principle, whose goal is the use of less information for the computation of all the observed data. Cluster analysis does not make use of category (or class) labels. The presence of category labels distinguishes supervised classification from cluster analysis (for more information on this see, for instance, Duda and Hart[15]). By employing a training set as a trainable classifier, one is able to partition the data space into decision regions. Each region will correspond to a class of the training set. We may say that the classifier learns the training classes. The boundary of a decision region is a *decision boundary*. The training classes separable by hyperplanes are said to be linearly separable. Two linearly separable classes in $\mathbb{R}^2$ are depicted in Fig. 1. The classes shown in Fig. 2 are however not linearly separable.

Decision regions can be obtained by using the training classes and an adequate training procedure (see, for instance, Duda and Hart[15]). After the classifier has been trained it is able to recognize an unknown object. If the unknown object x belongs to the decision region R_i, then x is from the category i, i.e., it is similar to the training vectors from the class A_i. Trainable classifiers may be used for a variety of applications. For instance,

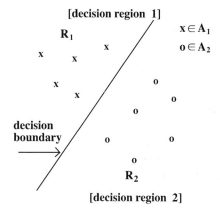

FIGURE 1 Two linearly separable classes in $\mathbb{R}^2$.

we may use them to determine if a complex chemical system is in a normal or abnormal state. The system may be characterized by several parameters which might include pressure and temperature. Two possible behaviors of the system are represented by certain states forming two training classes. Other applications may concern the analysis of drugs, food, water, or materials based on their chemical composition. Using a trainable classifer we may decide whether a novel object belongs within a predefined category. It is also conceivable that such an object cannot be assigned to any existing category.

Most real-world classes do not have sharp boundaries. Between the complete membership or nonmembership of an object in a class we may observe an infinity of intermediate situations. These intermediate situations are said to be fuzzy. Fuzzy set theory as developed by Zadeh[66] permits an object to belong to a cluster with a grade of membership that lies within the interval $[0,1]$. A class (or a cluster) of objects may be represented as a fuzzy set. Because real-world classes are more fuzzy than

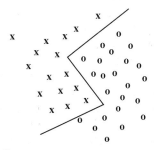

FIGURE 2 Two nonlinearly separable classes in $\mathbb{R}^2$.

crisp, fuzzy set theory provides us with a useful tool for clustering data (see Bezdek[7] or Dumitrescu[21]). The use of fuzzy methods can also help in the design of classifiers able to learn from fuzzy, i.e., ambiguous, incomplete, or contradictory, data (see Bezdek[8] and Dumitrescu[19, 20]). Fuzzy classification techniques represent a useful and powerful tool in analytical chemistry and related areas. Blaffert[10] and Otto and Bandemer,[49-51] for instance, have considered fuzzy pattern recognition methods in chemistry. Many applications of fuzzy classification techniques in analytical chemistry have been reported. However, we may regard the field as in its infancy with many potential applications still waiting to be exploited.

Hierarchical clustering methods are of particular interest in chemical data analysis. Usually a hierarchical clustering procedure is more suitable than a one-level (nonhierarchical) method. Using hierarchical methods we may obtain more information concerning cluster data structure. Unfortunately, however, there are very few methods that enable us to construct a fuzzy hierarchy. A hierarchy may be obtained either divisively or agglomeratively. In work by Dumitrescu[17] a fuzzy divisive hierarchical clustering method has been proposed. This method has been used for mineral water classification,[22, 23, 38] the study of acrylonitrile selectivity,[42] the selection and optimal combination of solvent systems in thin-layer chromatography (TLC),[26, 57] the classification of Roman pottery,[55] and for the cross-classification of muds using their chemical and mineralogical characteristics.[25] The chemical elements have also been classified using their chemical and physical characteristics. The clustering structure that resulted even suggested a new arrangement for the chemical elements.[58]

II. FUZZY PARTITION OF A FUZZY CLASS

Let us consider now a data set $X = \{x^1, x^2, \ldots, x^P\}$ and suppose that each object in X is characterized by the values of s features (or characteristics). x_k^j is the value of the kth characteristic with respect to the jth object. For instance, x_k^j would be the amount of a certain chemical element (labeled k) in the sample j or the amount of the substance k in the TLC solvent system j. The chemical significance of the sample vectors $x^1, \ldots, x^P$ changes according to each particular problem. Objects to be classified may arise from a great variety of sources including chemical experimental strategies, materials, observations and experimental results, technological parameters, or the parameters of a chemical process. If we intend to compare (classify) processes (techniques, experiments, ...), x_k^j will represent the value of the kth parameter in the process (technology, experiment, ...) j. As usual, we shall consider object x^j to be a column vector (or a point) in s-dimensional Euclidean vector space.

The objects in the data set may have hybrid characteristics that would assign them to a number of different classes. In such cases the data clusters may partially overlap. We may detect overlapping clusters if we allow each point to be a member of each class, i.e., to have a fuzzy membership degree. The membership degree is a number lying between 0 and 1. The existence of stray points (outliers), nonseparable classes, and bridges between classes (see Fig. 3) together represent major difficulties in proceeding with the classification process. These sources of ambiguity and error for most clustering techniques can be accommodated by representing clusters as fuzzy sets. Fuzzy set theory represents a natural framework for dealing with classification problems. Within this framework the clustering structure of a data set X may be represented as a fuzzy partition of X.

A fuzzy set on X is a mapping $A: X \rightarrow [0, 1]$. The value $A(x^j) \in [0, 1]$ is the *membership degree* of the object x^j in the fuzzy class A. Let A, B be fuzzy sets on X. Then $P = \{A, B\}$ is a (binary) *fuzzy partition* of X if the following conditions are fulfilled:

(i) $$A \cap B = \varnothing,$$

(ii) $$A \cap B = X.$$

P is a *partition of unity* if the condition

$$A(x) + B(x) = 1$$

holds for each x from X. A fuzzy partition P is a natural generalization of the classical notion of partition if P is also a partition of unity. A fuzzy partition and a partition of unity need not always be equivalent. Let us consider, for instance, intersection and union that are defined using the t-norm T_0

$$T_0(a, b) = \min(a, b), \qquad \forall a, b \in [0, 1],$$

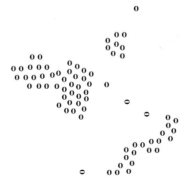

FIGURE 3 Overlapping classes, stray points, and bridges between classes.

and its dual conorm S_0

$$S_0(a,b) = \max(a,b), \qquad \forall a,b \in [0,1].$$

In this case we have

$$(A \overset{\circ}{\cup} B)(x) = S_0(A(x), B(x)) = \max(A(x), B(x))$$

and

$$(A \underset{\circ}{\cap} B)(x) = T_0(A(x), B(x)) = \min(A(x), B(x))$$

for each x from X. If these definitions are used, the equivalence between a fuzzy partition and a partition of unity will not hold.

In this chapter we shall admit the set operations for fuzzy sets are defined with the t-norm T_∞,

$$T_\infty(a,b) = \max(a + b - 1, 0),$$

and its dual conorm S_∞,

$$S_\infty(a,b) = \min(a + b, 1),$$

for each $a, b \in [0,1]$. For the present, we shall consider the following definitions for the union and intersection of fuzzy sets,

$$(A \cup B) = \min(A(x) + B(x), 1),$$
$$(A \cap B)(x) = \max(A(x) + B(x) - 1, 0),$$

for each x within X. Dumitrescu[18] has demonstrated that these definitions are unique for a binary fuzzy partition to be equivalent to a partition of unity. Therefore, for the considered definitions we have

$$\left. \begin{array}{r} A \cup B = X \\ A \cap B = \varnothing \end{array} \right\} \quad \Leftrightarrow \quad A(x) + B(x) = 1, \qquad \forall x \in X,$$

and these are the unique definitions for which this equivalence holds.

Let us now consider the fuzzy sets $A_1, A_2, \ldots, A_n$ on X with $n > 2$. We know that $\{A_1, A_2, \ldots, A_n\}$ is a *disjoint family* (Butnariu[11]) of fuzzy sets if and only if the condition

$$\left(\bigcup_{i=1}^{i} A_i \right) \cap A_{j+1} = \varnothing,$$

holds for $j = 1, \ldots, n - 1$. $P = \{A_1, A_2, \ldots, A_n\}$ is a *fuzzy n-partition* of $X(C)$ if the following requirements are fulfilled:

(1) P is a disjoint family of fuzzy sets on X; and

(2) $\bigcup_{i=1}^{n} A_i = X(C)$.

It is now easily seen that $P = \{A_1, A_2, \ldots, A_n\}$ is a fuzzy partition of X if and only if P is a partition of unity, i.e.,

$$\sum_{i=1}^{n} A_i(x) = 1$$

for each $x \in X$. Each A_i will be called a module of the fuzzy partition P. We shall denote by $F_n(C)$ the family of all fuzzy n-partitions of the fuzzy set C. It follows that $P = \{A_1, \ldots, A_n\}$ is a fuzzy partition of C if and only if the equality

$$\sum_{i=1}^{n} A_i(x) = C(x)$$

holds for each x from X.

If the data set X is comprised of two classes, the cluster structure of X may be described by two disjoint fuzzy sets A_1 and A_2 whose union is X. Each fuzzy set corresponds to a class (or cluster) of samples in X. The disjointness condition of A_1, A_2 is a minimal separation condition for the respective classes. It is also possible for the classes $A_i, i = 1, 2$, to have a cluster substructure. This cluster substructure may be described by a fuzzy partition of A_i. If there are $n > 2$ classes in X, the cluster structure of X may be described by a fuzzy n-partition $P = \{A_1, A_2, \ldots, A_n\}$ of X. In this section we shall concentrate on methods to obtain the subcluster structure of a fuzzy set C. It is seen that the cluster structure of the entire data set X may be obtained from a subcluster detection procedure by putting $C = X$ in such a procedure. The search for cluster substructure in a fuzzy class appears to be very necessary in the following circumstances:

(1) A one-level data classification may be insufficient for data characterization. Data structure knowledge may require a more sophisticated analysis.

(2) The problem itself may impose on us a search for the structure of hierarchically organized data.

(3) There is a need to design a hierarchical clustering procedure. Such a procedure is especially useful when the number of clusters in the data is unknown.

(4) There is a need to design a fuzzy training algorithm by a hierarchical procedure.

III. CLUSTER SUBSTRUCTURE IN A FUZZY CLASS

Let $X = \{x^1, x^2, \ldots, x^p\}$ be a data set. In what follows we shall consider each element x^j of X as an s-dimensional vector from $\mathbb{R}^s$. Let C be a fuzzy set corresponding to a class of objects from X. We shall assume that the cluster substructure of the class C is given by the fuzzy partition $P = \{A_1, \ldots, A_n\}$ of C. Subclass A_i may be represented by a prototype L^i. These prototypes could be (i) geometric prototypes (points from $\mathbb{R}^s$, lines in $\mathbb{R}^s$, hyperplanes in $\mathbb{R}^s$, linear varieties in $\mathbb{R}^s$, convex combinations of linear varieties, curves in $\mathbb{R}^s$, spherical shells in $\mathbb{R}^s$, any curved surfaces in $\mathbb{R}^s$) or (ii) set prototypes.

A. The Generalized Fuzzy n-Means Algorithm

If the classes are compact clusters of spherical or ellipsoidal shape, they may be represented by the points (vectors) $L^1, \ldots, L^n$ in $\mathbb{R}^s$. If we let $L = \{L^1, L^2, \ldots, L^n\}$ we may consider L as the representation of the fuzzy partition P. The *inadequacy* $I(A_i, L^i)$ between the fuzzy class A_i and its prototype may be defined as

$$I(A_i, L^i) = \sum_{j=1}^{p} D_i(x^j, L^i),$$

where $D_i(x^j, L^i)$ is the *dissimilarity* between the data point x^j and the prototype L^i. For the dissimilarity D_i we may choose the squared distance to the prototype of the fuzzy class A_i. If d is a norm-induced metric on X, i.e.,

$$d(x, y) = \|x - y\|,$$

and d is generated by a scalar (or inner) product, we obtain

$$d^2(x, y) = \|x - y\|^2 = (x - y, x - y).$$

In $\mathbb{R}^s$ the scalar product has the form

$$(x, y) = x^T M y = \sum_{ij} x_i M_{ij} y_j,$$

where M is a symmetric positive definite matrix and T denotes the transposition.

Using d we may also define a distance with respect to fuzzy class A_i. This distance is denoted by d_i and may be defined as

$$d_i(x, y) = \min(A_i(x), A_i(y)) d(x, y).$$

If the prototype L^i is from X, then we may assume that the prototype has the greatest membership in A_i, that is,

$$A_i(L^i) = \max_j A_i(x^j).$$

If L^i does not belong to X, we may extend A_i to $\mathbb{R}^s$ and put $A_i(x) = 1$, if $x \in \mathbb{R}^s$ and $x \notin X$. Inadequacy $I(A_i, L^i)$ is thus

$$I\left(A_i, L^i\right) = \sum_{j=1}^{p} d_i^2(x^j, L^i).$$

With the previous assumptions we may write

$$I\left(A_i, L^i\right) = \sum_{j=1}^{p} A_i^2(x^j)\|x^j - L^i\|^2.$$

Therefore, we have obtained

$$I\left(A_i, L^i\right) = \sum_{j=1}^{p} A_i^2(x^j)(x^j - L^i)^T M(x^j - L^i).$$

The inadequacy $J(P, L)$ between the fuzzy partition P and its representation L may be defined as

$$J(P, L) = \sum_{i=1}^{n} I\left(A_i, L^i\right).$$

We have obtained the criterion (or objective) function

$$J: F_n(C) \times \mathbb{R}^{sn} \to \mathbb{R},$$

which may be also written as follows:

$$J(P, L) = \sum_{i=1}^{n} \sum_{j=1}^{p} A_i^2(x^j)d^2(x^j, L^i).$$

If we consider the distance d as generated by a scalar product induced norm we have

$$J(P, L) = \sum_{i=1}^{n} \sum_{j=1}^{p} A_i^2(x^j)\|x^j - L^i\|^2,$$

and eventually we may write

$$J(P, L) = \sum_{i=1}^{n} \sum_{j=1}^{p} A_i^2(x^j)(x^j - L^i)^T M(x^j - L^i).$$

The optimal fuzzy n-partition and its representation could be obtained as a minimum of criterion function. To minimize the objective function an iterative method is at hand. This method is based on the successive minimization of the functions $J(P, \cdot)$, where partition P is fixed, and $J(\cdot, L)$, where the representation L is fixed.

The optimal prototypes have to satisfy the extremum condition

$$\frac{\partial J(P, L)}{\partial L^i} = 0, \qquad i = 1, \ldots, n.$$

From this optimum condition we obtain the result

$$\sum_{j=1}^{p} A_i^2(x^j) \frac{\partial}{\partial L^i} (x^j - L^i)^T M(x^j - L^i) = 0, \qquad i = 1, \ldots, n.$$

If A is a symmetric matrix, we have the following derivation rule:

$$\frac{\partial}{\partial x} x^T A x = 2 A x.$$

Using this rule, the minimum condition becomes

$$\sum_{j=1}^{p} A_i^2(x^j) 2 M(x^j - L^i) = 0, \qquad i = 1, \ldots, n.$$

Matrix M is nonsingular (since it is positive definite). Multiplying this equality on the left by M^{-1}, we find that

$$\sum_{j=1}^{p} A_i^2(x^j) x^j = \sum_{j=1}^{p} A_i^2(x^j) L^i, \qquad i = 1, \ldots, n.$$

From this equality we find that the representative point of the class A_i is given by

$$L^i = \frac{\sum_{j=1}^{P} A_i^2(x^j) x^j}{\sum_{j=1}^{p} A_i^2(x^j)}, \qquad i = 1, \ldots, n. \tag{1}$$

We have found that the prototype of each fuzzy class A_i is its center of mass (or the mean vector) of A_i. The component k of the vector L^i is L_k^i, where

$$L_k^i = \frac{\sum_{j=1}^{P} A_i^2(x^j) x_k^j}{\sum_{j=1}^{p} A_i^2(x^j)}.$$

The minimum point of the function $J(\cdot, L)$ may be obtained by using the Lagrange multiplier method. Thus we obtain[16, 17]

$$A_i(x^j) = \left(C(x^j) \right) \bigg/ \left(\sum_{k=1}^{n} \frac{d^2(x^j, L^i)}{d^2(x^j, L^k)} \right) \tag{2}$$

where $i = 1, 2, \ldots, n$ and $j = 1, 2, \ldots, p$. A difficulty appears when a prototype coincides with a data point. If, for instance, $L^k = x^j$, then $d(L^k, x^j) = 0$ and formula (2) for membership degrees no longer holds. We may avoid this difficulty by adding a small positive value to each distance between points and prototypes. For instance, the squared distance $d^2(x^j, L^k)$ in the expression for $A_i(x^j)$ could be replaced by

$$\hat{d}^2(x^j, L^k) = \left[d(x^j, L^k) + a \right]^2$$

or by

$$\bar{d}^2(x^j, L^k) = d^2(x^j, L^k) + a,$$

where a is a small positive number.

We may now give an iterative procedure for obtaining the cluster substructure of the fuzzy class C. This procedure, developed by Dumitrescu,[17] will be called the *generalized fuzzy n-means* (GFNM) procedure. Essentially, the GFNM algorithm consists of iterations based on formulas (1) and (2). The iterative process starts with an arbitrary initialization of the prototypes. The process stops when two successive partitions are arbitrarily close enough. To measure the distance between two fuzzy partitions we associate to each partition P a matrix Q having n lines and p columns. Q is called the *matrix representation* of the fuzzy partition P and is defined as

$$Q_{ij} = A_i(x^i), \qquad i = 1, \ldots, n; j = 1, \ldots, p.$$

Considering Q_1 and Q_2 as the matrix representations of the fuzzy partitions P^1 and P^2, we may define the distance between P^1 and P^2 as

$$d(P^1, P^2) = \| Q_1 - Q_2 \|.$$

The matrix norm may be calculated by using one of the following formulas:

$$\| Q \| = \max_{i, j} |Q_{ij}|$$

or

$$\| Q \| = \max_{i} \sum_{j=1}^{p} |Q_{ij}|.$$

Let P^r be the fuzzy partition at the rth iteration. The iterative process stops at the rth iteration if

$$d(P^r, P^{r+1}) = \| Q_r - Q_{r-1} \| < \varepsilon,$$

where ε is the admissible error. A typical value for ε is 10^{-5}.

The iterative procedure can be described in the terms of the generalized fuzzy n-means (GFNM) algorithm which has the following steps:

S1. Choose the number n of subclusters in the fuzzy class C. Choose P^1 to be an arbitrary fuzzy n-partition of C. Let Q_1 be the matrix associated with P^1.

S2. Compute the prototypes $L^i, i = 1, \ldots, n$, using

$$L^i = \frac{\sum_{j=1}^{p} A_i^2(x^j) x^j}{\sum_{j=1}^{p} A_i^2(x^j)}, \qquad i = 1, \ldots, n.$$

S3. Compute a new fuzzy partition P^2 of C using

$$A_i(x^j) = \left(C(x^j) \right) \Big/ \left(\sum_{k=1}^{n} \frac{\hat{d}^2(x^j, L^i)}{\hat{d}^L(x^j, L^k)} \right),$$

for $i = 1, \ldots, n$; $j = 1, \ldots, p$.

S4. Let Q_2 be the matrix representation of P^2. If $\|Q_1 - Q_2\| < \varepsilon$, then stop. Otherwise set $P^1 := P^2$, $Q^1 := Q^2$. Go to step S2.

Remarks. (i) In the expression of $A_i(x^j)$ we may replace $\hat{d}^2$ by $\bar{d}^2$, (ii) If we regard $C = X$, this procedure reduces to the well-known *fuzzy n-means* FNM algorithm of Dunn and Bezdek (see Bezdek[7]). In this latter case the cluster structure of the data set X is obtained. The memberships at step S3 are thus computed from the equation

$$A_i(x^j) = 1 \Big/ \left(\sum_{k=1}^{n} \frac{\hat{d}^2(x^j, L^i)}{\hat{d}^2(x^j, L^k)} \right), \qquad i = 1, \ldots, n; \ j = 1, \ldots, p.$$

B. Hard *n*-Means Algorithm

The FNM procedure may be notified to detect hard cluster structure in the data set. In this case clusters will be described by classical (hard) subsets of X. In the iterative process the classes are modified according to the rule

$$x^j \text{ is assigned to class } i \Leftrightarrow d^2(x^j, L^i) = \min_k d^2(x^j, L^k).$$

The prototype of the class A_i is

$$L^i = \frac{\sum_{x \in A_i} x}{p_i},$$

where p_i is the number of points in the class A_i. The iterative process stops when the partitions obtained at two successive steps coincide.

C. Adaptive Distances in Fuzzy Clustering

GFNM is a one-level (or horizontal) clustering procedure. It may be used to detect spherical or ellipsoidal clusters. The shape of the clusters detected by this procedure is influenced by the matrix M. If $M = I$ (I

denotes the identity matrix), then the detected classes tend to be spherical. If $M \neq I$, the detected classes will be approximate ellipsoids. These ellipsoids may be more or less elongated, according to the form of matrix M. Therefore the geometry of the detected clusters is imposed by the matrix M. Gustafson and Kessel[31] suggested characterizing each cluster A_i by a different matrix M_i. The distance of x^j to the prototype L^i was defined as

$$d^2(x^j, L^i) = (x^j - L^i)^T M_i (x^j - L^i),$$

where M_i is a symmetric and positive definite matrix. This approach enables us to adapt the metric to each class. Usually the clusters in a data set are of unequal size. For the case of a small cluster situated near a larger one, assume that these clusters are described by the fuzzy sets A_1 and A_2. The points having an equal membership to A_1 and A_2 are then situated on a hyperplane. The hyperplane of equal membership is half the distance between the prototypes L^1 and L^2. If the matrix M in the scalar product is the unit matrix, the hyperplane will be normal to the prototype line.

If the clusters are close it is possible that the hyperplane of equal membership will intersect the greater cluster. Some points of the greater cluster will be captured by its neighbor, as shown in Fig. 4. This represents a pathological situation that may be avoided by using a data-dependent (or adaptive) distance.[17, 21] With an adaptive metric the apparent sizes of clusters become equal. An adaptive distance may be induced by the radius or by the diameter of each fuzzy class A_i. The diameter δ_i of the fuzzy class A_i is defined as

$$\delta_i = \max_{x, y \in X} d_i(x, y).$$

We may also write

$$\delta_i = \max_{x, y} \min(A_i(x), A_i(y)) d(x, y).$$

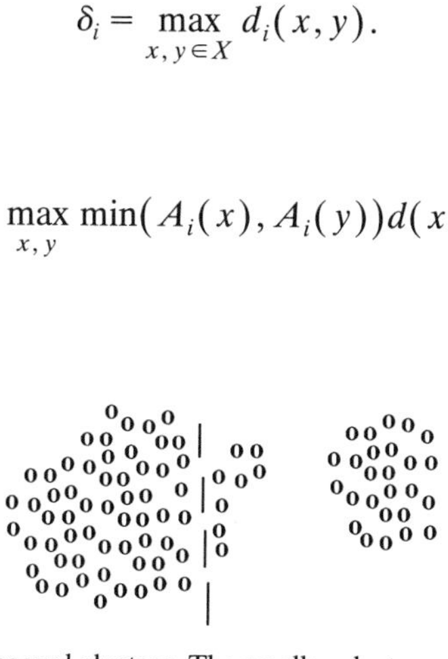

FIGURE 4 Two unequal clusters. The smaller cluster captures some points.

The *diameter induced adaptive distance* with respect to A_i is defined as

$$d_{\text{ia}}(x, y) = \frac{d_i(x, y)}{\delta_i}.$$

The diameter δ_{ia} of A_i with respect to this distance is

$$\delta_{\text{ia}} = \max_{x, y \in X} d_{\text{ia}}(x, y) = 1.$$

Therefore all clusters will have the same diameter with respect to respective adaptive distances.

Adaptive distances may be used for clustering purposes. The considered dissimilarity is the square of the adaptive distances, i.e.,

$$D_i(x^j, L^i) = d_{\text{ia}}^2(x^j, L^i).$$

This dissimilarity may also be written as

$$D_i(x^j, L^i) = A_i^2(x^j) \frac{d^2(x^j, L^i)}{\delta_i^2}.$$

From the dissimilarity the criterion function becomes

$$J(P, L) = \sum_{i=1}^{n} \sum_{j=1}^{p} A_i^2(x^j) \frac{d^2(x^j, L^i)}{\delta_i^2}.$$

If we now set

$$d_a(x^j, L^i) = \frac{d(x^j, L^i)}{\delta_i},$$

we find that

$$J(P, L) = \sum_{i=1}^{n} \sum_{j=1}^{p} A_i^2(x^j) d_a^2(x^j, L^i).$$

Using this criterion function we can obtain the membership degrees thus:

$$A_i(x^j) = (C(x^j)) \Bigg/ \left(\sum_{k=1}^{n} \frac{d_a^2(x^j, L^i)}{d_a^2(x^j, L^k)} \right), \qquad i = 1, \ldots, n; j = 1, \ldots, p.$$

Hence, the use of the adaptive distance requires the replacement of $d(x^j, L^i)$ by $d_a(x^j, L^i)$ in the expression for $A_i(x^j)$. The expression for the prototype L^i however remains unchanged.

A two-stage clustering process serves to illustrate the use of the adaptive distance:

(1) First compute the fuzzy partition by using the GFNM algorithm and the usual distance. The diameters of the classes in this partition are computed.
(2) Use the adaptive distances induced by diameters obtained from step 1 to compute a new fuzzy partition and its prototypes.

We may also adopt a different procedure in which the adaptive distances are computed at each iteration in the GFNM algorithm. At the first iteration the diameters are all equal: $\delta_i = 1$, $i = 1, 2, \ldots, n$. The diameter of the fuzzy class A_i obtained at iteration k induces an adaptive distance. The adaptive distances are used to compute the fuzzy classes at iteration $k + 1$. For a relatively large data set the computation of diameters may involve the storage of a large distance matrix. We may avoid the computational difficulties in such cases by using a simpler adaptive distance. We have supposed the classes are approximately spherical (or ellipsoidal). The mean of a fuzzy class may thus be considered as an approximation of the geometric center of the class. On this basis we may define the radius r_i of the fuzzy class A_i as

$$r_i = \max_{x \in X} d_i(x, L^i) = \max_x A_i(x) d(x, L^i).$$

The membership degrees are then computed using the distance d'_a, where

$$d'_a(x^j, L^i) = \frac{d(x^j, L^i)}{r_i}.$$

D. Linear Clusters

Sometimes data sets may contain linear clusters. Now each fuzzy class A_i may be represented by a direction u^i and a point v^i. The points of the cluster A_i are supposed to fall on the line

$$y = v^i + tu^i, \qquad t \in \mathbb{R}.$$

This is a line passing through v^i and having direction u^i. If $L^i = (v^i, u^i)$ is the linear prototype of A_i, the inadequacy between the fuzzy partition $P = \{A_1, \ldots, A_n\}$ and its representation $L = \{L^1, \ldots, L^n\}$ may be written[7, 21] as

$$J(P, L) = \sum_{i=1}^{n} \sum_{j=1}^{p} A_i^2(x^j) d^2(x^j, L^i)$$

$$= \sum_{i=1}^{n} \sum_{j=1}^{p} A_i^2(x^j) \left[\|x^j - v^i\|^2 - (x^j - v^i, u^i)^2 \right].$$

Minimizing function J, we obtain for $A_i(x^j)$ the same expression as for point prototypes and v^i will be the mean [given by Eq. (1)] of the fuzzy class A_i: $v^i = m^i$. Direction u^i is the eigenvector corresponding to the largest eigenvalue of the matrix S_i:

$$S_i = M\left[\sum_{j=1}^{p} A_i^2(x^j)(x^j - m^i)(x^j - m^i)^T\right]M.$$

S_i is the *scatter matrix* of the fuzzy class A_i. The linear cluster substructure of a fuzzy class may be obtained by an iterative procedure similar to that of the GFNM algorithm. If $C = X$, the procedure reduces to the fuzzy n-lines algorithm of Bezdek *et al.*[9]

E. Clusters with a Degree of Linearity

We may suppose a class A_i is not purely linear. The *linearity degree* of a cluster may be expressed by a number a, $0 \le a \le 1$. Let $L^i = (v^i, u^i)$ be the prototype of the fuzzy class A_i. The distance of x^j to L^i is then

$$d(x^j, L^i) = \left[\|x^j - v^i\|^2 - a(x^j - v^i, u^i)^{-2}\right]^{1/2}.$$

If $a = 1$, the cluster is completely linear and $d(x^j, L^i)$ is the distance of x^j to the line of direction u^i passing through the point v^i. If $a = 0$, the cluster is nonlinear and $d(x^j, L^i)$ is the distance of x^j to the point prototype v^i. The criterion function obtained is

$$J_a(P, L) = \sum_{i=1}^{n}\sum_{j=1}^{p} A_i^2(x^j)d^2(x^j, L^i)$$

$$= \sum_{i=1}^{n}\sum_{j=1}^{p} A_i^2(x^j)\|x^j - v^i\|^2 - a\sum_{i=1}^{n}\sum_{j=1}^{p} A_i^2(x^j)(x^j - v^i, u^i)^2.$$

The optimal fuzzy partition is given by

$$A_i(x^j) = \left(C(x^j)\right)\bigg/\left(\sum_{k=1}^{n}\frac{d_{ji}^2}{d_{jk}^2}\right), \qquad i = 1,\ldots,n; j = 1,\ldots,p,$$

where

$$d_{jk}^2 = \|x^j - v^i\|^2 - a(x^j - v^i, u^i)^2$$

$$= (x^j - v^i)^T M(x^j - v^i) - a(x^j - v^i)^T Mu^i.$$

The optimal prototypes are those corresponding to the purely linear case, i.e., for $a = 1$.

F. Principal Components of a Fuzzy Class

The principal components of any fuzzy class C are the most important directions along which this cluster is spread out. Let L be the line through v with the direction u:

$$L = \{ y \in \mathbb{R}^s \mid y = v + tu, \, t \in \mathbb{R} \}.$$

The proximity between the fuzzy class C and the line L may be given by the number $W(L, C)$, defined as

$$W(L, C) = \sum_{j=1}^{p} d_c^2(x^j, L)$$

where d_c denotes distance with respect to C:

$$d_c(x, y) = \min(C(x), C(y)) d(x, y).$$

We are interested in finding the closest line to the cloud C and so need to minimize $W(L, C)$. The best line minimizing $W(L, C)$ passes through m_c— the mean vector of the class C shown by Dumitrescu[16] to be given by

$$m_c = \frac{\sum_{j=1}^{p} C^2(x^j) x^j}{\sum_{j=1}^{p} C^2(x^j)}.$$

The principal directions of the fuzzy class C are eigenvectors of S_c, the scatter matrix of C.

$$S_c = M \left[\sum_{j=1}^{p} C^2(x^j)(x^j - m_c)(x^j - m_c)^T \right] M.$$

If $u^1, \ldots, u^s$ are the eigenvectors of S_c and $\lambda_1, \ldots, \lambda_s$ the corresponding eigenvalues, i.e.,

$$S_c u^i = \lambda_i u^i, \qquad i = 1, \ldots, s,$$

and we suppose that $\lambda_1 \geq \lambda_2 \geq \cdots \lambda_s$, the vector u^1 gives the most important direction along which the fuzzy cluster C is spread out. The principal components $u^1, u^2, \ldots, u^s$ of C are mutually orthogonal. The problem now is to find the most representative directions of the class C. The ratio

$$r_k = \frac{\sum_{j=1}^{k} \lambda_j}{\sum_{i=1}^{s} \lambda_i}$$

represents a measure of the goodness of fit to the cloud C of the first k principal components. The k-dimensional subspace spanned by the first k principal components represents a linear skeleton of the cluster. This skeleton may be considered as a linear representation of the fuzzy class C.

It gives a global geometric characterization of the class. The information given by principal components may be used for classification purposes and for dimensionality reduction of a data set having varied characteristics.

G. Cluster Validity

Cluster validity is the evaluation of (fuzzy) partitions to establish which partition best explains the unknown cluster structure in a data set. An effective cluster validity analysis would be able to determine whether (i) there exists any cluster structure in the data, (ii) the data are random, and (iii) the identified clusters are "real," i.e., they represent relationships among data and are not artificially generated by the classification procedure. The most important algorithmic parameter is the cluster number. A slight error in the cluster number choice will render accurate detection of the clustering structure impossible even if this structure is minimal. The search for the cluster number is thus of central importance in cluster analysis. For ranking fuzzy partitions, use can be made of cluster validity functionals (see Bezdek[7]). A validity functional is a function which assigns to a fuzzy partition P a number expressing the quality n of the clustering provided by P.

Let $X = \{x^1, x^2, \ldots, x^p\}$ be a data set and let $P = \{A_1, A_2, \ldots, A_n\}$ be a fuzzy partition of X. To measure how close P is from a hard (classical) partition, Bezdek[6] has introduced the *partition coefficient* of P, defined as

$$C(P) = \frac{1}{p} \sum_{i=1}^{n} \sum_{j=1}^{p} A_i^2(x^j).$$

The partition coefficient has the following properties (see Bezdek[6]):

(i) $\dfrac{1}{n} \le C(P) \le 1, \qquad \forall P \in F_n(X);$

(ii) $C(P) = \dfrac{1}{n} \quad \Leftrightarrow \quad A_i(X^j) = \dfrac{1}{n}, \qquad \forall ij;$

(iii) $C(P) = 1 \quad \Leftrightarrow \quad P$ is a hard partition of X.

The partition coefficient is a measure of the quality of a fuzzy partition. The closer $C(P)$ is to 1, the better the fuzzy partition P will be. The outputs of a fuzzy clustering algorithm for several different values of n may be compared by means of the partition coefficient. The best partition (and the best n) is that associated with the highest partition coefficient value.

Unfortunately, the partition coefficient is sensitive to the values of n independently of any structure in the data. We cannot therefore select the best fuzzy partition with $C(P)$. A validity functional that is the dual of the partition coefficient needs to be used. Such a functional is the *classification entropy* (see Bezdek[7] and Roubens[56]), defined as

$$E(P) = -\frac{1}{p} \sum_{i=1}^{n} \sum_{j=1}^{p} A_i(x^j) \log_2 \left(A_i(x^j) \right).$$

This classification entropy has the following properties:

(i) $\qquad\qquad 0 \leq E(P) \leq \log_2 n;$

(ii) $\qquad E(P) = \log_2 n \quad \Leftrightarrow \quad A_i(x^j) = \frac{1}{n}, \qquad \forall i, j;$

(iii) $\qquad E(P) = 0 \quad \Leftrightarrow \quad P$ is a hard partition of X.

The classification entropy may be interpreted as measuring the ambiguity associated with a fuzzy partition. A good cluster identification would be indicated by an entropy value close to 0. Various transformations of the validity functionals C and E have been proposed to mitigate their monotonic tendencies. Here we consider a normalization that reduces the validity functional's range of variation to $[0, 1]$. Normalized expressions of C and E assume the forms

$$C'(p) = \frac{n}{n-1} \left[1 - C(P) \right],$$

$$E'(P) = \frac{E(P)}{\log_2 n}.$$

IV. FUZZY DIVISIVE HIERARCHICAL CLUSTERING

In many cases the chemical data to be analyzed have a built-in hierarchical structure. Such a structure cannot be discovered by a one-level (horizontal) clustering procedure. Some information in the data set can be lost in using such an inadequate classification method. We are thus led to seek algorithms able to detect an inner cluster hierarchy of a data set. Hierarchical clustering is also useful in the search for the optimal number of clusters. The problem of choosing the optimal number of clusters in a data set is known as the cluster validity problem. It is the principal problem in all clustering techniques. As already mentioned, in one-level fuzzy clustering, validity functionals are used to solve this problem. Unfortunately, however, validity functionals do not provide a complete solution of the cluster validity problem and the method is time-consuming. In what follows we consider a fuzzy clustering procedure based on a hierarchical data decomposition. This approach seems to be more robust than the fuzzy clustering methods based on validity functionals. The possibility of arriving at a wrong classification is drastically reduced vis-à-vis the one-level classification approach. Especially for large data sets having many clusters, the hierarchical scheme is simpler and less time-consuming. Hierarchical clustering procedures are able to detect the intrinsic hierarchical structure in data. In this regard some natural connections between clusters may be emphasized. A data hierarchy may be constructed agglomeratively or divisively. We use a fuzzy divisive scheme to construct a fuzzy hierarchy in a data set. In our method,[17] classes are subdivided as long as necessary to yield a final objective classification. A decomposition criterion is applied to retain only real clusters in the hierarchy. In this way no a priori knowledge concerning the optimal number of clusters is required.

A. Polarization Degree of a Fuzzy Partition

For a fuzzy partition $\{A_1, A_2\}$ of X, where X has no structure or X is a compact cluster, the membership degrees for A_1 and A_2 tend to be equal. By contrast, if A_1 and A_2 are well separated, the memberships $A_1(x)$ and $A_2(x)$ tend to polarize toward the extreme values 0 and 1 for each x. The presence of distinct clusters induces a polarization tendency in the membership degrees, whereas the absence of a cluster structure induces uniformity in the membership degrees. We may conjecture that polarization of a membership degree is associated with the presence of a cluster structure in a fuzzy partition. Hence structuredness implies polarization. The polarization degree of a binary fuzzy partition may thus be considered as measuring the quality of the partition. For the fuzzy partition $P = \{C_1, C_2\}$ of a fuzzy class C, the polarization degree of C may be defined[17, 21] as

$$R(P) = \frac{\sum_{x \in X} \max(C_1(x), C_2(x))}{\sum_{x \in X} C(x)}.$$

It is then easy to prove that $\frac{1}{2} \leq R(P) \leq 1$ for each binary fuzzy partition.

In what follows we show that a fuzzy partition $P = \{A_1, A_2\}$ describes *real clusters* if and only if for each class $A_i, i = 1, 2$, there exists an x such that $A_i(x) \geq \frac{1}{2}$ and $R(P) \geq t$, where t is an appropriate threshold in the interval $(0.5, 1)$. We may interpret t as our confidence limit that the detected clusters are real. We consider now a data set $X = \{x^1, x^2, x^3, x^4\}$ and a fuzzy partition $P = \{A_1, A_2\}$ of X having the memberships specified as follows:

	x^1	x^2	x^3	x^4
$A_1(x^j)$	0.9	0.2	0.7	0.6
$A_2(x^j)$	0.1	0.8	0.3	0.4

The degree of polarization of this partition is

$$R(P) = \frac{0.9 + 0.8 + 0.7 + 0.6}{4} = 0.75.$$

A_1 and A_2 describe real clusters with a degree confidence limit of $t = 0.75$. Consider the fuzzy class C on X having the membership degrees

	x^1	x^2	x^3	x^4
$C(x^j)$	1	0.5	0.7	0.8

and a fuzzy partition $Q = \{C_1, C_2\}$ of C defined as

	x^1	x^2	x^3	x^4
$C_1(x^j)$	0.5	0.4	0.1	0.4
$C_2(x^j)$	0.5	0.1	0.6	0.4

The polarization degree of Q will be

$$R(Q) = \frac{0.5 + 0.4 + 0.6 + 0.4}{3} \approx 0.63$$

We may assume with a confidence limit of $t = 0.63$ that C_1 and C_2 represent real clusters.

B. Fuzzy Divisive Hierarchical Clustering

Next we discuss a fuzzy hierarchical clustering procedure proposed by Dumitrescu.[17-21] The procedure starts with the computation of a binary fuzzy partition $\{A_1, A_2\}$ of the data set X. To do this, the FNM algorithm or other appropriate procedure may be used. If A_1 and A_2 do not describe real clusters, we may conjecture there is no structure in X or that the data comprise a single compact cluster. The process then ends. If the fuzzy classes A_1 and A_2 correspond to real clusters, we set $P^1 = \{A_1, A_2\}$. Assume the cluster structure of each class A_i, $i = 1, 2$, is given by a binary fuzzy partition of A_i. We may compute the cluster substructure of A_i using the GFNM algorithm or one of its relatives. There are now two possibilities:

(1) The binary fuzzy partition of A_i does not represent real clusters. In this case the class A_i will be marked to avoid subsequent attempts to split it. The marked class will be allocated to a new fuzzy partition P^2. It is clear that P^2 is a fuzzy partition of the data set X.

(2) The binary fuzzy partition of A_i represents real clusters. The two atoms (modules) of this partition are then allocated to the fuzzy partition P^2 of X.

For the unmarked classes of P^2 we follow the same procedure. In this way a new fuzzy partition P^3 of X is obtained. The partitioning procedure is stopped when all the modules of the current partition P^k have been marked. Using this procedure we can detect a hierarchical structure in the data set X. The optimal number of clusters in X and the corresponding final classes are also obtained.

The fuzzy hierarchy obtained is richer in information than a hierarchy based on classical sets. Specimens having nearly equal membership degrees in different clusters are of special interest for the scientist. Such specimens may be problematic and may cause the scientist to search for an explanation of their behavior. Another possibility is to search for those characteristics that are responsible for unexpected behavior of some of the points in the data set. If we delete some characteristics or add new characteristics, we generally obtain a different clustering. To interpret the obtained classification it is useful to seek for the corresponding classical hierarchy. This may be done using a hierarchical assignment rule. We let C be a fuzzy class at a decomposition level k, and denote C' as the corresponding classical set. The fuzzy partition $\{C_1, C_2\}$ of C is assumed to describe real clusters. The classical set corresponding to C_1 is C_i', $i = 1, 2$. These classical sets are obtained according to the rules

$$x \in C_1' \quad \Leftrightarrow \quad x \in C' \quad \text{and} \quad C_1(x) \geq C_2(x)$$

and

$$x \in C_2' \quad \Leftrightarrow \quad x \in C' \quad \text{and} \quad C_1(x) < C_2(x).$$

It is easy to see that $\{C_1', C_2'\}$ is a hard partition of the classical set C'. Thus, from these assignment rules we finally obtain the classical (or hard) hierarchy corresponding to the respective fuzzy hierarchy. Accordingly, the fuzzy divisive hierarchical clustering (FDHC) procedure may be used to obtain the optimal cluster structure of the data set and a hierarchical relationship between clusters and subclusters. The method is especially useful when the number of clusters is unknown. In most cases the number of clusters to be expected in the data set is unknown.

V. FUZZY CROSS-CLASSIFICATION

Chemical data can sometimes be very complex. Each object (or sample) in the data set is then described by a large number of features (characteristics, attributes). Usually the characteristics are not of equal importance for the clusters in the data set. Each cluster may be characterized in terms of its relevant features, the latter being those that are responsible for holding together the points of the cluster. The objects in a class will have comparable values of the common relevant characteristics.

Knowledge of the cluster structure of a data set is complete only if we are able to detect the clusters as well as the corresponding feature classes. Simultaneous classification of the objects and their characteristics is necessary in many of the practical problems encountered in analytical chemistry and related fields. Such problems include the following:

(1) The need to give a complete characterization of the experimental results.
(2) The design of very large classifiers. In this case not all the features are relevant for all the objects. In case of a divisive hierarchical classifer, each node of the classifier will contain a set of objects as well as the characteristics that differentiate these objects.
(3) The need to manage an experiment such that insignificant or inconvenient classes are eliminated from the possible results.
(4) To obtain a synthetic characterization of certain experiments according to the classes of experimental conditions.
(5) Giving suitable classes of analytical methods for a set of analyses.
(6) Obtaining the modular structure of very complex chemical systems.
(7) Process control and model identification. Simultaneous classification may be useful for control model identification and for extraction of the rules of control from data.

A. One-Level Cross-Classification

Let $X = \{x^1, x^2, \ldots, x^p\}$ and let x^j from $\mathbb{R}^s$ be a data set as represented in the following table:

	y^1	y^2	$\cdots$	y^s
x^1	x_1^1	x_2^1	$\cdots$	x_s^1
x^2	x_1^2	x_2^2	$\cdots$	x_s^2
$\vdots$	$\vdots$	$\vdots$	$\vdots$	$\vdots$
x^p	x_1^p	x_2^p	$\cdots$	x_s^p

Each column i, $i = 1, 2, \ldots, s$, in this table represents the values of the i-th characteristic, and y^i denotes the corresponding p component vector. If now $Y = \{y^1, y^2, \ldots, y^s\}$ is the characteristics set where y^k is from $\mathbb{R}^p$, then y_j^k is the value of the k-th set of characteristics with respect to the object x^j. As indicated, some practical problems require simultaneous classification (or cross-classification) of the objects and their corresponding characteristics. Therefore we need a fuzzy partition of X compatible with a fuzzy partition of the set Y of characteristics.

If X is a set of materials (or substances) and Y is a set of analyses of these materials, we require a partition of X and a partition of Y such that each material in a given class i may be analyzed by using the analyses of the corresponding class i. Simultaneous clustering is the search for a pair (P, Q), where P is a fuzzy partition of data set X and Q is a fuzzy partition of characteristics set Y. Let us suppose that P and Q have the same number of atoms (modules), say n. A fuzzy procedure may be used to compute the partition $P = \{A_1, A_2, \ldots, A_n\}$ of X. The problem is then to find the corresponding fuzzy partition of the characteristics set Y. The value $y_i^{\prime k}$ of the characteristic k with respect to the fuzzy class A_i may be defined as the sum of the values $y_j^k, j = 1, 2, \ldots, p$, of this characteristic weighted with the membership of $x^j, j = 1, 2, \ldots, p$, in A_i. Hence, we have

$$y_i^{\prime k} = \sum_{j=1}^{p} A_i(x^j) y_j^k.$$

Let $Y' = \{y'^1, y'^2, \ldots, y'^s\}$ be the set of modified characteristics, where y'^k is a vector of dimension n and $Q = \{B_1, B_2, \ldots, B_n\}$ is the fuzzy partition of Y' that fits the partition P. The prototype of the fuzzy class B_i is v^i from $\mathbb{R}^n$. The fuzzy partition Q may be obtained by minimizing the criterion function,

$$J(Q, v) = \sum_{i=1}^{n} \sum_{k=1}^{s} B_i^2(y'^k) d^2(y'^k, v^i),$$

where d is the distance induced by a scalar product in $\mathbb{R}^n$:

$$d^2(y'^k - v^i) = \|y'^k - v^i\|^2 = (y'^k - v^i, y'^k - v^i).$$

Minimizing the function $J(\cdot, v)$, we obtain

$$B_i(y'^k) = 1 \bigg/ \left(\sum_{r=1}^{n} \frac{d^2(y'^k, v^i)}{d^2(y'^k, v^r)} \right), \tag{3}$$

where $i = 1, 2, \ldots, n$ and $k = 1, 2, \ldots, s$.

The optimum prototype, obtained by minimizing the function $J(Q, \cdot)$, is the mean of the fuzzy class B_i,

$$v^i = \frac{\sum_{k=1}^{s} B_i^2(y'^k) y'^k}{\sum_{k=1}^{s} B_i^2(y'^k)}, \tag{4}$$

where $i = 1, 2, \ldots, n$. The optimal fuzzy partition and its representation will be obtained iteratively as usual from formulas (3) and (4) for computing memberships and prototypes. The process may stop here or be continued by computing a new fuzzy partition P' of X induced by the partition Q. We may expect that new partition P' better matches the characteristics partition Q. The way in which the elements of Y have been defined enables us to conjecture that the optimal fuzzy partition Q of Y will be correlated strongly with the fuzzy partition P (or P') of X. Our procedure is therefore able to obtain matching fuzzy partitions of objects and characteristics; further details are given in Dumitrescu.[21]

B. Hierarchical Cross-Classification

Divisive hierarchical simultaneous clustering procedures build a fuzzy hierarchy of objects and a fuzzy hierarchy of characteristics.[21] Each node of the corresponding tree is labeled by a pair (C, D), where C is a fuzzy class of objects and D is a fuzzy class of characteristics. At the first level a binary fuzzy partition of data set X and the corresponding binary partition of characteristics set Y are computed. The classes that emerge are subdivided until no pair of real clusters can be obtained.

The simultaneous clustering procedure for X and Y may generalize to a pair (C, D). This procedure may be described as follows:

P1. Compute a binary partition $P = \{A_1, A_2\}$ of the fuzzy set C using the GFNM algorithm (or other appropriate procedure).

P2. Compute the characteristic $y_i'^k$ associated with A_i:

$$y_i'^k = \sum_{j=1}^{p} A_i(x^j) x_k^j, \qquad k = 1, \ldots, s; i = 1, 2.$$

The new characteristics may be represented by the following table:

	y'^1	$\cdots$	y'^s
A_1	$y_1'^1$	$\cdots$	$y_s'^1$
A_2	$y_1'^2$	$\cdots$	$y_s'^2$

P3. Let us denote

$$Y' = \{y'^n, y^n, \ldots, y'^s\}.$$

The fuzzy set D induces a fuzzy set D' on the set Y'. We may define D' naturally as

$$D'(y'^k) = D(y^k), \qquad k = 1, \ldots, s.$$

P4. A fuzzy partition $Q = \{B_1, B_2\}$ of the fuzzy set D' is obtained using GFNM procedure. Thus we have

$$B_i(y'^k) = \left(D'(y'^k)\right) \bigg/ \left(\sum_{r=1}^{s} \frac{d^2(y'^k, v^i)}{d^2(y'^k, v^1)}\right),$$

and v^1 is the mean of the fuzzy class B_i.

P5. At the next classification level the fuzzy classes A_1 and A_2 of the partition P have to be split. We may compute a binary fuzzy partition of A_i, $i = 1, 2$, as usual using GFNM algorithm. To obtain a stronger correlation between the partition of objects and characteristics we may re-characterize the objects using the classes B_1 and B_2 of characteristics. An object j will be now described by a two-component vector x'^j. We may put:

$$x'^j = \begin{pmatrix} x_1'^j \\ x_2'^j \end{pmatrix},$$

where $x_v'^j$ $i = 1, 2$, is an average of the features of the object j with respect to the class B_i of characteristics. We may define $x_1'^j$ as

$$x_i'^j = \sum_{k=1}^{s} B_i(y'^k)x_k^j,$$

where $i = 1, 2$ and $j = 1, 2, \ldots, p$.

The objects $x'^j, j = 1, 2, \ldots, p$, may now be represented as

	B_1	B_2
x'^1	$x_1'^1$	$x_2'^1$
x'^2	$x_1'^2$	$x_2'^2$
$\vdots$	$\vdots$	$\vdots$
x'^p	$x_1'^p$	$x_2'^p$

Let us denote $X' = \{x'^1, x'^2, \ldots, x'^p\}$. The class A_i induces a fuzzy class A_i' on the universe X' and we have $A_i'(x'^j) = A_i(x^j)$. Instead of computing a binary fuzzy partition of A_i, we calculate a binary fuzzy partition of A_i'. For this the GFNM algorithm will be used.

We are now able to state the hierarchical cross-classification procedure. The root node of the partition tree corresponds to the pair (X, Y). At the first level a fuzzy partition $P = \{A_1, A_2\}$ of X is computed. Let $Y' = \{y'^1, y'^2, \ldots, y'^2\}$ be the characteristics set induced by A_1 and A_2. We then have

$$y_i'^k = \sum_{j=1}^{p} A_i(x^j)x_k^j.$$

A fuzzy partition $\{B_1, B_2\}$ of the set Y' is calculated. In this way we have built the first level of the clustering tree (see Fig. 5). If $\{A_1, A_2\}$ and $\{B_1, B_2\}$ correspond to real clusters, the process continues. Using bidimensional characteristics induced by B_1 and B_2, binary fuzzy partitions of $A_i, i = 1, 2$, are computed. A pair (C, D) corresponds to a terminal node of the partition tree if the successors of C or D do not correspond to real clusters. A more realistic procedure is to stop the partitioning process only for a class $(C$ or $D)$ that has no real successors. The other fuzzy class may be further divided.

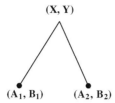

FIGURE 5   The first decomposition level of hierarchical cross-classification.

VI. FUZZY HIERARCHICAL CLASSIFICATION TECHNIQUES IN ANALYTICAL CHEMISTRY

Classification techniques have been used in analytical chemistry for a wide range of applications. These applications deal with the interpretation of analytical chemical data, automated recognition of substances from their spectra, detection of protein or nucleic acid structures and their classification, signal interpretation, problems of inference concerning the nature or provenance of complex materials, and so on. Some general analytical chemical problems may also be approached with various classification methods. Such general problems might be selection of the best analytical procedure for a given analysis or database searching and detection of the similarities among classes of objects representing experimental or observational results. Cluster analysis may be used to solve various multicriteria optimization problems related to chemical applications. Quality control, including that for air, water, food and drugs, also involves useful applications of classification methods.

In the extensive literature on the applications of classification methods to analytical chemistry many techniques have been described and several reviews have been published. A detailed discussion of all these techniques is not our intention here. Useful information may be found in the books of Jurs and Insenhour,[35] Malinowski and Howery,[43] Varmuza,[64] Kateman and Pijpers,[37] Massart and Kaufman,[44] Jolliffe,[34] Coomans and Broeckaert,[12] Sharaf et al.,[60] Massart et al.,[45] Willett,[65] and Zupan.[67] Some early reviews and interesting applications are to be found in Kowalski,[39] Kryger,[41] and Frank and Kowalski.[28] For other useful methods and applications, also see Allen et al.,[1] Auf der Heyde,[3] Derde and Massart,[13, 14] Forina and Armanino,[27] Holm and Sande,[32] Karpen et al.,[36] Kowalski and Wold,[40] Murray-Rust and Raftery,[47] Okada and Wipke,[48] Perkins and Barlow,[52] Perkins and Dean,[53] Pijpers,[54] Shenkin et al.,[61] Shenkin and McDonald,[62] and Swanson.[63] The books of Duda and Hart[15] and Jain and Dubes[33] are also recommended.

The advent of fuzzy set concepts and models gave a new flavor to classification techniques in analytical chemistry. Blaffert[10] and Otto and

Bandemer[49–51] considered the role of fuzzy set theory in analytical chemistry. The applications they described focused on pattern recognition problems,[10, 50] the calibration of analytical methods,[49, 50] quality control,[51] and component identification and mixture evaluation.[4] Gordon and Somorjai[30] applied a fuzzy clustering technique to the detection of similarities among protein substructures. A molecular dynamics trajectory of a protein fragment was analyzed. In the following subsections, some applications based on the hierarchical fuzzy clustering techniques presented in this chapter are reviewed.

A. Selectivity Control in Acrylonitrile Electroreduction

A large number of experimental data were accumulated during the elaboration of a new electrochemical proprionitrile synthesis, founded on the nondimerizing electroreduction of acrylonitrile. To extract the most significant information from the over 2000 electroreduction experiments, carried out under very different conditions, a powerful mathematical technique was needed. The fuzzy hierarchical clustering method fulfills the goal of critically evaluating the analytical results of the proprionitrile electrosynthesis. The results obtained are consistent with the experimental facts. The objects to be classified are gas-chromatographic data, represented as six-component vectors. These components represent the concentration of the electroreduction products: proprionitrile, adiponitrile, succinonitrile, methylglutaronitrile, bis(cyanoethyl)ether, and tricyanohexanes. The concentration vectors are normalized for reproducible experimental conditions (a current density of 70 mA/cm^2 and a duration of 4 h).

A data set consisting of 112 concentration vectors was considered. The fuzzy divisive hierarchical clustering (FDHC) procedure yielded eight final clusters. In Fig. 6 the corresponding fuzzy hierarchy, which has five decomposition levels, is presented (Lowy et al.[42]). To obtain more sharply defined classes (intermediate and final), the contribution of the linear prototypes (of linearity degree a) was successively modified. The composition of the final clusters has not been affected by the screening of the mathematical conditions of the classification. Only a slight change in the membership degrees was observed. A careful analysis of the obtained clusters reveals that the assignment of a concentration vector to a given cluster is determined mainly by its proprionitrile content. Consequently, the clusters attained within the FDHC procedure are in agreement with the physical reality of the nondimerizing acrylonitrile electroreduction.

The hierarchical clustering procedure outlined here permits us to extract the most significant information in the data set. All 14 clusters obtained at 5 successive hierarchical levels were homogenous and reflected the experimental conditions. In several cases, careful checking of the membership value of a given datum point in its assigned hard cluster

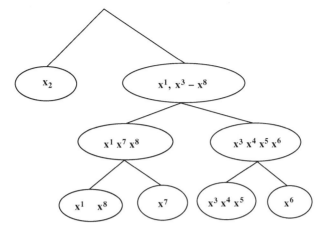

FIGURE 6   Partition tree of archeological artifacts (obsidian).

against membership values in other clusters may reveal similarities between experimental results obtained under completely different conditions. Whenever a datum point is found to have close membership values in several final clusters, the reason it was assigned to one particular class should be sought. By doing so, the main feature that holds together each cluster can be assessed. In Lowy et al.[42] the hope is expressed that the FHDC procedure may prove to be a valuable technique to monitor the progress of electrochemical reactions occurring during the nondimerizing electroreduction of acrylonitrile and to ascertain the success of the selective orientation of the process toward proprionitrile formation.

B. Classification of Mineral Waters

Dumitrescu and Kekedy[22, 23] used the FHDC procedure for the classification of mineral waters. Each sample was characterized by the analytical chemical data for eight major components: CO_2, HCO_2^-, Cl^-, Ca^{2+}, Mg^{2+}, Fe^{2+}, Na^{2+}, and mineralization. Tentatively, classifications were made with a smaller number of components. The components present in greater concentration were successively omitted so as to investigate the possible influence on the classification of components present in smaller concentrations. Thus fuzzy hierarchical clustering was performed by considering successively only seven components (mineralization omitted), six constituents (mineralization and HCO_3^- content omitted), or only five constituents (the previous and CO_2 content omitted), respectively. With eight components the FHDC method yielded three clusters. To characterize the three distinct classes of water found, the total hardness (in German degrees) was also taken into account. Note that total hardness was not a characteristic for clustering purposes.

We observed that the three detected clusters corresponded to groups of mineral waters with distinct characteristics as follows:

Cluster 1. Mineralization 1000–2000 mg/liter; total hardness less than 50 German degrees; CO_2 and HCO_3^- content small. The cluster is very compact, the degree of membership of waters to this cluster being between 0.995 and 0.965.

Cluster 2. Increased mineralization (3000–6000 g/liter); total hardness between 50 and 100 German degrees. The content of CO_2 and HCO_3^- is higher than in cluster 1. The cluster is less compact. The membership degree in this cluster ranges between 0.544 and 0.992. There are two waters that on the basis of their HCO_3^- content only would belong to cluster 1, but their high mineralization as well as their high CO_2 content put them in this cluster.

Cluster 3. Mineralization higher than 6000 mg/liter; total hardness greater than 100 degrees; high CO_2 and HCO_3^- content as well.

Gradual reduction of the number of components considered in clustering produced no significant changes. With only seven or six components, three clusters were again obtained. Cluster 1 remained unchanged, but small modifications appeared in clusters 2 and 3 associated with a decrease of the membership degrees. Fuzzy hierarchical clustering with five components yielded a single cluster. This indicates that components present in small amounts do not influence the classification. Finally, we mention that the use of clustering methods to characterize waters has been emphasized by Scarminio *et al.*[59] and Bartels *et al.*[5]

C. Provenance of Archaeological Artifacts

The FHDC algorithm was used[24] to make inferences on the provenance of obsidian artifacts (vitreous lavas) from an archaeological site at Iclod (in the district of Cluj, Romania). Spectrographical data were used to characterize eight samples, some of them having a known provenance. The samples x^1 and x^2 were obsidians from Melos and Sardinia. Samples x^3, x^4, x^5, and x^6 corresponded to obsidians from Iclod and the samples x^7 and x^8 corresponded to an obsidian from Hungary. At the first classification level the sample x^2 (Sardinia) formed a class. At the second classification level the samples x^3, x^4, x^5, and x^6 (i.e., those from Iclod) were clustered together. Samples x^1, x^7, and x^8 were in a separate class. This classification scheme (see Fig. 6) enabled us to conclude that the obsidian from Iclod did not originate from the same source as the samples x^7

and x^8. Therefore the origin of the Iclod obsidian would be in the Transylvanian Mountains.

D. Optimal Choice of Solvent Systems

In Dumitrescu et al.[26] and Sârbu et al[57] the FHDC approach was used for the optimal choice of sets of solvents in thin-layer chromatography (TLC). This approach proved its capabilities in finding the best system and the optimal combination of two or more TLC systems according to their positions in the obtained fuzzy hierarchy. Seven solvent systems recommended in the literature for the TLC of carotenoids were considered. Each system was characterized by the R_f values of 11 carotenoids. Using the FDHC clustering scheme, we grouped the solvent systems into three clusters in a fuzzy hierarchy with two decomposition levels (see Fig. 7). With this fuzzy clustering hierarchy we can combine effective systems that are sufficiently dissimilar. In this way the quantity of redundant information will be minimal. The quality of a system with respect to a fuzzy class may be expressed by its membership degree in the respective class. If a system has a large membership degree in one class, it will be far removed from the other classes. Therefore, it seems reasonable to combine systems from two classes that are as far apart as possible and for which each system has the greatest membership in its own class. The tree structure of hierarchical classification allows us to observe the relationship between detected classes. To obtain an optimal combination of two chromatographic systems from a binary fuzzy partition tree, one could choose for instance one system from class A_1 and the other from class A_{21}.

The FDHC algorithm may be successfully used not only to compare the merits of different TLC systems, but also to detect the inner structure of each system. The fuzzy cluster structure of a system may supply a more rational basis for evaluating TLC systems. The FDHC procedure has been used to detect the cluster structure of each system. TLC systems have been compared by using the partition coefficient $C(P)$ of the final fuzzy partition P. Since $C(P)$ depends on the number of classes, the normalized

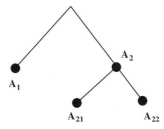

FIGURE 7 Fuzzy partition of TLC solvent systems.

partition coefficient $C'(P)$ has been used,

$$C'(P) = \frac{n}{n-1}[1 - C(P)],$$

where n is the number of clusters in the final fuzzy partitions. Possible values for $C'(P)$ are between 0 and 1. With $C(P)$ or $C'(P)$ it is possible to select the system with the best separation characteristics. A system having a uniform distribution of the values is better than a structural system. Therefore, a system displaying a lower partition coefficient is to be considered as the optimal choice. The FHDC method permits a rational classification and selection procedure for separating systems in TLC. The algorithm yields not only a fuzzy partition, but also additional information such as the partition tree and membership values.

E. Classification of Roman Pottery

Roman pottery (terra sigillata) has been produced in central Italy since the first century B.C. and traded throughout the Roman world. Aruga et al.[2] analyzed 48 shards of terra sigillata by atomic absorption spectrometry with electrothermal atomization in a graphite furnace (GFAAS) and by inductively coupled plasma atomic emission spectrometry (ICP-AES) to determine the seven major elements K, Mg, Ca, Ti, Mn, Fe, and Al. Hierarchical agglomerative clustering based on Euclidean distance was performed on the normalized data. To verify the class assignments and to solve doubtful attributions, a supervised classification algorithm, namely, the soft independent modeling of class analogy (SIMCA), was applied.

Pop et al.[55] also analyzed the same set of 48 shards. The essentials of the classification results of Aruga et al. were again obtained. These authors correctly considered some points misclassified by their agglomerative clustering procedure. All these points however have been correctly placed by the FDHC procedure. The relationship between some classes has also been elucidated in the FDHC approach. The use of a fuzzy class to describe clusters also permitted us to correct some other doubtful assignments. The FDHC procedure seems to provide better results than the hierarchical agglomerative procedure. Because the FDHC procedure corrected the error observed within the agglomerative classification method, no more supervisory intervention in the clustering process was necessary. It is also important to emphasize that the fuzzy membership degrees of the objects allow us to perform a more realistic analysis of the objects situated at the border or in the overlapping area of two neighboring classes. Principal component analysis (see, for instance, Varmuza[64] or Duda and Hart[15]) has been used to reduce the data dimension. Divisive hierarchical classifications obtained for the reduced data are practical as is the case with the classification of the original data set.

F. Cross-Classification of Therapeutic Muds

Muds have been used empirically, but nonetheless efficiently, since antiquity. However, it was not until 1931 that the International Society of Medical Hydrology established certain criteria for the classification of muds and their therapeutic usage. Forty mud and peloid samples were collected from eight different locations during the summer of 1990. Their physicochemical characteristics were determined[46] to evaluate their possible use for therapeutic applications. Samples were selected that represented deposits that are either currently used or may be used for medical purposes. Samples of five Greek muds (Krinides, Pikrolimni, Lisbori, Thermi, and Kyllini) were taken. The Pikrolimni and Thermi samples had a pH 9.8 and 8.4, respectively, which suggests their potential as beautifying muds. The high montmorillonite content of the Lisbori and Thermi samples exhibited poor mud characteristics due to their low content of clay minerals and high content of silica. Samples of various well-known beautifying muds, such as black mud from the Dead Sea and Argilla Solare from Italy, as well as a matured mud from Boario Therme in Italy, were also examined for comparative purposes. Their composition substantiates the fact that desirable mud characteristics are a high amount of clay minerals, a high pH, and a low amount of free silica, feldspar, and carbonates.

The fuzzy hierarchical cross-classification algorithm was used to classify eight mud samples.[25] Each sample was characterized as a vector with 23 components representing chemical analysis. The fuzzy partition tree obtained by using simultaneous classification of muds and their characteristics is shown in Fig. 8. There are six final fuzzy classes in this hierarchy. The classical partition corresponding to the final fuzzy classes of the muds is A_{111} Krinides; A_{1121}, Lisbori; A_{1122}, Argilla Solare; A_{121}, Prikolimni;

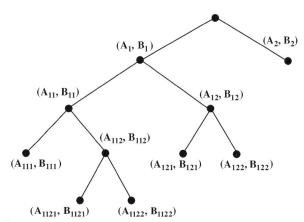

FIGURE 8 Fuzzy cross-classification hierarchy of therapeutic muds.

A_{122}, Thermi, Kyllini, Boaris; A_2, black mud. The corresponding classes of characteristics are

B_{111}	organic, K_2O, MgO, Fe_2O_3
B_{1121}	Na^+, Cl^-, SO_4^{2-}, $MgCO_3$, Fe_2O_3*, Na_2O, CaO
B_{1122}	K^+, Ca^{2+}, Mg^{2+}, HCO_3^-, NO_3^-, CO_3^{2-}
B_{121}	$CaCO_3$, CEC, Al_2O_3
B_{122}	pH
B_2	ash, SiO_2

(The asterisk denotes the chemical analysis of the aluminosilicate residue.)

We immediately observe that the black mud sample (Cluster A_2) is different in comparison to all the others. The corresponding characteristics class is $B_2 = \{ash, SiO_2\}$. From the composition table (see Mitrakis and Sikalidis[46]), it is apparent that the two characteristics have the lowest values for the class A_2. This fact is somewhat surprising. In reality, a characteristic is representative for individuals in a certain class if its value with respect to this class is strongly different from the other values. The remarkable individuality of the black mud sample is determined by its low content of ash and SiO_2. The low content of ash is strongly correlated to the relative high content of HCO_3^- and CO_3^{2-}. By contrast, for the Krinides sample (cluster A_{111}), the corresponding characteristics class is B_{111}. We may thus infer that the singularity of this sample is due to its high organic content, a high amount of K_2O, and a low amount of MgO and Fe_2CO_3. The positions of the Pikrolimni (class A_{121}) and Thermi (class A_{122}) muds between the Argilla Solare and the Boario (within the partition tree) confirm the conclusion concerning their potential as beautifying muds. The position of the Lisbori sample (class A_{1121}) is between Argilla Solare and Krinides, a position that appears to reflect the poorest mud characteristics. The position of Kyllini (class A_{122}) is much closer to the best muds (Argilla Solare, Boario, and black mud) and so indicates good mud characteristics. This conclusion is contrary to that obtained by subjective visual examination only.[46]

The fuzzy cross-classification algorithm produces both a fuzzy partition and a fuzzy partition of characteristics compatible with the former. The advantages of this algorithm include the ability to observe not only the fuzzy classes obtained and their relationship, but also the characteristics corresponding to each final class of objects. Each object class may be well described using the corresponding characteristics. These are the characteristics that have contributed to the separation of the respective fuzzy class. Fuzzy divisive hierarchical cross-classification of therapeutic muds based on their physicochemical characteristics allowed an objective interpretation of their origin and maturation and helped in their classification. It also permitted quantitative and qualitative identification of the components that influence the physicochemical properties and therapeutic potential of the analyzed muds.

REFERENCES

1. F. H. Allen, H. J. Doyle, and R. Taylor, *Acta Crystallogr. Sect. B* **4.7**, 50 (1991).
2. R. Aruga, P. Mirti, and A. Casoli, *Anal. Chim. Acta* **276**, 197 (1993).
3. T. P. E. Auf der Heyde, *J. Chem. Educ.* **67**, 461 (1990).
4. H. Bandermer and M. Otto, *Mikrochim. Acta (Wein)* **2**, 93 (1986).
5. J. Barteles, T. Janse, and F. Pijpers, *Anal. Chim. Acta* **177**, 35 (1985).
6. J. C. Bezdek, *J. Cybernetics* **3**, 58 (1974).
7. J. C. Bezdek, *Pattern Recognition with Fuzzy Objective Function Algorithms.* Plenum, New York, 1981.
8. J. C. Bezdek, *J. Intell. Fuzzy Systems* **1**, 1 (1993).
9. J. C. Bezdek, C. Coray, R. Gunderson, and J. Watson, *SIAM J. Appl. Math.* **40**, 339 (1981).
10. T. Blaffert, *Anal. Chim. Acta* **161**, 135 (1984).
11. D. Butnariu, *J. Math. Anal. Appl.* **93**, 436 (983).
12. D. Coomans and I. Broeckaert, *Potential Pattern Recognition.* Wiley, New York, 1986.
13. M. P. Derde and D. L. Massart, *Fresenius Z. Anal. Chim* **313**, 484 (1982).
14. M. P. Derde and D. L. Massart, *Anal. Chim. Acta* **191**, 1 (1986).
15. R. Duda and P. Hart, *Pattern Classification and Scene Analysis.* Wiley, New York, 1973.
16. D. Dumitrescu, *Studia Univ. Babes-Bolyia, Ser. Math.* **33**, 48 (1988).
17. D. Dumitrescu, *Fuzzy Sets and Systems* **28**, 47 (1988).
18. D. Dumitrescu, *Fuzzy Sets and Systems* **47**, 193 (1992).
19. D. Dumitrescu, *Fuzzy Sets and Systems* **56**, 155 (1993).
20. D. Dumitrescu, *Fuzzy Sets and Systems* **67**, 277 (1994).
21. D. Dumitrescu, *Fuzzy Sets and Their Applications to Clustering and Training.* Unpublished manuscript.
22. D. Dumitrescu and L. Kekedy, *Rev. Chim. (Bucharest)* **38**, 47 (1987).
23. D. Dumitrescu and L. Kekedy, *Studia Univ. Babes-Bolyai, Chemia* **32**, 68 (1987).
24. D. Dumitrescu and Gh. Lazarovici, in *Archaeometry in Romania* (P. T. Frangopol and V. Morariu, eds.), p. 87. I.A.P. Press, Bucharest, 1990.
25. D. Dumitrescu, H. Pop, and C. Sârbu, *J. Chem. Inf. Comput. Sci.* **35**, 851 (1995).
26. D. Dumitrescu, C. Sârbu, and H. Pop, *Anal. Lett.* **27**, 1031 (1994).
27. M. Forina and C. Armanino, *Ann. Chim. (Rome)* **72**, 127 (1982).
28. I. E. Frank and B. R. Kowalski, *Anal. Chem.* **54**, 232 (1982).
29. D. Gautheret, F. Mayor, and R. Cedergren, *J. Mol. Biol.* **229**, 1049 (1993).
30. H. I. Gordon and R. L. Somorjai, *Proteins: Struct., Funct., Genetics* **14**, 249 (1992).
31. D. E. Gustafson and W. Kessel, in *Advances in Fuzzy Set Theory and Applications* (M. Gupta, R. Rogade, and R. Yager, eds.), p. 605. North-Holland, Amsterdam, 1979.
32. L. Holm and C. Sander, *J. Mol. Biol.* **233**, 123 (1993).
33. A. K. Jain and R. C. Dubes, *Algorithms for Clustering Data.* Prentice-Hall, Englewood Cliffs, NJ, 1988.
34. I. J. Jolliffe, *Principal Component Analysis.* Springer-Verlag, New York/Berlin, 1986.
35. R. C. Jurs and T. C. Isenhour, *Chemical Applications of Pattern Recognition.* Wiley, New York, 1975.
36. M. E. Karpen, D. J. Tobias, and C. L. Brooks, III, *Biochemistry* **32**, 412 (1993).
37. C. Kateman and F. W. Pijpers, *Quality Control in Analytical Chemistry.* Wiley, New York, 1981.

38. L. Kekedy and D. Dumitrescu, *Rev. Chim.* (*Bucharest*) **38**, 428 (1987).
39. B. R. Kowalski, *Anal. Chem.* **47**, 1152 (1975).
40. B. R. Kowalski and S. Wold, *in Handbook of Statistics* (P. R. Krishnaiah and L. N. Kanal, eds.), Vol. 2, p. 673. North-Holland, Amsterdam, 1982.
41. L. Kryger, *Talanta* **28**, 871 (1981).
42. D. A. Lowy, D. Dumitrescu, H. Pop, and L. Oniciu, *Proceedings of the 8th International Forum on Process Analytical Chemistry* (IFPAC-SM), Houston, Texas, 1994.
43. E. R. Malinowski and D. G. Howery, *Factor Analysis in Chemistry.* Wiley, New York, 1980.
44. D. L. Massart and L. Kaufman, *The Interpretation of Analytical Chemical Data by the Use of Cluster Analysis.* Wiley, New York, 1983.
45. D. L. Massart, B. G. Vandeginste, S. N. Deming, Y. Michotte, and L. Kaufman, *Chemometrics: A Textbook.* Elsevier, Amsterdam, 1988.
46. M. Mitrakis and C. Sikalidis, *Chimica Cronica* **3**, 171 (1993).
47. P. Murray-Rust and J. Raftery, *J. Mol. Graphics* **3**, 50 (1985).
48. T. Okada and T. Wipke, *Tetrahedron Comp. Meth.* **2**, 249 (1989).
49. M. Otto and H. Bandemer, *Chemom. Intell. Lab. Syst.* **1**, 71 (1986).
50. M. Otto and H. Bandemer, *Anal. Chim. Acta* **184**, 21 (1986).
51. M. Otto and H. Bandemer, *Anal. Chim. Acta* **191**, 193 (1989).
52. T. D. J. Perkins and D. J. Barlow, *J. Mol. Graphics* **8**, 156 (1990).
53. T. D. Perkins and P. M. Dean, *J. Computer-Aided Mol. Design* **7**, 155 (1993).
54. F. Pijpers, *Analyst* **109**, 229 (1984).
55. H. Pop, D. Dumitrescu, and C. Sârbu, *Anal. Chim. Acta* **310**, 269 (1995).
56. M. Roubens, *Eur. J. Oper. Res.* **10**, 294 (1982).
57. C. Sârbu, D. Dumitrescu, and H. Pop, *Rev. Chim.* (*Bucharest*) **44**, 450 (1993).
58. C. Sârbu, H. Pop, O. Horowitz and D. Dumitrescu, *J. Chem. Inf. Comput. Sci*, **36**, 465 (1996).
59. I. S. Scarminio, R. E. Burns, and E. A. Zagalto, *Energ. Nucl. Agric.* **4**, 99 (1982).
60. M. A. Sharaf, D. L. Illman, and B. R. Kowalski, *Chemometrics.* Wiley, New York, 1986.
61. P. S. Shenkin, B. Erman, and L. D. Mastrandrea, *Proteins: Struct., Funct., Genetics* **11**, 297 (1991).
62. P. Shenkin and D. Q. McDonald, *J. Comput. Chem.* **15**, 899 (1994).
63. R. M. Swanson, *J. Chem. Educ.* **67**, 206 (1990).
64. K. Varmuza, *Pattern Recognition in Chemistry.* Springer-Verlag, Berlin, 1980.
65. P. Willet, *Similarity and Clustering in Chemical Information Systems.* Research Studies Press, Letchworth, UK, 1987.
66. L. A. Zadeh, *Inform. Control* **8**, 338 (1965).
67. J. Zupan, *Algorithms for Chemists.* Wiley, New York, 1989.

Index